The ANATOMY ASSIGNMENT

LEVEL 1

STEPHANIE A. LANOUE

Kendall Hunt
publishing company

Cover image © Shutterstock, Inc.

www.kendallhunt.com
Send all inquiries to:
4050 Westmark Drive
Dubuque, IA 52004-1840

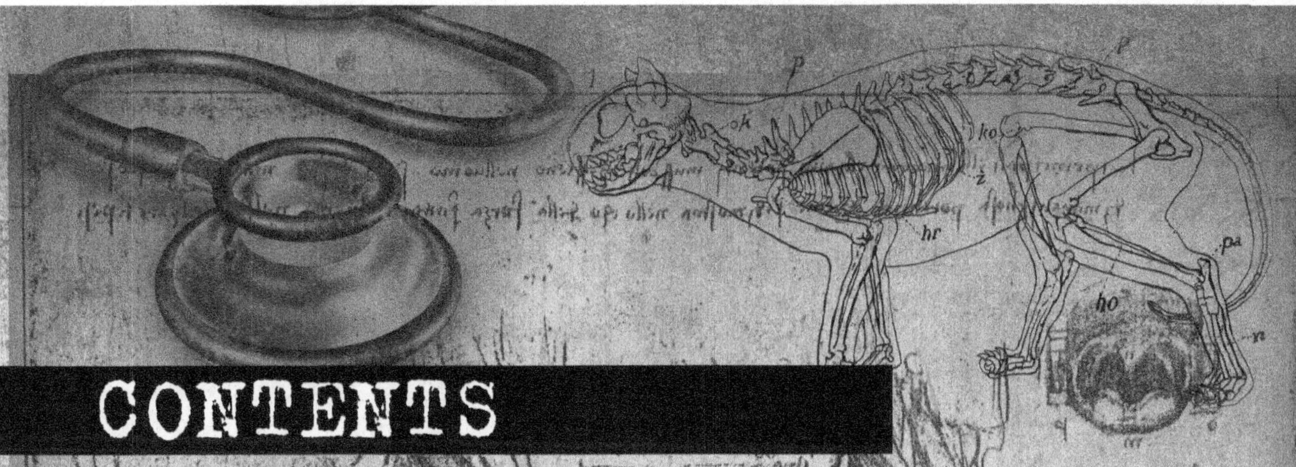

CONTENTS

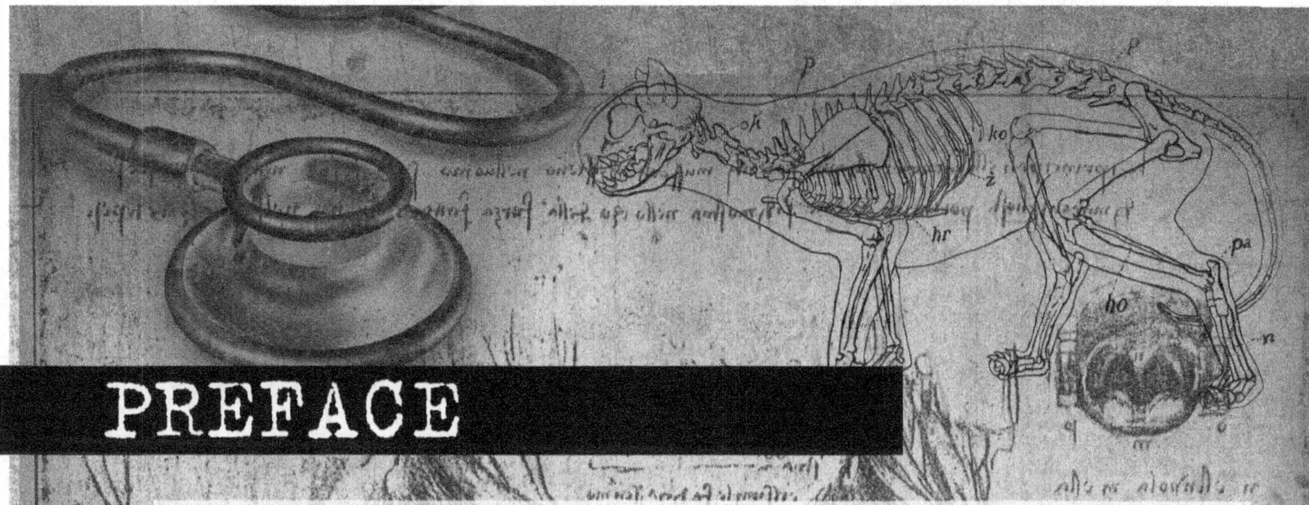

PREFACE

When I surveyed current Anatomy and Physiology college textbooks, I was struck with one huge observation. They all appeared to be very similar. At best, reference books that appear more like an encyclopedia than an inviting book that a college student would like to read and study.

While high-gloss images, consistent formatting, and sound content are not to be understated, they certainly don't make a textbook stand out as unique in today's market. Nor do they necessarily provide an engaging learning experience for the typical community college science student. Students want to know how what they are learning is relevant to their lives. They also want to be "hooked" or interested in what they are studying.

The Anatomy Assignment – Level I is a fresh approach to studying college human anatomy and physiology. While it maintains solid content, it reads more like a novel and less like a reference book so students are both drawn into the storyline as well as the science. Students are also encouraged to write their ideas and answers to thought-provoking questions directly in the textbook. Research supports that students learn most effectively by writing. Writing also allows for additional practice, which also benefits most students.

Speaking of the textbook's storyline, loose yourself in awe as you have fun following two college students taking anatomy and physiology for the first time. Their professor, who expects more of them than just book knowledge and rote memorization, gives them the ultimate challenge that she refers to as, "The Anatomy Assignment." Come along as the two main characters follow their instructor's requirements and journey through "live" field experiences, case studies, and the latest discoveries involving the human body. Each chapter places the students in a unique setting such as a hospital radiology department, a morgue, the local crime lab, etc.

Who knew learning could be so fun at the college level? Are you ready to get started? Carpe diem! And, let me know how you like the textbook. The author may be contacted at: salanoue@lit.edu.

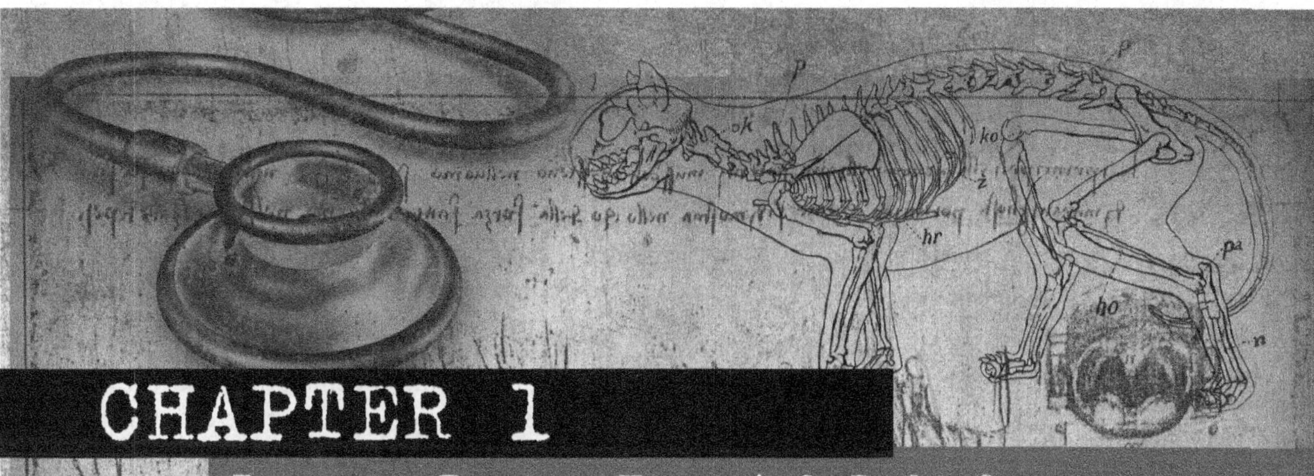

CHAPTER 1

Lessons From a Hospital Radiology Department: Orientation to the Human Body

KEY WORDS

The study of Anatomy & Physiology involves many new vocabulary words. It is helpful to gain familiarity with these new key words, just as you would a foreign language.

anterior	inferior	positive feedback	supine
deep	intermediate	posterior	transverse
distal	lateral	prone	ventral
dorsal	medial	proximal	visceral serosa
frontal	negative feedback	superficial	
homeostasis	parietal serosa	superior	

STUDENT STUDY GUIDE/OUTCOMES

It is helpful to have an idea of what you need to learn before proceeding. Here is a study guide to assist you:

1. Explain homeostasis in your own words. Give a supporting example of how your body maintains homeostasis.
2. Discuss basic anatomical positioning and directions along which the body is divided by the sagittal, frontal, and transverse planes.
3. Use a diagram while describing the directional terms *ventral, dorsal, anterior, posterior, cephalic, supine, prone, superior, inferior, medial, lateral, proximal, distal, superficial,* and *deep.*
4. Using a diagram, identify the regional terms (e.g., acromial, coxal) of the human body.
5. Identify the four abdominopelvic quadrants of the body: right upper quadrant, right lower quadrant, left upper quadrant and left lower quadrant.
6. Identify the nine abdominopelvic regions of the body: right hypochondriac region, right lumbar region, right iliac region, epigastric region, umbilical region, hypogastric region, left hypochondriac region, left lumbar region, and left iliac region.

Grace and Will, on special assignment from their college Anatomy & Physiology professor, entered the bustling hospital lobby and briskly made their way to the information desk to inquire about the location of the hospital's radiology department. Grace, a diligent and conscientious redhead, hated to be late for appointments and was running too close for comfort. It wasn't easy to get an appointment with the head radiologist, Dr. Monroe, and she didn't want to make a bad impression by being late.

Somewhat out of breath she managed to ask for directions, grab Will's arm, and head for the elevator. As they scurried off, the volunteer behind the desk gave a resounding repeat chorus, "That's 526-B." As they approached the elevator, the doors opened and fortunately they were able to board right away. When they arrived at the radiology department, they checked in. After a brief wait, an employee came out and escorted them to see Dr. Monroe.

"Ah-h-h, there you are! Welcome!" he belted with a grin. "I understand that you are on a very special assignment! Are you ready to see our imaging equipment and learn more about taking scans of the body?" Both Grace and Will eagerly nodded their heads.

HOMEOSTASIS

Will half-jokingly interjected, "Before we officially get started, there is one thing I have always wanted to know. Why is it always so cold in hospitals? Especially, imaging and testing centers? How are patients supposed to stay warm?" Walking forward Dr. Monroe chuckled then replied, "**Homeostasis** my friend… despite the temperature in this room, your body fights to maintain a stable internal temperature of 98.6° F. It does this by shivering in response to the cold. This helps to warm the body so that the internal temperature will stay fairly constant." Will continued, "Like when your body produces sweat to cool off when it's too hot?" "Exactly!" replied Dr. Monroe.

Then Dr. Monroe added, "Homeostasis is the ability to maintain a constant internal environment in response to environmental changes. All of your body's systems work together to maintain homeostasis inside your body. Homeostasis is achieved by making sure the temperature, pH (acidity), and oxygen levels as well as many other factors are set just right for your cells to survive.

Grace asked, "How does your body know how to regulate itself?"

Dr. Monroe answered, "Great question Grace! Well, there are both negative and positive feedback loops. **Negative feedback** is a process that happens when your systems need to slow down or completely stop a process that is happening. For example, when you eat food travels into your stomach, and digestion begins. You don't need your stomach working if you aren't eating. The digestive system works with a series of hormones and nervous impulses to stop and start the secretion of acids in your stomach. Another example of negative feedback occurs when your body's temperature begins to rise and a negative feedback response works to counteract and stop the rise in temperature. Sweating is a good example of negative feedback.

IMAGE 1.1 Sweating is a good example of Negative Feedback

Will added, "Oh wow, that makes sense."

Dr. Monroe continued, "**Positive feedback** is the opposite of negative feedback in that it encourages a physiological process or amplifies the action of a system. Positive feedback is a cyclic process that can continue to amplify your body's response to a stimulus until a negative feedback response takes over. An example of positive feedback also can happen in your stomach. Your stomach normally secretes a compound called pepsinogen that is an inactive enzyme. As your body converts pepsinogen to the enzyme pepsin, it triggers a process that helps convert other pepsinogen molecules to pepsin. This cascade effect occurs and soon your stomach has enough pepsin molecules to digest proteins.

Will questioned, "Can reptiles regulate their body temperature like we do?"

Dr. Monroe answered, "We are homeothermic organisms, which means we regulate our own body temperature. Other species like reptiles are not homeothermic. If our body gets too cold, a series of actions are taken to warm our body. Sensors throughout your nervous system can recognize when the temperature drops and might trigger your muscular system to start shivering. The constant contractions of your muscles allow heat to be generated. Your nervous and endocrine systems may also contract the blood vessels of your circulatory system to keep blood in the core of your body and not the extremities like fingers.

Dr. Monroe continued walking and gestured toward some equipment in the distance, "As you might imagine, imaging techniques have greatly accelerated progress in medical science. In most cases, it allows us to look inside the body without having to do exploratory surgery first." Dr. Monroe opened the door to a large room and excitedly exclaimed, "Let's check out the Computerized Tomography (CT) scanner first."

CT SCAN

"It was formerly called 'computerized axial tomography (CAT)' scanning and was developed in 1972. The patient is moved through this circular machine that omits low-intensity X-rays on one side of the circle while receiving them via a detector on the other side. It is basically a medical imaging procedure that uses a more sophisticated application of X-rays to show cross-sectional images or 'slices' through the body. These slices would be similar to slicing a loaf of bread. Most of the slices are about as thin as a coin, however. It allows an exceptional 3D look through the body especially when examining aneurysms, kidney stones, and tumors. In some cases, contrast media may be administered through an IV in the patient's arm to further enhance the images. Here are some images produced by the scanner."

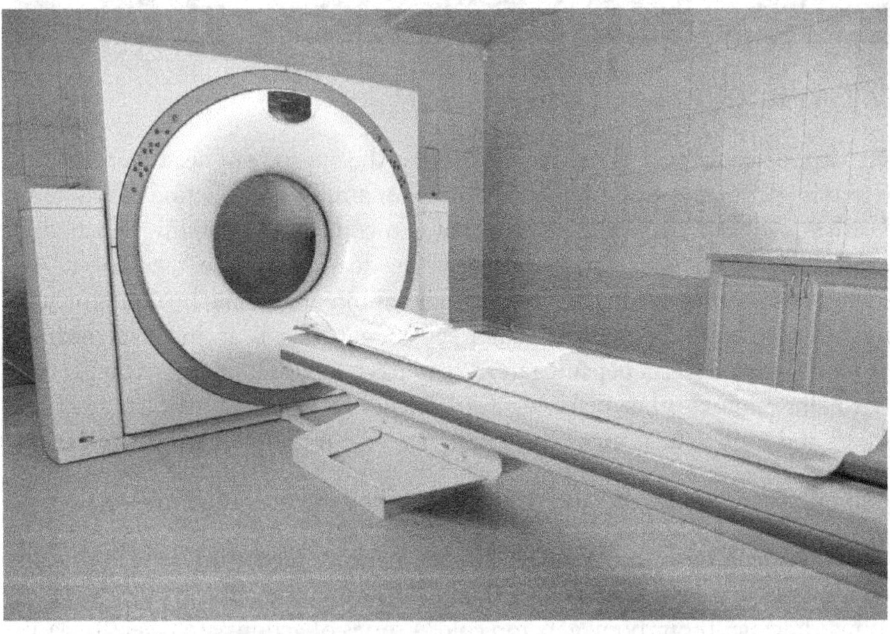

© JupiterImages LLC

IMAGE 1.2 Computer Tomography (CT) Machine

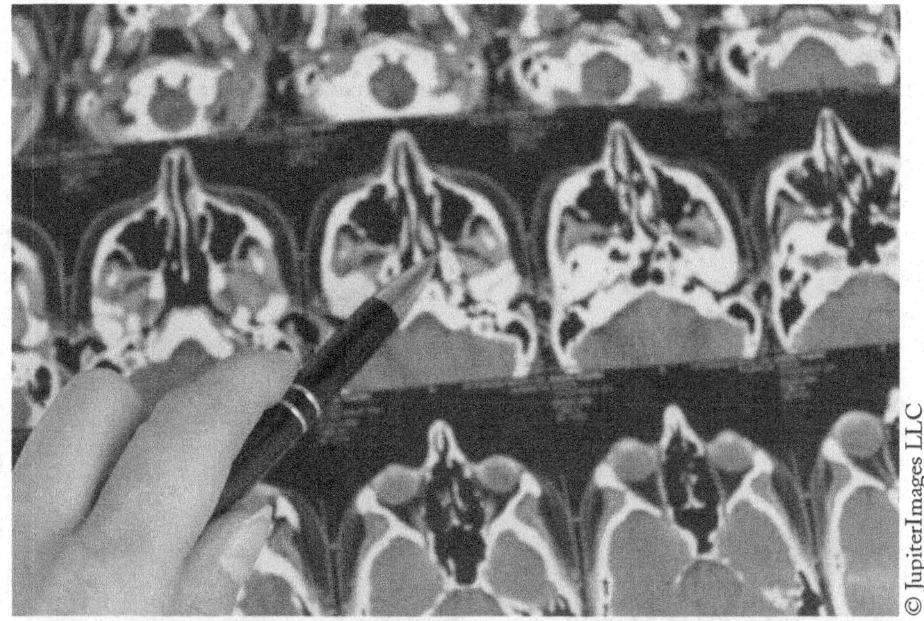

IMAGE 1.3 Computer Tomography (CT) Images
of the Human Skull

"Generally, the human body can be studied by taking cross-sections or plane views through it. It may be a **transverse plane** (or **horizontal plane**; as in a horizontal image through the torso), a **frontal plane** (also known as the **coronal plane**; or an image that separates the person's front from their back) or a **median plane** (also known as a mid-sagittal plane; in effect an image that cuts a person into two halves vertically at their midline such as along the nose, chin, and navel)."

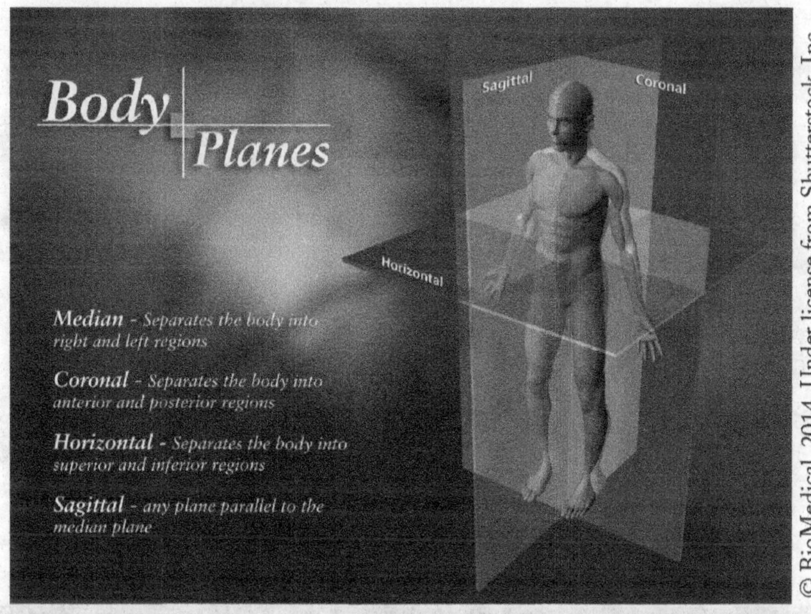

IMAGE 1.4 Planes of the Body

Dr. Monroe further explained, "Before we scan a patient, we must position them correctly. If we ask a patient to lie down on their back and position them face up in the CT scanner, they are in a **supine** position. If we have them lie face down, they are **prone**. Aside from scanning, if a doctor is generally observing a patient, they may place them in basic **anatomical position** which is standing face forward, arms downward but palms up (supine), with feet close together and flat on the floor."

IMAGE 1.5 Supine Position

IMAGE 1.6 Prone Position

Self-Check – Answer these questions now before proceeding:

1. In your own words, what is homeostasis? Give an example.

    ____

    ____

    ____

2. Explain the difference between a transverse, frontal, and medial plane. What other words can be used to describe transverse, frontal, and medial?

    ____

    ____

    ____

3. Distinguish between the terms *supine* and *prone*.

    ____

    ____

    ____

4. Refer back to image 1.3, the CT images of the skull shown to Grace and Will. What plane do these images appear to represent? Explain your answer.

    ____

    ____

    ____

After seeing the CT scanner, Will was curious about other ways to image the body and questioned further, "Hey doc, can we see some more machines and images?" "Absolutely!" answered Dr. Monroe as he walked into another suite adjacent to the room they were in.

RADIOGRAPHY OR X-RAY

"Over half of all clinical imaging is Radiography or X-ray. Although clinical equipment wasn't put in place until 1903, X-rays were actually discovered in 1895. X-rays penetrate the body and darken X-ray film to capture an image called a radiograph. X-rays are best absorbed by dense tissues such as bones, teeth, and solid mass tumors. Hollow organs can be imaged using X-ray if filled with radiopaque substances like barium sulfate. Blood vessels can also be examined if substances are injected for better visualization as in angiography of the aorta and branching vessels off the aorta, shown here."

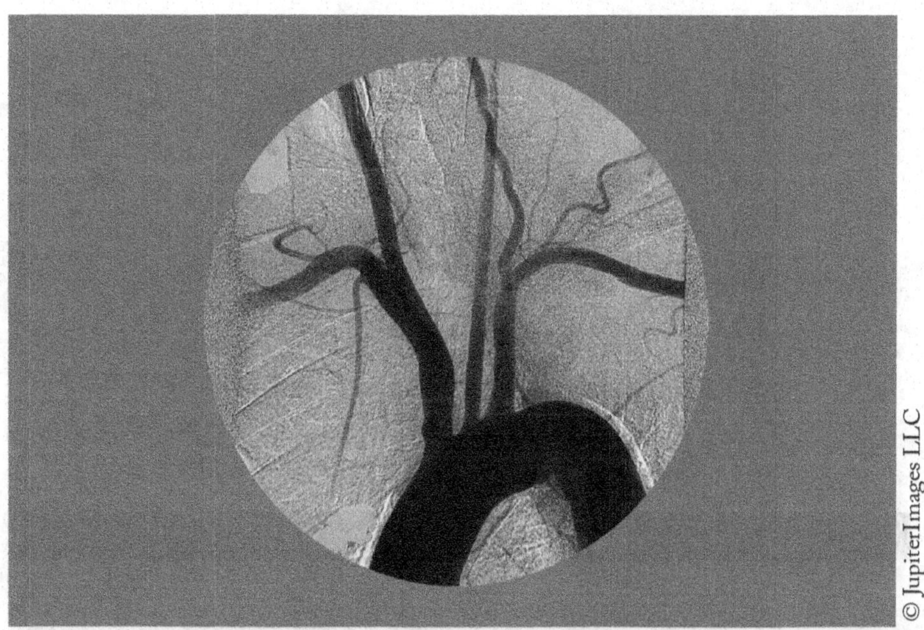

IMAGE 1.7 Angiogram of the Aorta and Arteries Branching Off the Aorta

Dr. Monroe explained, "The other imaging areas are currently in use by patients and technologists, so I am going to give you a brochure on the additional imaging techniques we have available."

MRI (MAGNETIC RESONANCE IMAGING)

An imaging method generally regarded as superior to CT scanning when it comes to visualizing soft tissues. It was developed in the 1970s and works on the principle of magnetism. The patient passes through a large cylinder comprised of a large electromagnet. Our body tissues contain the common element *hydrogen* that aligns itself with the magnetic field. The technologist, in turn, sends out radio waves causing the hydrogen to align in a different position. By turning off the radio waves, the hydrogen in our tissues realigns itself to the magnetic field giving off varying rates of energy. This energy is captured and analyzed producing an image of the body. A new variation of this technique called Functional MRI or fMRI has been developed. It can visualize moment-to-moment changes in actual tissue function. For example, it can show shifting patterns of actual brain activity, such an increased blood flow to an area during a specific task. It is regarded as even more accurate than PET scans of the brain.

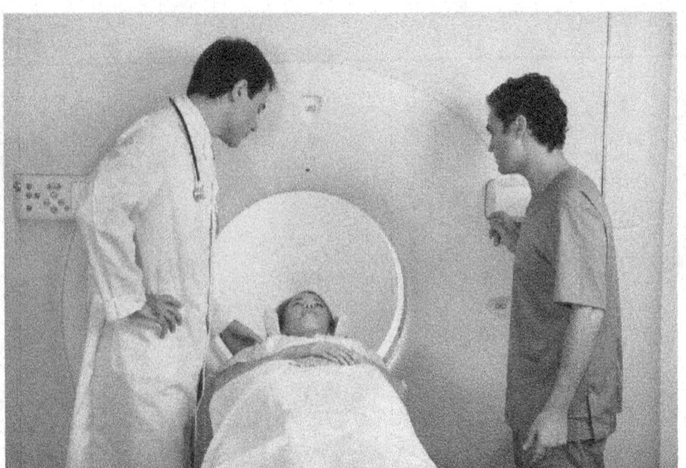

© Tyler Olson, 2012.
Under license from Shutterstock, Inc.

IMAGE 1.8 MRI Machine

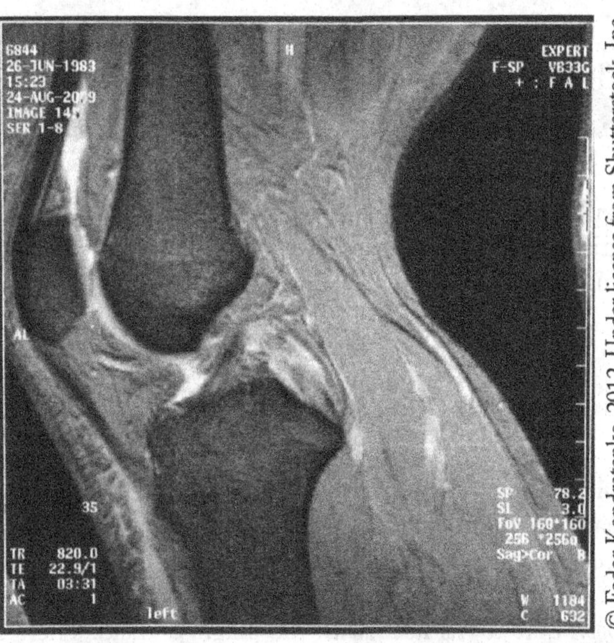

© Fedor Kondratenko, 2012. Under license from Shutterstock, Inc.

IMAGE 1.9 MRI Image of the Knee

PET (POSITRON EMISSION TOMOGRAPHY)

PET scans are used to see how your body uses certain substances such as glucose. Your doctor can track anything unusual about a substance as it moves through your body and how it is being used. For example, using glucose tracking, a PET scan can reveal cancerous tumors. This is because cancerous tissue uses more glucose than normal body tissue and will show up as a bright area on the PET image.

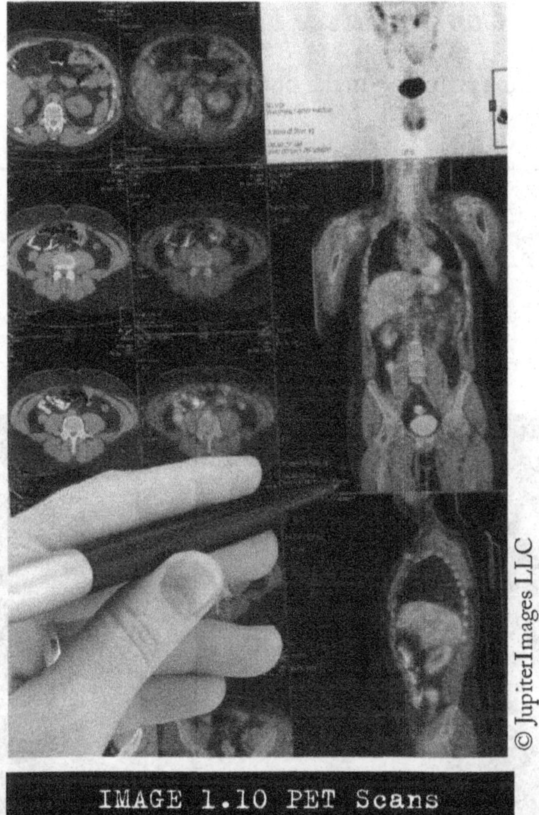

IMAGE 1.10 PET Scans

© JupiterImages LLC

SONOGRAPHY

One of the most common types of imaging, it does not produce X-rays or harmful radiation, making it very popular among obstetricians examining developing fetuses. Sonography has been around since the 1950s, however, in order to get decent images, computer technology needed to evolve. It works by using a handheld transducer on the skin which produces high-frequency ultrasound waves that echo back from the internal organs.

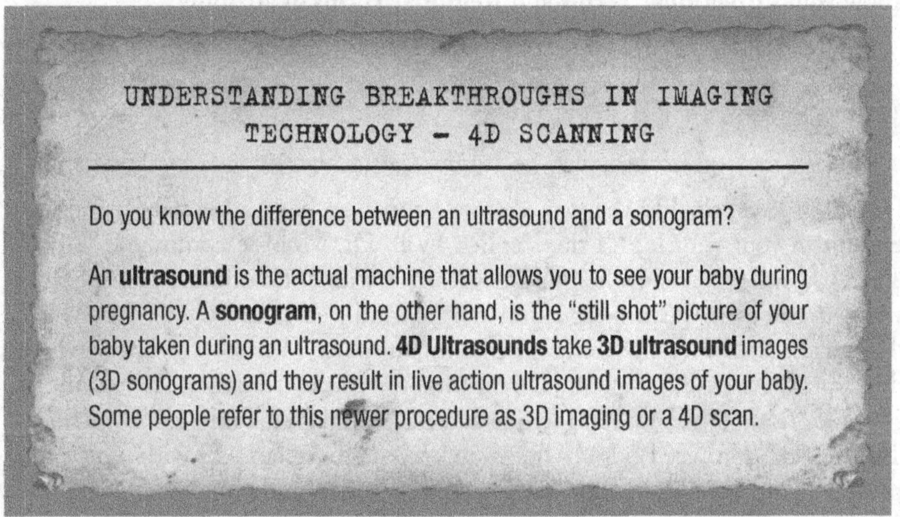

UNDERSTANDING BREAKTHROUGHS IN IMAGING TECHNOLOGY – 4D SCANNING

Do you know the difference between an ultrasound and a sonogram?

An **ultrasound** is the actual machine that allows you to see your baby during pregnancy. A **sonogram**, on the other hand, is the "still shot" picture of your baby taken during an ultrasound. **4D Ultrasounds** take **3D ultrasound** images (3D sonograms) and they result in live action ultrasound images of your baby. Some people refer to this newer procedure as 3D imaging or a 4D scan.

Self-Check – Continued see p18

5. What features can you make out in the 4D scan below? List them here.

____

____

____

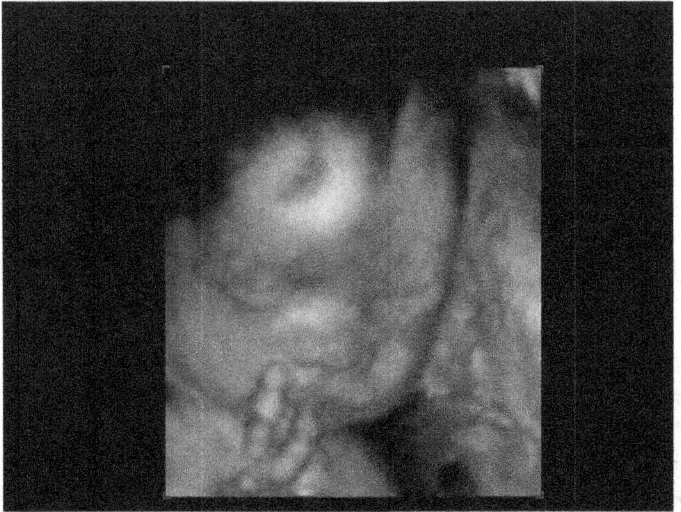

IMAGE 1.11 4D Ultrasound Image of a Developing Fetus

With the students' time growing to a close with Dr. Monroe, Will asked, "Is there any additional advice you could give us for learning more about the general anatomy of the body?" Dr. Monroe explained, "It is also helpful to know your **Directional Terms** and **Regional Terms** of anatomy."

DIRECTIONAL TERMS

Dr. Monroe asked, "By the way, do you happen to have a Global Positioning System in your vehicle?" "I do," replied Will. Dr. Monroe continued, "Good. Then you understand how helpful navigation instrumentation can be—especially when it comes to helping you find north, south, east, and west when you're driving. Basically, it's the same on the human body. We need to know which direction we are talking about. **Directional terms** describe the location of one structure relative to another. Here is some material (see Table 1.1 Directional Terms of the Human Body) to help you become more familiar with what I am talking about. I recommend reading each description and visualizing it."

TABLE 1.1 Directional Terms of the Human Body

Directional Term	Meaning	Example
superior	above	The head is **superior** to the abdomen.
inferior	below	The navel is **inferior** to the chin.
ventral or **anterior**	front or belly	The sternum is **ventral** to the spinal column.
dorsal or **posterior**	back or spine	The spinal column is **dorsal** to the sternum.
medial	toward the median plane	The navel is **medial** to the hips.
lateral	away from the median plane (on the side)	The ears are **lateral** to the nose.
intermediate	between two areas	The nose is **intermediate** to the forehead and the chin.
proximal	closer to the attachment point of a limb *NOTE: this term is reserved for the limbs*	The elbow is **proximal** to the wrist.
distal	farther from the attachment point of a limb *NOTE: this term is reserved for the limbs*	The wrist is **distal** from the elbow.
superficial	close to the surface of the body	The skin is **superficial** to the muscles.
deep	farther from the surface of the body	The organs are **deep** to the skin.

Self-Check – Continued

6. Review image 1.9. MRI of the Knee. What **directional term** best describes the aspect or view of the knee? Explain your answer.

____

____

7. Which two of the Directional Terms are typically reserved for limbs? Name the terms. Give an original example of how each may be used.

____

____

True/False – Circle true if you believe the statement to be true; circle false if you believe the statement to be false but give the correct answer.

8. The knee is distal to the ankle.   True   False

____

9. The nose is medial to the ears.   True   False

____

10. The ears are lateral to the nose.   True   False

____

11. The stomach is inferior to the heart.   True   False

____

12. The spinal column is posterior to the lungs.   True   False

____

13. The chin is superior to the forehead.   True   False

14. The skin is deep to the muscles.   True   False

REGIONAL TERMS

Regional terms describe the general landmarks of the body at the surface level. They are significant when performing a physical examination as well as many other clinical applications.

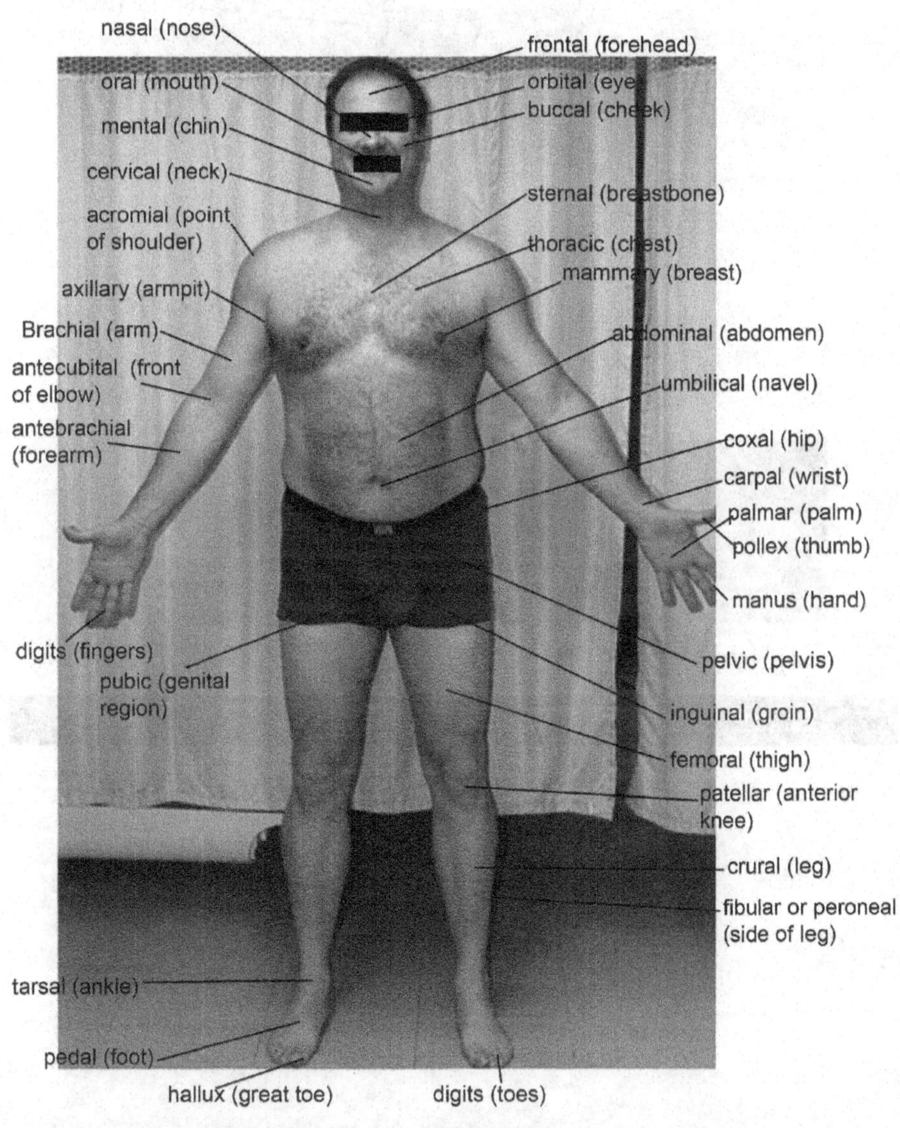

IMAGE 1.12A Regional Terms of the Human Body

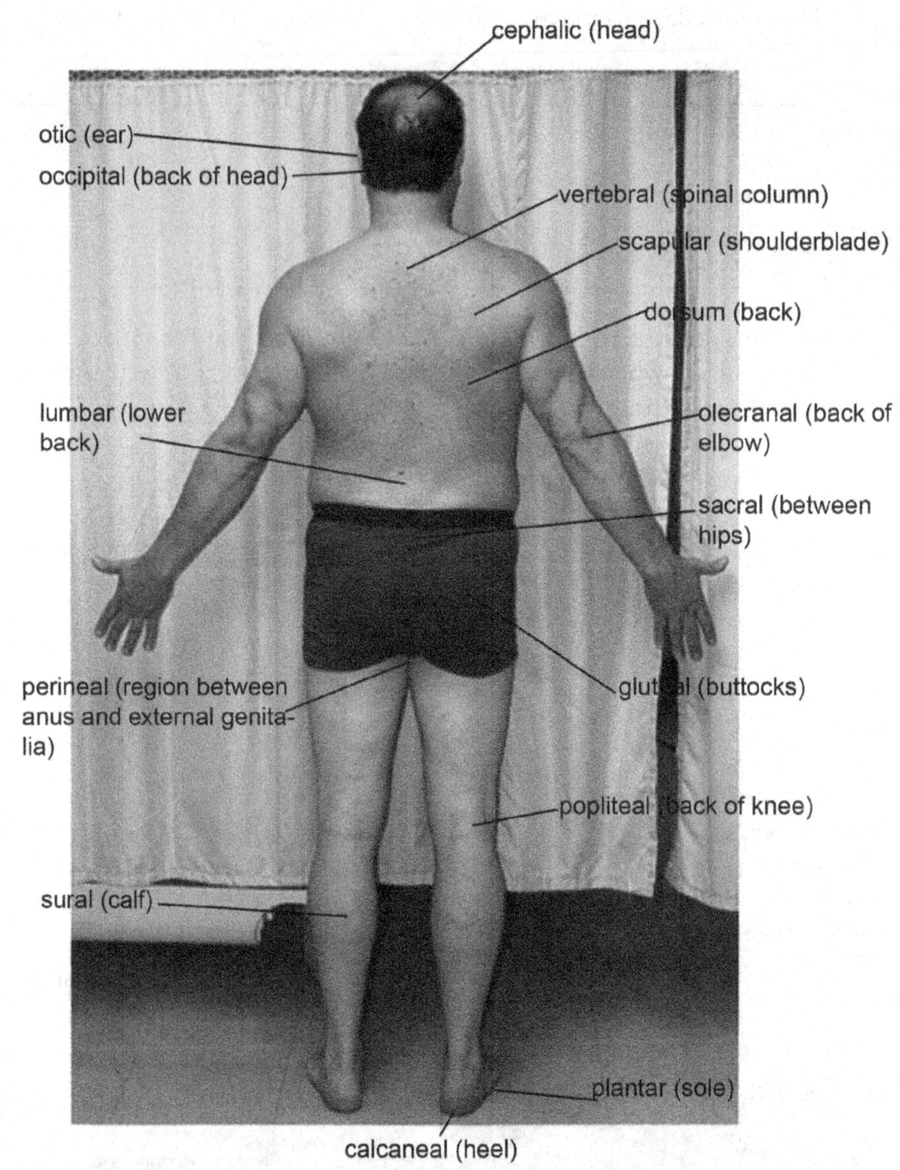

IMAGE 1.12B Regional Terms of the Human Body

PRONUNCIATION OF REGIONAL TERMS

Pronunciation can be a stumbling block when you are first learning about Anatomy & Physiology. Here are some helpful tips for getting it right.

TABLE 1.2 Regional Terms and Their Pronunciations

Anterior Terms	Posterior Terms
Abdominal (ab-dom-uh-nl) – abdomen	Acromial (a-cro-mi-al) – point of shoulder
Acromial (a-cro-mi-al) – point of shoulder	Brachial (brey-kee-uhl) – arm
Antecubital (an-te-cu-bi-tal) – front of elbow	Calcaneal (cal-ca-ne-al) – heel
Axillary (ak-suh-ler-ree) – armpit	Cephalic (suh-fal-ik) – head
Brachial (brey-kee-uhl) – arm	Dorsum (dawr-suhm) – back
Buccal (buhk-uhl) – cheek	Femoral (fem-er-uhl) – thigh
Carpal (kahr-puhl) – wrist	Gluteal (gloo-tee-uhl) – buttock
Cervical (sur-vi-kuhl) – neck	Lumbar (lum-bahr) – loin or lower back
Coxal (cox-al) – hip	Manus (mey-nuhs) – hand
Crural (kroor-uhl) – lower leg near shin	Otic (oh-tic) – ear
Digital (dij-i-tl) – fingers and/or toes	Occipital (ok-sip-i-tl) – base of skull
Femoral (fem-er-uhl) – thigh	Olecranal (o-lec-ra-nal) – back of elbow
Fibular (fib-u-lar) / Peroneal (per-uh-nee-uhl) – side of leg	Perineal (per-i-ne-al) – region between the anus and external genitalia
Frontal (fruhn-tl) – forehead	Plantar (plan-ter) – sole of foot
Hallux (hal-uhks) – great toe	Popliteal (pop-lit-ee-uhl) – back of knee
Inguinal (ing-gwuh-nl) – groin	Sacral (sey-kruhl) – between the hips
Mammary (mam-uh-ree) – breast	Scapular (skap-yuh-ler) – shoulder blade
Mental (men-tl) – chin	Sural (soor-uhl) – calf
Nasal (ney-zuhl) – nose	Vertebral (vur-tuh-bruhl) – spinal column
Oral (awr-uhl) – mouth	
Orbital (awr-bi-tl) – eye	
Palmar (pal-mer) – palm	
Pedal (ped-l) - foot	
Petellar (pa-tel-lar) – anterior knee	
Pelvic (pel-vik) – pelvis	
Pollex (pol-eks) – thumb	
Pubic (pyoo-bik) – genital region	
Sternal (stur-nl) – breastbone	
Tarsal (tahr-suhl) – ankle	
Thoracic (thaw-ras-ik) – chest	
Umbilical (uhm-bil-i-kuhl) – navel	

Self-Check – Continued

15. Practice labeling the diagram.

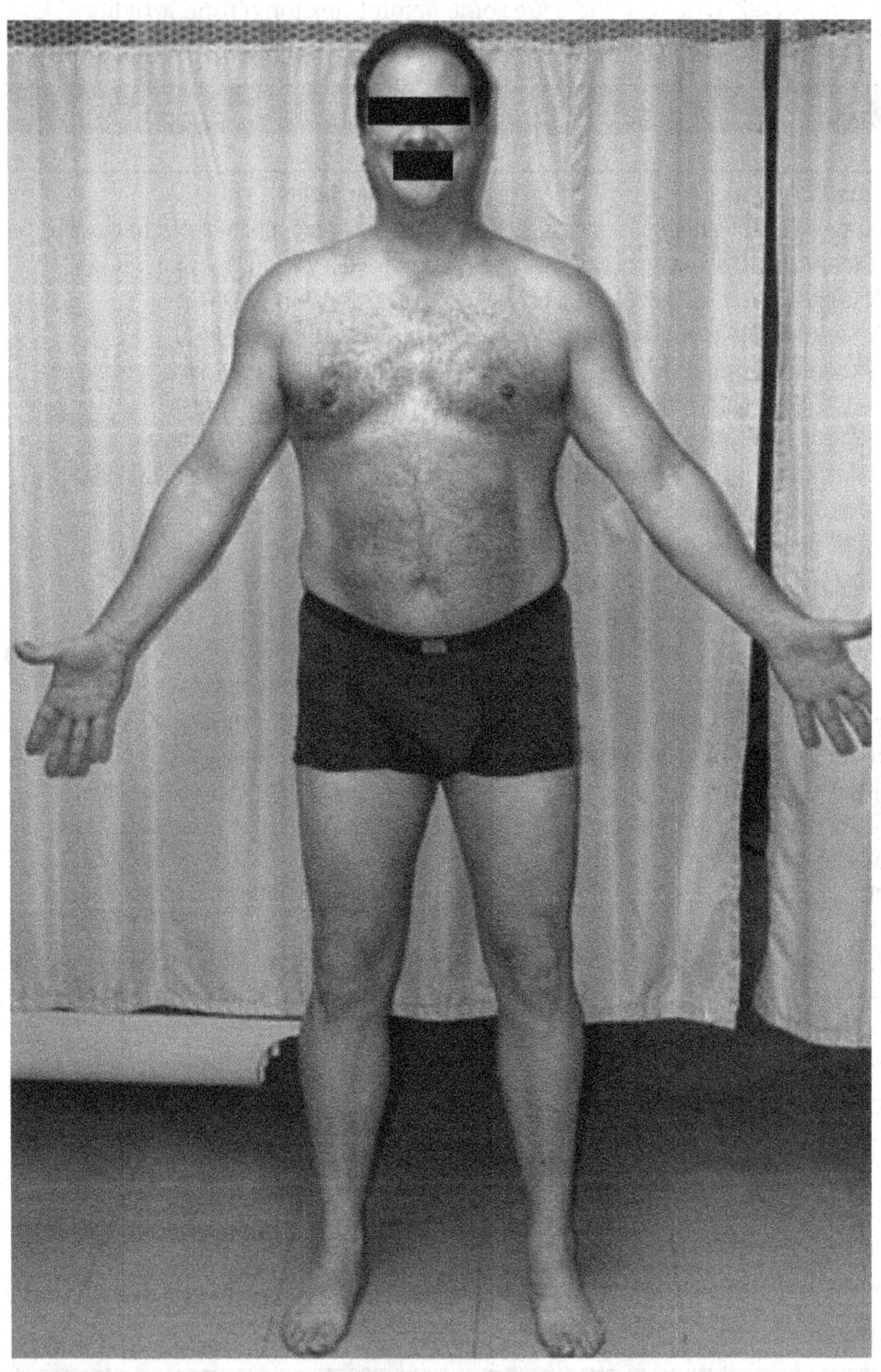

IMAGE 1.13 Practice Diagram–Regional Terms of the Body

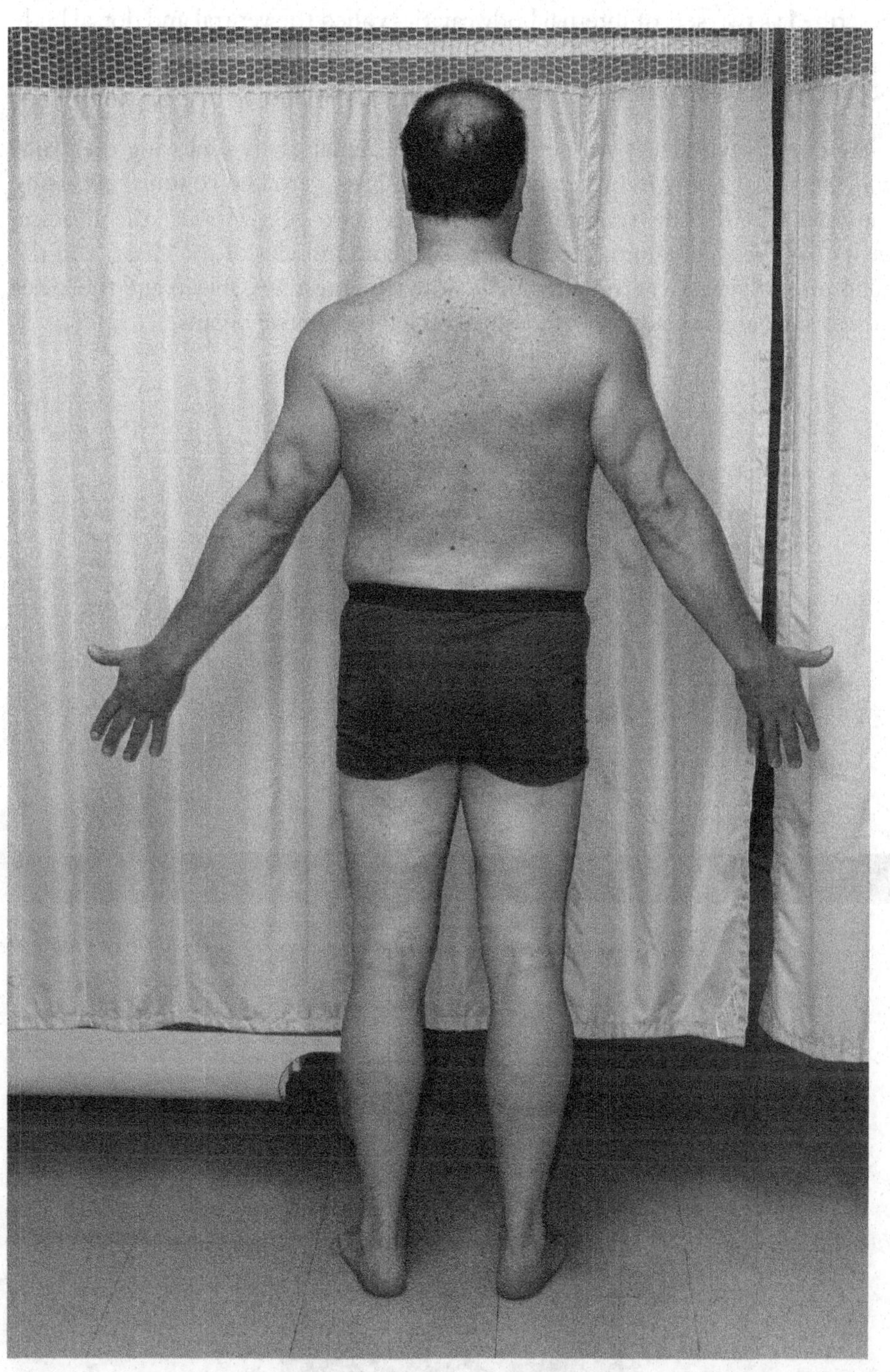

IMAGE 1.14 Practice Diagram—Regional Terms of the Body

BODY CAVITIES AND MEMBRANES

Dr. Monroe continued, "It is also helpful to understand that the body is typically described as two sets of internal body cavities called the ventral and dorsal body cavities. These cavities are closed to the outside and provide different degrees of protection to the organs contained within them.

"The dorsal cavity has two subdivisions: the cranial cavity (encasing the skull) and the spinal cavity (which runs within the bony vertebral column) enclosing the spinal cord. The ventral cavity also has two subdivisions: the thoracic cavity (or chest containing heart, lungs, ribs and muscles of the chest) and the abdominopelvic cavity (containing the stomach, intestines, spleen and the lower organs such as the bladder, rectum and some reproductive organs)."

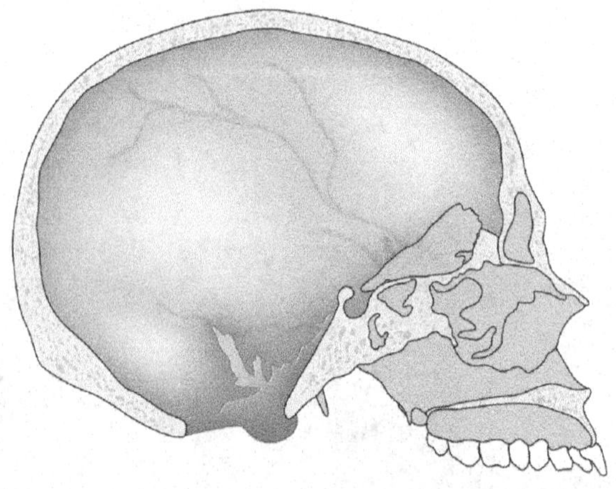

© Blamb, 2014. Under license from Shutterstock, Inc.

IMAGE 1.15 Example of the Cranial Cavity

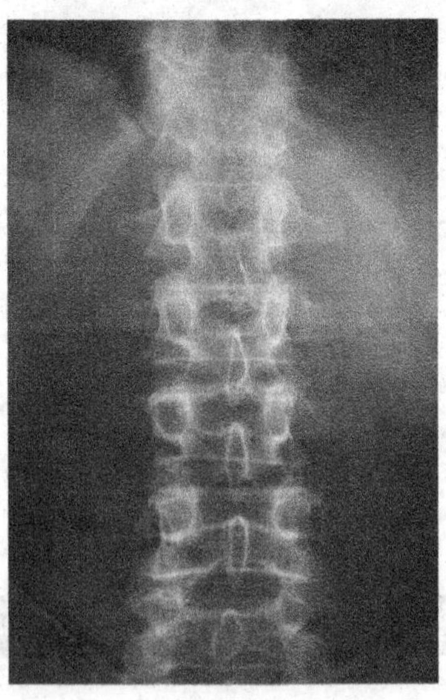

© George N, 2014. Under license from Shutterstock, Inc.

IMAGE 1.16 Example of Spinal Cavity

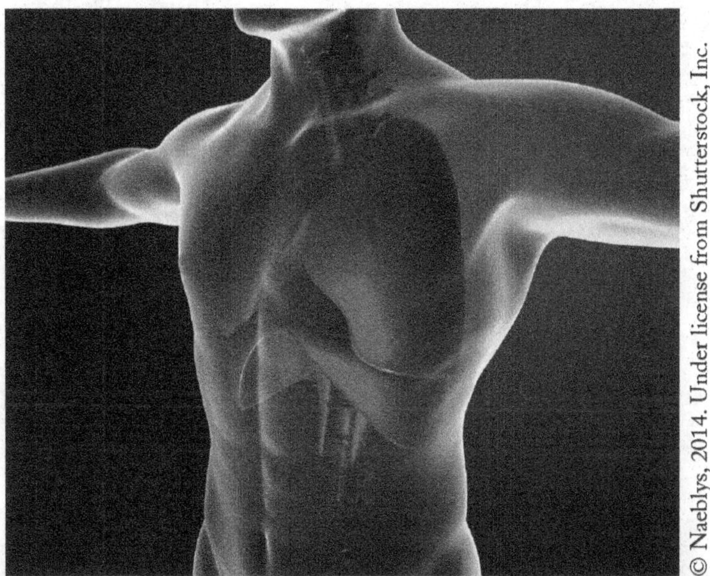

IMAGE 1.17 Example of Thoracic Cavity

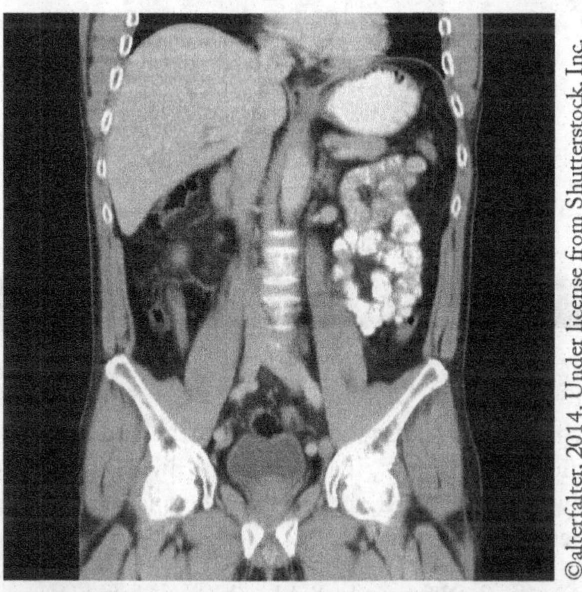

IMAGE 1.18 Example of Abdominopelvic Cavity

Dr. Monroe added, "Additionally, the walls of the ventral body cavity and the outer surfaces of the organs contain a thin, double-layered membrane called the serosa. The part of the membrane lining the cavity walls is called the **parietal serosa**. It folds in on itself to form the **visceral serosa** covering the organs in the cavity.

"Other body cavities include the oral (tongue and teeth) and digestive cavities (digestive organs), nasal cavity, orbital cavities, middle ear cavities (contain ear bones), and synovial cavities (joints)."

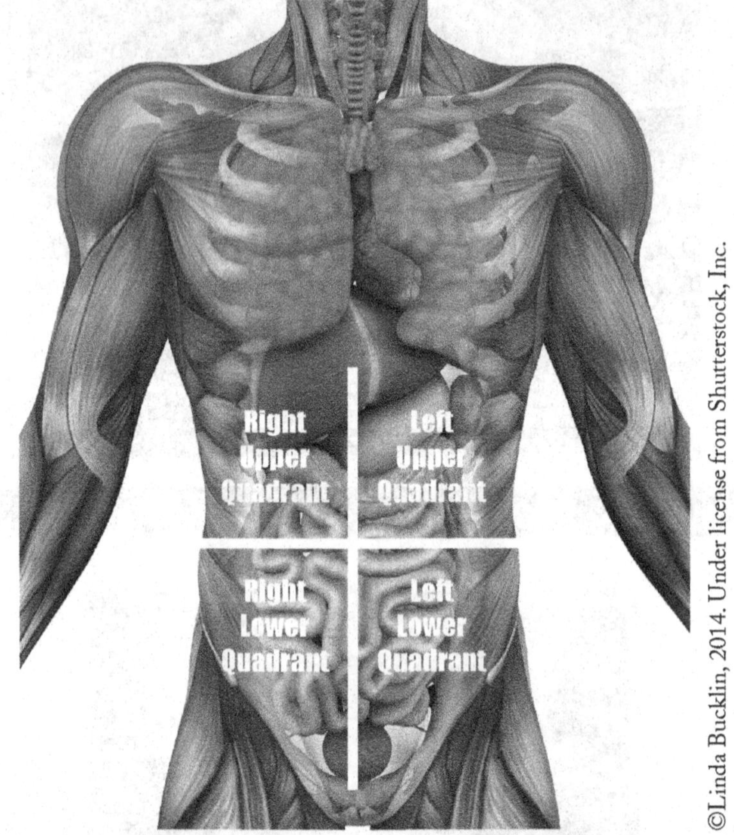

IMAGE 1.19 The Four Abdominopelvic Quadrants

Dr. Monroe continued, "Additionally, because the abdominal region is large and contains numerous organs, it is helpful to divide it into basic quadrants for closer study - there are four abdominopelvic quadrants: right upper, right lower, left upper and left lower.

"Another division method for study divides the body into nine regions: the right hypochondriac region, the epigastric region (located superior to the umbilical region), the left hypochondriac region, the right lumbar region, the umbilical region, the left lumbar region, the right iliac region, the hypogastric (pubic) region, and the left iliac (inguinal region)."

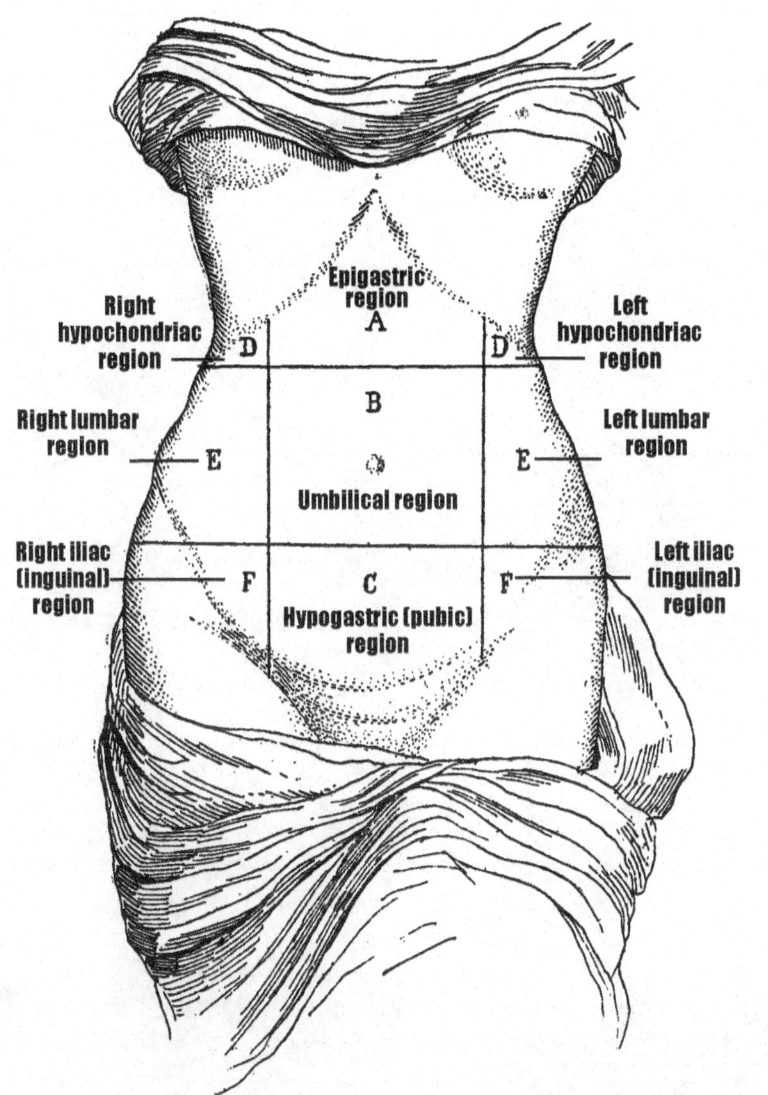

IMAGE 1.20 The Nine Abdominopelvic Regions

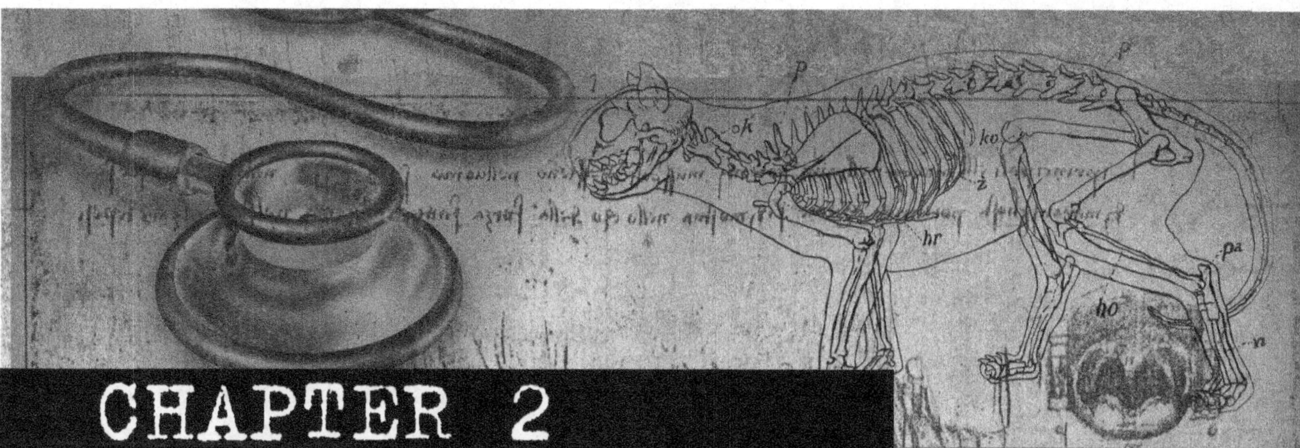

CHAPTER 2

A Crash Course in Biochemistry from the Local Medical Lab

KEY WORDS

The study of Anatomy & Physiology involves many new vocabulary words. It is helpful to gain familiarity with these new key words, just as you would a foreign language.

acid	compounds	enzyme	molecules	protein
amino acid	covalent bond	exchange reaction	monosaccharide	protons
anabolic	decomposition	exergonic	neutrons	quaternary
atoms	reaction	free radicals	nucleic acid	proteins
base	disaccharide	hydrogen bond	organic	secondary
biochemistry	electrolytes	inorganic	pH	proteins
carbohydrate	electrons	ionic bond	polypeptide chains	surface tension
catabolic	element	ions	polysaccharide	synthesis reaction
catalyst	endergonic	lipid	primary proteins	tertiary proteins

STUDENT STUDY GUIDE/OUTCOMES

It is helpful to have an idea of what you need to learn before proceeding. Here is a study guide to assist you:

1. Name the chemical elements of the body and their chemical symbols (abbreviations).
2. Identify the basic four organic compounds of life (carbohydrates, lipids, proteins, and nucleic acids). State their types and functions.
3. Explain the significance of checking your blood cholesterol levels. What is involved? How do you interpret the test(s)?
4. Explain how atoms and molecules are constructed.
5. Discuss the unique significance of ions, electrolytes, and free radicals in the human body.
6. Describe the three types of chemical bonds.
7. Explain the three types of chemical reactions.
8. Explain the difference between an acid and a base.
9. Describe the pH scale and what it measures.

Grace and Will, on special assignment from their college Anatomy & Physiology professor, entered the local medical laboratory to get more information on biochemistry as it relates to the human body. They had many questions and had often wondered things like, Why are trans fats so bad for you? How do I know what multivitamin to select? Why is too much potassium or cholesterol harmful? And, how do sports drinks really work? Maybe you have even wondered the same things? Biochemistry, or the study of molecules that make up living things, can help answer these questions. Biochemistry can also provide a foundation for sound decisions so that you can avoid worthless health fads, unnecessary purchases of products, and increase your understanding of medical tests.

Will introduced himself to the representative at the intake desk, "Hi, we're here for a 9:00 a.m. appointment with your Clinical Biochemist, Dr. Sheldon. I believe she is expecting us. My name is Will." The representative glanced up and said, "One moment please," as she walked into a back hallway. Then she continued, "Okay, come with me. I'll show you where Dr. Sheldon's lab is located. She is expecting you."

They entered a large laboratory with multiple projects, benches, and machinery. In the corner, they could see an attractive woman setting up some type of lab test involving blood. She smiled at them through an array of glassware and greeted them as they approached, "You are just in time! I'm getting ready to double-check a test result on a patient that we got a call on. Their doctor has asked us to re-run a specific test involving this patient's blood lipids (specifically known as lipoproteins). Do you know what lipids refer to?" "I think so," answered Grace in a non-confident voice.

Self-Check – Answer these questions now before proceeding:

1. What do you already know about lipids? What would you tell Grace about them? Write your response here.

____

____

____

____

ORGANIC MOLECULES OF LIFE

"Lipids are composed of carbon, hydrogen, and oxygen—with a high ratio of hydrogen to oxygen. Some people refer to them generally as fats, but there are actually five primary types of lipids in humans—fatty acids, triglycerides, phospholipids, eicosanoids, and steroids. Ever heard of any of those?" Dr. Sheldon asked. Grace shrugged and replied, "I have heard of some of them."

Dr. Sheldon continued, "Basically, a **lipid** is a fat, it is insoluble in water but can be dissolved in other lipids and in **organic** solvents such as alcohol and ether. Lipids are comprised of carbon, hydrogen and oxygen (in low amounts). Phosphorus is also found in complex lipids.

"Triglycerides, like most fats for example, are composed of two types of building blocks - fatty acids and glycerol. Fatty acids are linear chains of hydrogen and carbon atoms (hydrocarbons) with an organic acid group (-COOH) at one end. Glycerol is a modified sugar (a sugar alcohol). Consequently, triglycerides provide the body's most efficient form of stored energy and when oxidized they yield large amounts of energy.

"Fatty acids with only single covalent bonds between their carbon atoms are known as saturated fats. They form a solid fat at room temperature. They should also be consumed in moderate amounts. Fatty acids that contain one or more double-bonds between carbon atoms are said to be unsaturated (monounsaturated or polyunsaturated) fats. These are heart healthy fats (usually liquid at room temperature). Examples include avocados and olive oil. Lastly trans fats, common in many margarines and baked goods, have been solidified by adding hydrogen atoms to their double carbon bonds. These fats are regarded as unhealthy and should not be eaten."

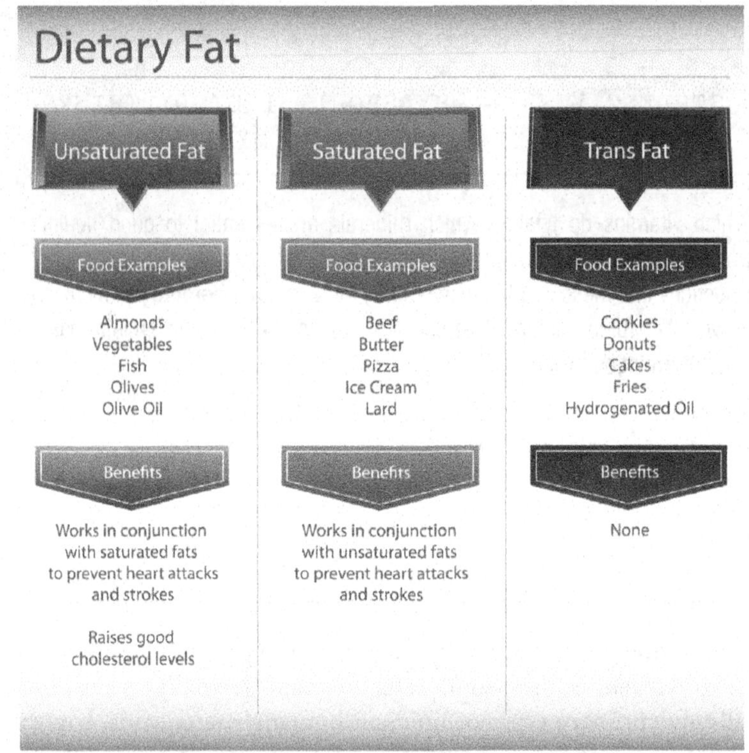

©John T Takai, 2014. Under license from Shutterstock, Inc.

IMAGE 2.1 Dietary Fats

Dr. Sheldon continued, "There are also modified triglycerides called phospholipids. A phospholipid is composed of two fatty acids, a glycerol unit, a phosphate group and a polar molecule (the molecule has a negative side and a positive side; like a water molecule). Triglycerides comprise cell membranes.

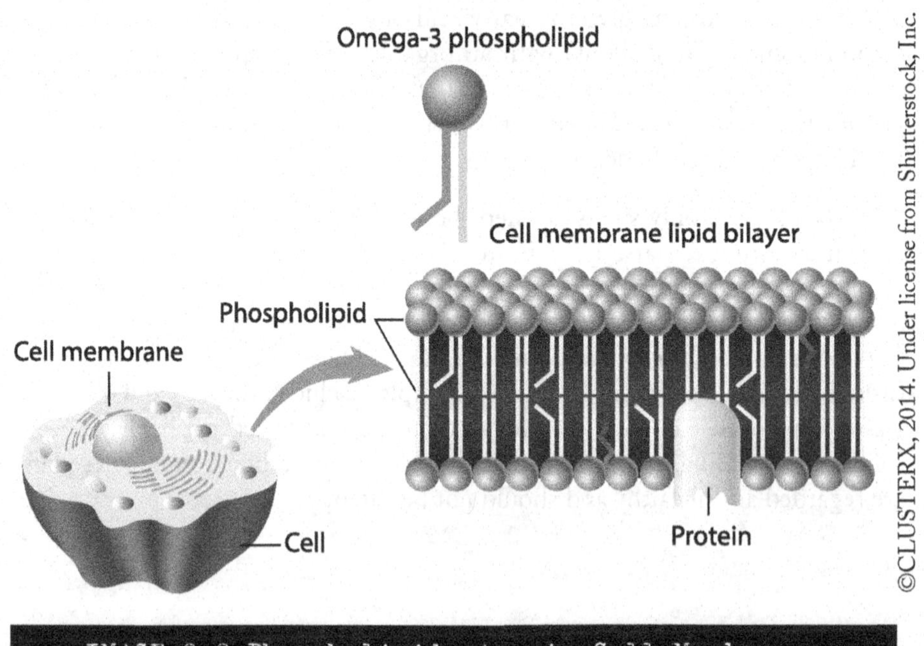

©CLUSTERX, 2014. Under license from Shutterstock, Inc.

IMAGE 2.2 Phospholipids Are in Cell Membranes

"The phosphate group and polar head region of the molecule is hydrophillic (attracted to water), while the fatty acid tail is hydrophobic (repelled by water). When placed in water, phospholipids will orient themselves into a bilayer in which the non-polar tail region faces the inner area of the bilayer. The polar head region faces outward and interacts with the water."

Steroids

Dr. Sheldon continued, "Steroids differ quite a bit from fats and oils. Steroids are basically flat molecules made of four interlocking hydrocarbon rings. They are fat-soluble and contain little oxygen like triglycerides. The single most important molecule in steroid chemistry is cholesterol. Steroids comprise steroid hormones (like estrogen and progesterone) and they are also vital to homeostasis. Without these steroid hormones, reproduction would be impossible and the lack of corticosteroids produced by the adrenal gland would be fatal."

Eicosanoids

Dr. Sheldon continued, "Eicosanoids are diverse lipids found in all cell membranes. Most important of these are the prostaglandins and their relatives, which play a role in various body processes including blood clotting, blood pressure, inflammation and labor contractions.

"There're actually three other main organic molecules of life besides lipids—**carbohydrates, proteins,** and **nucleic acids.**" Picking up a piece of paper she added, "I keep this brochure handy when I train new employees. It has a table that shows the four major organic molecules, their types, and functions. Check it out."

TABLE 2.1 Dr. Sheldon's Brochure Showing the Four Major Organic Molecules, Their Types, and Functions	
Organic Molecules	
I. Lipids	
Type	Function
Fatty acids	Source of energy; precursor to triglycerides
Phospholipids	Aids in fat digestion; major component of cell membranes
Triglycerides	Energy storage; thermal insulation; binding and cushioning organs
Eicosanoids (prostaglandins)	Chemical messengers between cells
Steroids	Hormones; chemical messengers between cells; cholesterol
Other (bile acids, fat-soluble vitamins A, D, K, & E)	Various functions including aid in fat digestion and nutrient absorption; component of cell membranes and precursor of other steroids; nutrients for the body involved in blood clotting, wound healing, vision, and calcium absorption, respectively

II. Carbohydrates	
Type	Function
Monosaccharides	*NOTE: Simplest of all sugars*
Glucose	Blood sugar; energy source for most cells
Galactose	Converted to glucose and metabolized
Fructose	Fruit sugar; converted to glucose and metabolized
Disaccharides	*NOTE: Two simple sugars put together*
Sucrose	Cane sugar; digested to glucose and fructose
Lactose	Milk sugar; digested to glucose and galactose; important in infant nutrition
Maltose	Malt sugar; product of starch digestion; further digested to glucose
Polysaccharides	*NOTE: two or more sugars strung together*
Cellulose	Polysaccharide in plants; dietary fiber
Starch	Energy storage in plants
Glycogen	Energy storage in animal cells (liver, brain, muscle, etc.)

III. Proteins	
Type	Function
Amino acids	Building blocks of proteins (example, alanine); 20 known
Enzymes	Biological catalysts (example, amylase digests starch)

IV. Nucleic Acids	
Type	Function
Deoxyribonucleic acid (DNA)	Constitutes all genes, gives instructions for making all of the body's proteins, and transfers hereditary information from cell to cell during cell growth and division (mitosis) and from generation to generation when humans reproduce (meiosis)
Ribonucleic acid (RNA)	Carries out instructions and synthesizes proteins, assembles amino acids in the right order to produce each protein dictated by DNA
Other (nucleotides)	Energy-transfer molecule made in a cell's mitochondria known as adenosine triphosphate (ATP); made of carbon and nitrogen, a monosaccharide, and one or more phosphate groups

CARBOHYDRATES

Dr. Sheldon continued, "So, we talked about lipids but let's discuss carbohydrates now. Do either of you know the three elements found in carbohydrates?"

Grace replied, "Well, if I remember correctly from high school chemistry, carbohydrates contain carbon (C), hydrogen (H), and oxygen (O)."

Dr. Sheldon replied, "Correct!" She continued, "Do either of you know the three forms of carbohydrates?"

Will answered, "I think the three forms of carbohydrates are monosaccharides, disaccharides, and polysaccharides."

Dr. Sheldon added, "Right again!" She questioned further, "So what is a monosaccharide?"

Grace replied with uncertainty, "Um-m-m, the simplest form of sugar?"

Monosaccharides

Dr. Sheldon continued, "Yes, correct. **Monosaccharides** are the simplest form of sugar and are usually colorless, water-soluble, crystalline solids. Some monosaccharides have a sweet taste. Chemically, monosaccharides are classified by the number of carbon atoms they contain: triose (3 carbons), tetrose (4 carbons), pentose (5 carbons), hexose (6 carbons), heptose (7 carbons), octose (8 carbons), etc. Specific examples of monosaccharides include glucose (a 6 carbon sugar) and fructose (also a 6 carbon sugar)."

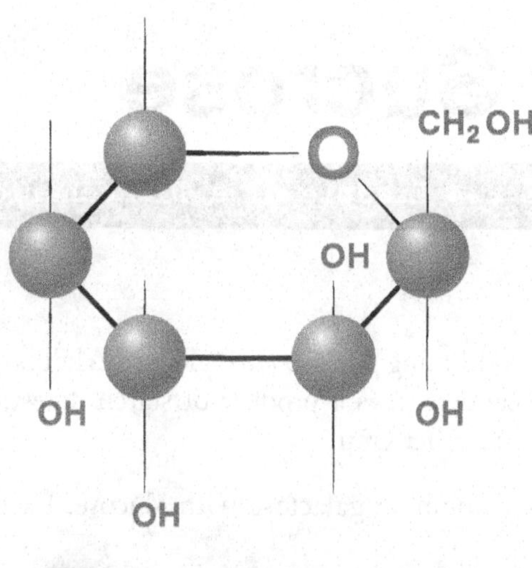

©lyricsaima, 2014. Under license from Shutterstock, Inc.

IMAGE 2.3 Example of a Monosaccharide

Will asked, "Isn't fructose also known as fruit sugar?"

Dr. Sheldon replied, "Yes, it is." Then she added, "Furthermore, monosaccharides are the building blocks of disaccharides such as sucrose and polysaccharides (such as cellulose and starch)."

Disaccharides

Dr. Sheldon continued, "**Disaccharides** are sugars or carbohydrates made by linking two monosaccharides. This is a list of some disaccharides, including the monosaccharides they are made from and foods containing them. Sucrose, maltose, and lactose are the most familiar disaccharides, but there are others.

"Sucrose is formed by combining glucose with fructose. Sucrose is table sugar. It is purified from sugar cane or sugar beets.

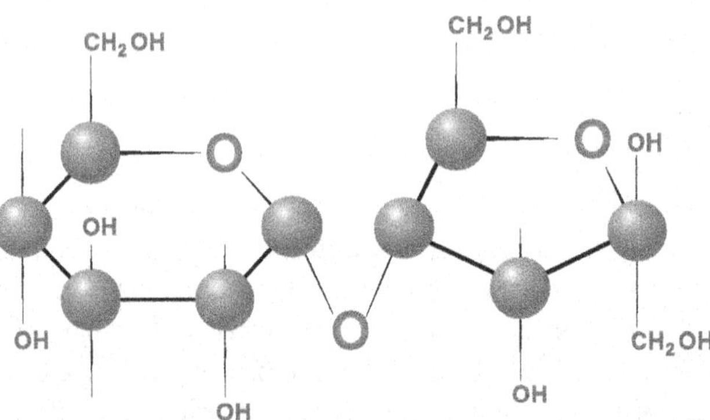

IMAGE 2.4 Example of a Disaccharide

©lyricsaima, 2014. Under license from Shutterstock, Inc.

"Maltose is made by combining glucose with glucose. Maltose is a sugar found in some cereals and candies. It is a product of starch digestions and may be purified from barley and other grains.

"Lactose is made by combining galactose with glucose. Lactose is the sugar found in milk."

Polysaccharides

Dr. Sheldon continued, "And then there are **polysaccharides** which consist of a number of monosaccharides joined by special bonds known as glycosidic bonds. They have some special properties including not sweet in taste, insoluble in water, they do not form crystals, and they can be extracted to form a white powder."

Cellulose

Functions of Carbohydrates

Dr. Sheldon summarized, "Overall, the main functions of carbohydrates are they serve as a substrate for respiration (glucose is essential for cardiac tissues), they are intermediate in respiration (e.g. glyceraldehydes), they serve as energy stores (e.g. starch, glycogen), they are important structurally (e.g. cellulose, chitin in arthropod exoskeletons and fungal walls), and they are important for recognition of molecules outside a cell (e.g., attached to proteins or lipids on cell surface membrane)."

PROTEINS

Dr. Sheldon spoke up, "Now that we have talked about carbohydrates, let's discuss proteins. Do either of you know what makes up a protein?"

Grace answered, "If I remember correctly, I think it is **amino acids**."

"Right you are!" praised Dr. Sheldon. She continued, "So then how would you define a protein?"

Grace continued, "A protein is a large macromolecule made up of long chains of amino acids. I know we need protein(s) for our bodies to function properly. In fact, they are involved in virtually all cell functions."

Will jumped in, "Yeah, and we need to eat foods like eggs, beans, peas, nuts, and meat to put protein into our diets."

Dr. Sheldon continued, "Some proteins are also involved in structural support, while others are involved in bodily movement, or in defense against germs. Proteins vary in structure as well as function. They are constructed from a set of 20 amino acids and have distinct three-dimensional shapes. Here is a list of several types of proteins and their functions."

Protein Functions

"Antibodies - are specialized proteins involved in defending the body from antigens (foreign invaders). They can travel through the blood stream and are utilized by the immune system to identify and defend against bacteria, viruses and other foreign intruders. One way antibodies counteract antigens is by immobilizing them so that they can be destroyed by white blood cells.

"Contractile Proteins - are responsible for movement. Examples include actin and myosin. These proteins are involved in muscle contraction and movement.

"**Enzymes** - are proteins that facilitate biochemical reactions. They are often referred to as **catalysts** because they speed up chemical reactions. Examples include the enzymes lactase and pepsin. Lactase breaks down the sugar lactose found in milk. Pepsin is a digestive enzyme that works in the stomach to break down proteins in food.

"Hormonal Proteins - are messenger proteins which help to coordinate certain bodily activities. Examples include insulin, oxytocin, and somatotropin. Insulin regulates glucose metabolism by controlling the blood-sugar concentration. Oxytocin stimulates contractions in females during childbirth. Somatotropin is a growth hormone that stimulates protein production in muscle cells.

"Structural Proteins - are fibrous and stringy and provide support. Examples include keratin, collagen, and elastin. Keratins strengthen protective coverings such as hair, quills, feathers, horns, and beaks. Collagens and elastin provide support for connective tissues such as tendons and ligaments.

"Storage Proteins - store amino acids. Examples include ovalbumin and casein. Ovalbumin is found in egg whites and casein is a milk-based protein.

"Transport Proteins - are carrier proteins which move molecules from one place to another around the body. Examples include hemoglobin and cytochromes. Hemoglobin transports oxygen through the blood. Cytochromes operate in the electron transport chain as electron carrier proteins."

Protein Structure

Dr. Sheldon explained further, "There are four levels of protein structure. These levels are distinguished from one another by the degree of complexity in their polypeptide chain (linked amino acids). The four, progressively complex levels of protein structure are **primary** (resembles a strand of 'amino acid' beads; is the backbone of a protein molecule), **secondary** (resembles an alpha helix like a slinky toy or coils of a telephone cord; or a beta helix where they are not coiled but form a pleated, ribbon-like structure), **tertiary** (resemble a compact ball-like or globular molecule), and **quaternary** (the most complex structure – like tangled pasta; when two or more chains aggregate) structure. A single protein molecule may contain one or more of these protein structure types. The structure of a protein determines its function. For example, collagen has a super-coiled helical shape. It is long, stringy, strong, and resembles a rope. This structure is great for providing support.

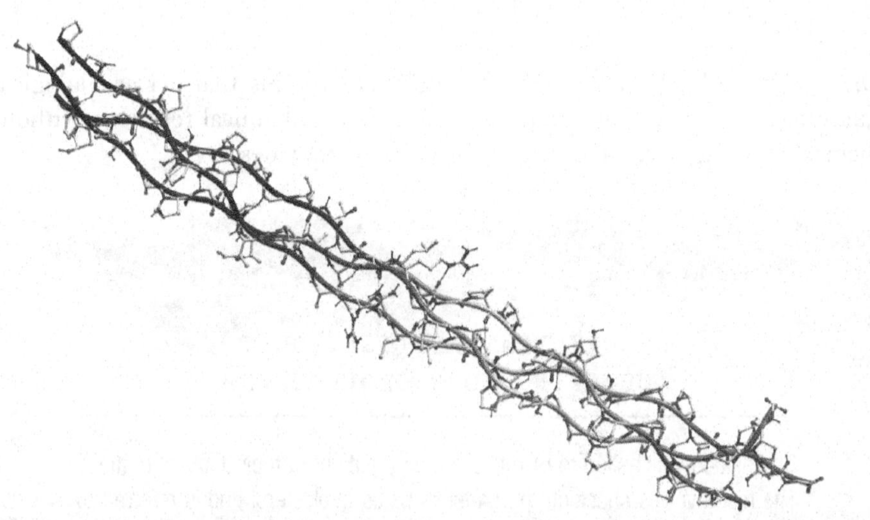

©Leonid Andronov, 2014. Under license from Shutterstock, Inc.

IMAGE 2.6 Collagen is an Example of a Secondary Protein

Hemoglobin, on the other hand, is a globular protein that is folded and compact. Its spherical shape is useful for maneuvering through blood vessels."

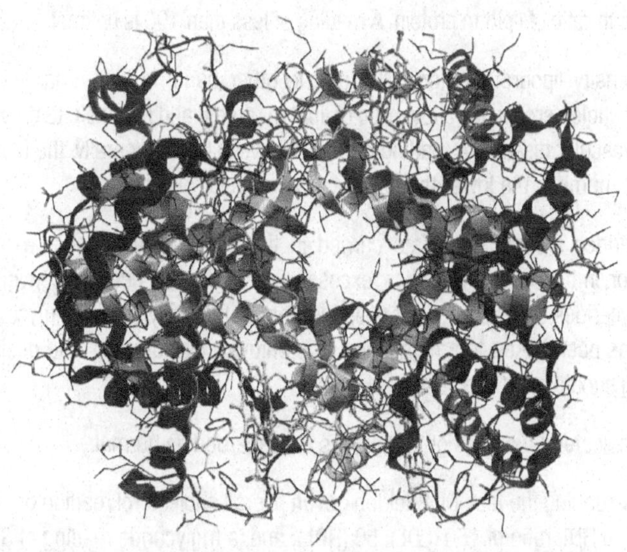

©Leonid Andronov, 2014. Under license from Shutterstock, Inc.

IMAGE 2.7 Hemoglobin is an Example of a Tertiary Protein

Protein Synthesis

"Proteins are synthesized in the body through a process called translation. Translation occurs in the cytoplasm and involves the translation of genetic codes assembled during DNA transcription, into proteins. Cell structures called ribosomes help translate these genetic codes into **polypeptide chains** that undergo several modifications before becoming fully functioning proteins."

Enzymes

Dr. Sheldon added, "There are also globular proteins that act as biological catalysts (substances that speed up or slow down chemical reactions without themselves being used up in the process) called enzymes."

CLINICAL APPLICATION: CHECKING YOUR CHOLESTEROL

Dr. Sheldon was asked to re-run a test on a patient at their doctor's request. The test involves checking the patient's blood cholesterol and is referred to as a lipid profile. Checking the cholesterol level actually entails measuring droplets of lipoproteins (a complex of several things including cholesterol, fat, phospholipids, and protein) in the blood. Doctors like to get an overall picture of your low-density lipoproteins (LDL), high-density lipoproteins (HDL), triglycerides, and total cholesterol (compilation of all).

Low-density lipoproteins (or LDL) is regarded as "bad" cholesterol because it can build up in the walls of your arteries and contribute to cardiovascular disease. It has a high ratio of lipid to protein. A reading of less than 100 is optimal.

High-density lipoproteins (or HDL), on the other hand, is well-regarded as "good" cholesterol. It has a low ratio of lipid to protein and may help to prevent cardiovascular disease. A reading of 60 or above is best. Generally, the higher the HDL number, the lower the risk of cardiovascular disease.

Triglycerides are generally fats carried in the blood by the foods we eat. However, in the case of diabetics, excess sugar in the body can be converted to triglycerides and stored in fat cells throughout the body. A high triglyceride level has been linked to a higher risk of cardiovascular disease. The desired reading should be 200 or less.

Total cholesterol readings of 200 or less are regarded as normal.

After re-running the test, Dr. Sheldon got an overall cholesterol reading of 190; an LDL to HDL ratio of 150 (LDL): 50 (HDL); and, a triglyceride reading of 350.

Nucleic Acids (DNA and RNA)

Dr. Sheldon continued, "Now that we have discussed lipids, carbohydrates and proteins - there is a fourth macromolecule we need to mention... do either of you know what it is?"

Will and Grace both nodded their heads to indicate no.

Dr. Sheldon added, "Nucleic acids which are some of the largest molecules in the body - specifically DNA (deoxyribonucleic acid) and RNA (ribonucleic acid). You will study more about nucleic acids in the next chapter."

Self-Check – Continued

2. What are the four main organic molecules?

    ____

    ____

    ____

    ____

3. Discuss the types and functions of carbohydrates.

    ____

    ____

    ____

    ____

    ____

4. Which carbohydrates are you most familiar with? Name some that you have heard of before.

____

____

____

5. Discuss the types and functions of proteins.

____

____

____

6. Discuss the types and functions of nucleic acids.

____

____

____

7. How would you explain the results of this blood lipid test to the patient? What would you advise?

____

____

____

See information in the last paragraph of the Cliical Application box page 36.

Dr. Sheldon questioned again, "I mentioned that lipids were made out of carbon, hydrogen, and oxygen. What do you know about substances like that?" Grace shyly answered again, "I don't know too much I guess."

Gesturing toward a poster hanging on the wall, Dr. Sheldon added, "There are 91 naturally occurring elements here on earth. Elements are often abbreviated by their chemical symbol. An **element** is the simplest form of matter that retains unique properties. Water, for example, can be broken down into two pure elements—hydrogen (H) and oxygen (O)—that each have unique properties of their own. Of the 91 elements, only 24 of these play an essential role in the normal physiology of the body. This table shows these elements according to their abundance in the body. Six of them account for 98.5% of the body's weight: oxygen, carbon, hydrogen, nitrogen, calcium, and phosphorus. The next 0.8% consists of another 6 elements: sulfur, potassium, sodium, chlorine, magnesium, and iron. And the remaining 12 account for 0.7% of your body's weight and are referred to as **trace elements** because no one of them accounts for more than 0.2%."

TABLE 2.2 Dr. Sheldon's Poster Showing Elements of the Human Body

Name	Chemical Symbol	Percentage of Body Weight (%)
Major Elements (Total 98.5%)		
Oxygen	O	65
Carbon	C	18
Hydrogen	H	10
Nitrogen	N	3
Calcium	Ca	1.5
Phosphorus	P	1.0
Lesser Elements (Total 0.8%)		
Sulfur	S	0.25
Potassium	K	0.20
Sodium	Na	0.15
Chlorine	Cl	0.15
Magnesium	Mg	0.05
Iron	Fe	0.006
Trace Elements (Total 0.7%)		
Chromium	Cr	Less than 0.2
Cobalt	Co	Less than 0.2
Copper	Cu	Less than 0.2
Fluorine	F	Less than 0.2

Iodine	I	Less than 0.2
Manganese	Mn	Less than 0.2
Molybdenum	Mo	Less than 0.2
Selenium	Se	Less than 0.2
Silicon	Si	Less than 0.2
Tin	Sn	Less than 0.2
Vanadium	V	Less than 0.2
Zinc	Zn	Less than 0.2

A BRIEF WORD ABOUT MINERALS

Dr. Sheldon further added, "Some of these elements are also classified as **minerals—inorganic** (does not contain carbon) elements extracted from soil by plants and passed up the food chain to humans and other living animals. Minerals make up about 4% of the human body by weight. Nearly three-quarters of this is calcium and phosphorus; the rest is chiefly chloride, magnesium, potassium, sodium, and sulfur. Minerals support body structures such as bones and teeth. Minerals like sulfur and phosphorus are also used as major components of proteins in the manufacturing of ATP and cell membranes. Lastly, minerals help enzymes and some hormones function."

Self-Check – Continued

8. Practice completing the Chemical Symbols for the Elements of the Human Body. Write your answers below in column #2 of the practice table.

Name	Symbol	Percentage of Body Weight (%)
Major Elements (Total 98.5%)		
Oxygen		65
Carbon		18
Hydrogen		10
Nitrogen		3
Calcium		1.5
Phosphorus		1.0
Lesser Elements (Total 0.8%)		
Sulfur		0.25
Potassium		0.20
Sodium		0.15
Chlorine		0.15
Magnesium		0.05
Iron		0.006

Trace Elements (Total 0.7%)		
Chromium		Less than 0.2
Cobalt		Less than 0.2
Copper		Less than 0.2
Fluorine		Less than 0.2
Iodine		Less than 0.2
Manganese		Less than 0.2
Molybdenum		Less than 0.2
Selenium		Less than 0.2
Silicon		Less than 0.2
Tin		Less than 0.2
Vanadium		Less than 0.2
Zinc		Less than 0.2

COMPOUNDS

Dr. Sheldon smiled and summarized, "If you put two or more elements together, you form a **compound**. A compound is different from just an element because it is made up of two different elements. For example, the element sodium and the element chloride combine to form the compound salt."

Sodium + Chloride → Salt (sodium chloride)

A BRIEF WORD ABOUT ATOMS AND MOLECULES

Grace chimed in, "When I was in high school, I remember studying physical chemistry and we learned about atoms. Isn't sodium also made up of atoms?" Dr. Sheldon answered, "Yes it is. **Atoms** are the smallest particles of any element. Do you remember how they are constructed?" Grace continued, "Let's see… I believe the center of an atom has a nucleus made out of **protons** that have a positive charge and **neutrons** that have no charge. Around that nucleus are clouds of **electrons** that carry a negative charge. Typically the number of protons equals the number of electrons in a stable atom." Dr. Sheldon happily replied, "Right! And if you have a collection of sodium atoms, you create the element *sodium*." And Will interjected, "And if you have a bunch of chloride atoms, you create the element *chloride*!"

IMAGE 2.8 A Simple Model of an Atom Showing the Nucleus and Surrounding Electrons

Grace continued, "We also studied molecules. If I remember correctly, a **molecule** is two or more atoms united by a **chemical bond**. And a chemical bond is the 'glue' that holds molecules together."

> ### STUDY TIP – WHAT IS THE DIFFERENCE BETWEEN A COMPOUND AND A MOLECULE?
>
> A molecule is formed when two or more atoms join together chemically. A compound is a molecule that contains at least two different elements. All compounds are molecules but not all molecules are compounds.
>
> For example, molecular hydrogen (H_2), molecular oxygen (O_2) and molecular nitrogen (N_2) are not compounds because each is composed of a single element. Water (H_2O), carbon dioxide (CO_2) and methane (CH_4) are compounds because each is made from more than one element. The smallest bit of each of these substances would be referred to as a molecule. For example, a single molecule of molecular hydrogen is made from two atoms of hydrogen while a single molecule of water is made from two atoms of hydrogen and one atom of oxygen.

CHEMICAL BONDS AND REACTIONS

Dr. Sheldon continued, "There are three major types of chemical bonds (attractive forces between atoms) - **ionic**, **covalent** and **hydrogen**.

"An ionic bond is a chemical bond between atoms formed by the transfer of one or more electrons from one atom to another."

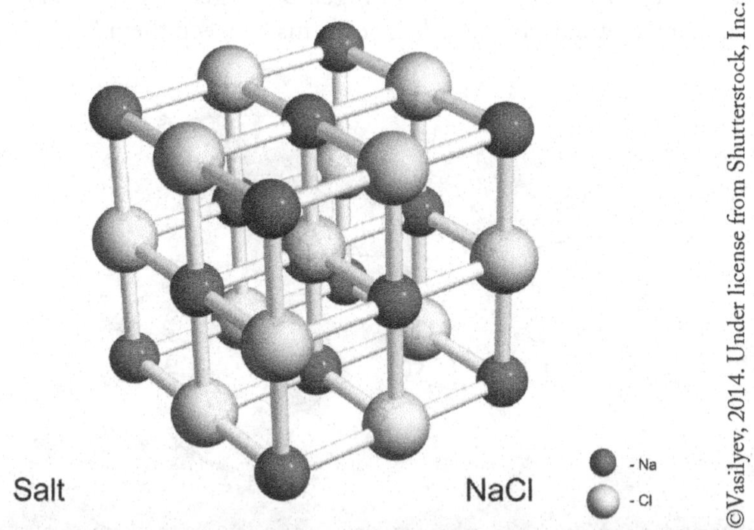

IMAGE 2.9 Sodium Chloride Is an Example of an Ionic Bond

Will added, "An example of ionic bonding would be the formation of table salt."

"Yes, that's right," replied Dr. Sheldon. She then continued, "Electrons do not have to be completely transferred for atoms to achieve stability. Instead, they may also be shared so that each atom is able to fill its outer electron shell at least part of the time. Electron sharing produces molecules in which the shared electrons occupy a single orbital common to both atoms - this is covalent bonding."

Grace added, "Methane (CH_4) is an example of a molecule that has a covalent bond."

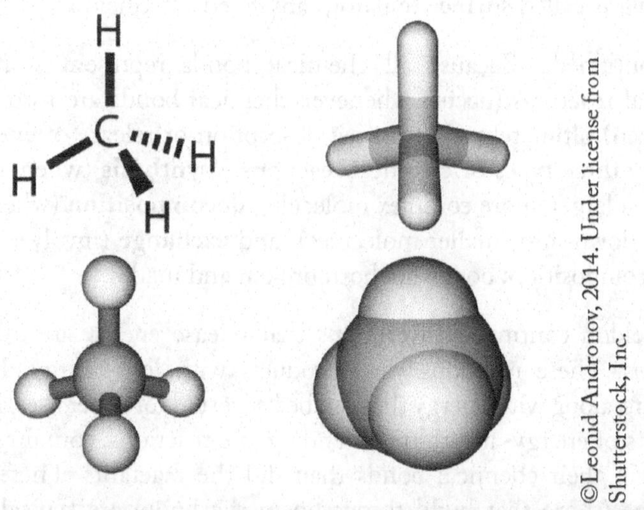

IMAGE 2.10 Methane Molecule Is an Example of a Covalent Bond

"Yes, you are right!" Dr. Sheldon replied gleefully. She then continued, "And hydrogen bonds form when a hydrogen atom, already covalently linked to one electronegative atom (usually nitrogen or oxygen) is attracted by another electron-hungry atom, so that a 'bridge' forms between them."

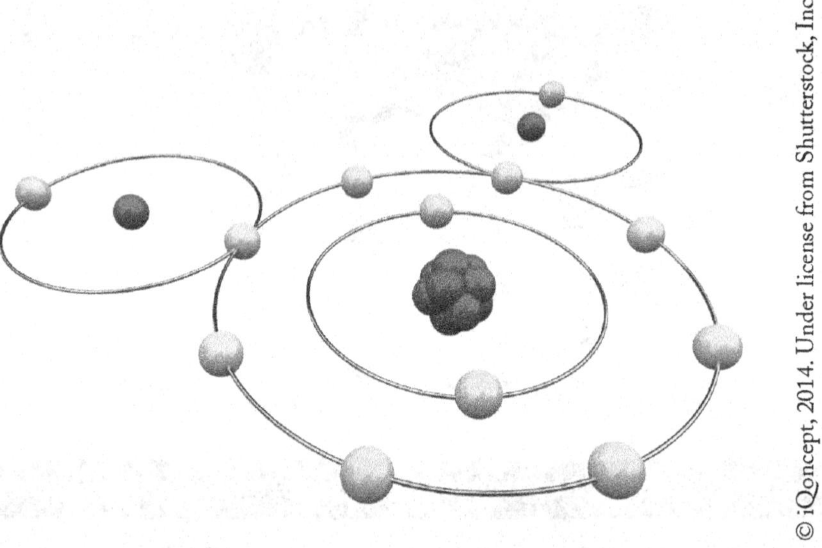

IMAGE 2.11 Hydrogen Bond Formed by Water

"Like water molecules?" questioned Will.

"Yes, like water molecules and have you ever noticed how water molecules cling together ?" questioned Dr. Sheldon.

"Oh - like water beading up on my just waxed car hood," exclaimed Will.

"Yes, that is called **surface tension**," answered Dr. Sheldon.

She continued, "Because all chemical bonds represent stored energy, all chemical reactions (occurs whenever chemical bonds are formed, rearranged or broken) ultimately result in net absorption or release of energy. There are basically three types of chemical reactions - **synthesis** (when atoms combine to form a larger, more complex molecule), **decomposition** (when a molecule is broken down into smaller molecules) and **exchange** (involves both synthesis and decomposition; bonds are both broken and made)."

Dr. Sheldon continued, "Reactions that release energy are called **exergonic** reactions. These reactions yield products with less energy than the initial reactants, along with energy that can be harvested for other uses. In contrast, the products of energy-absorbing, or endergonic reactions, contain more potential energy in their chemical bonds than did the reactants. Therefore, **anabolic** reactions (those that build things up in the body) are typically **endergonic** reactions. There are also **catabolic** reactions (those that tend to tear things apart or break things down) in the body."

Will commented, "And most chemical reactions are reversible. Theoretically, if chemical bonds can be made, they can be broken and vice versa."

Grace added, "And, there are certain factors that influence the rate of a reaction like temperature (higher temperatures proceed more quickly), concentration, particle size and the presence of catalysts."

IONS, ELECTROLYTES, AND FREE RADICALS

Dr. Sheldon nodded in agreement and then added, "There are some interesting variations of atoms and molecules that you have to be aware of in a lab setting or that you will hear more about when you study Anatomy & Physiology. Have either of you ever heard of ions, electrolytes, or free radicals?" Will assuredly replied, "I have heard of electrolytes in sports drinks and free radicals coming off of charred meat on the grill."

Dr. Sheldon chuckled and responded, "That's a good start! **Ions** are actually particles with unequal numbers of protons and electrons. Ions with opposite charges (+ and -) are often attracted to one another and will follow one another through the body. Sodium (Na+) and chloride (Cl-), which have opposite charges, demonstrate this ionic attraction for one another.

"**Electrolytes** are found in sports drinks as well as the human body. They are substances that ionize or give off ions in water like acids, bases, and salts. Electrolytes are capable of conducting electricity. The electrical current coming from the brain, for example, can be measured because electrolytes found in body fluids conduct electrical currents from that organ back to the skin's surface, which can then be measured using electrodes.

"Finally, **free radicals** are very unstable chemical particles that typically have an odd number of electrons and are thought of as destructive to your body's cells. Will, you mentioned burnt meat earlier as being a source of free radicals and you are correct. The burning of the meat and fat actually makes a small amount of a compound called polycyclic aromatic hydrocarbons, which is thought to be carcinogenic (or cancer-causing). Another common source of free radical formation is the preservative known as nitrate, which is found in processed meats like bacon, sausage, and hot dogs. Free radicals are also naturally produced by some reactions in the body including ATP reactions in the mitochondria of your cells. When we are healthy, our bodies have multiple ways to neutralize free radicals including the use of antioxidants. One theory of degenerative diseases like cancer involves the belief that it results from lifelong cellular damage by free radicals."

Will added, "So I am thinking that the formation of free radicals is kind of like when I drive my car and it burns gasoline by combining it with oxygen in the pistons of the engine. My car moves due to the released energy, but it also gives off exhaust fumes as a byproduct. Likewise, something very similar happens in the cells of your body. When oxygen combines with glucose (made from food) in your cells you make energy, but you also make 'free radicals' - your body's version of exhaust fumes."

Grace commented, "Yeah, if I heard you correctly, free radicals are unstable, destructive oxygen atoms. They are missing electrons. These radicals injure your body's healthy molecules by stealing electrons to replace their missing electrons and to balance themselves. In the process, the free radicals injure cells and leave damaged cells and tissues in their wake. This process is called oxidation and it's what makes iron rust and fruit turn brown."

Will commented, "Oh wow, like sliced apples that are left out."

Dr. Sheldon continued, "Exactly! And, in excess these free radicals can cause tissue and organ damage, they can combine with innocuous cholesterol into sticky plaque that clogs arteries, they can destroy cell membranes, the protective covering of cells. They can get into the DNA and inflict damage and thus impair the body's cell's ability to divide and repair itself. This can lead to a breakdown of the bodily systems. Human beings and animals have developed mechanisms that can protect against the formation of these dangerous byproducts of metabolism. The body produces antioxidants. When an antioxidant finds a free radical, it grabs hold of it, and escorts it out of the body before it can do any damage. Chemically it helps to neutralize the free radical by offering their own electrons and thus protect healthy tissue. Antioxidants are found in the food we eat. However, foods lose their antioxidant powers when they are processed and cooked. Stress, air pollution, cigarette smoke and chemicals can cause your body to produce even more free radicals. Additionally, radiation, ultraviolet light in sunshine, toxic chemicals, pesticides, excess saturated fats and even the processed foods itself create more free radicals. The best bet is to keep your antioxidant level high by eating foods high in the vitamins needed to create antioxidants. Another way to control the formation of free radicals is to take antioxidant supplements which can help restore the proper balance of antioxidants and free-radicals."

Self-Check – Continued

9. Write the chemical abbreviation for salt. (Hint: write the abbreviation for sodium immediately followed by the abbreviation for chloride.)

   ____

10. Explain why sodium chloride is not classified as an element.

   ____

   ____

11. In the discussion that Grace, Will, and Dr. Sheldon had in the paragraph above, it was mentioned that electrolytes are substances like acids and bases. What other example was given?

____

____

____

12. Why are free radicals thought of as destructive? Give some examples of known sources of free radicals.

____

____

____

ACIDS, BASES AND PH CONCENTRATION

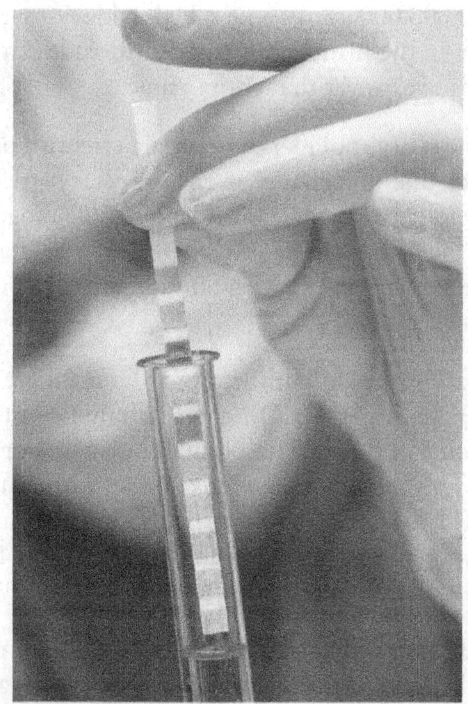

IMAGE 2.12 Measuring pH Concentration

"Hey, what are the lab technicians working on over there?" asked Will.

Dr. Sheldon replied, "Oh - pH work. You see, like salts - acids and bases are also electrolytes. That is to say that they ionize and disassociate in water and can then conduct an electrical current."

Dr. Sheldon continued, "An **acid** has a sour taste in general, can react with or dissolve metals, and release hydrogen ions (H+) in detectable amounts. An acid is a substance that donates hydrogen ions. Because of this, when an acid is dissolved in water, the balance between hydrogen ions and hydroxyl ions is shifted. Now there are more hydrogen ions than hydroxyl ions in the solution. This kind of solution is acidic."

Will explained, "For example, hydrochloric acid (HCl), an acid produced by stomach cells that aid digestion, disassociates into a proton and a chloride ion":

$$HCl ---> H+ \quad + Cl-$$

Dr. Sheldon continued, "A **base** is a substance that accepts hydrogen ions. When a base is dissolved in water, the balance between hydrogen ions and hydroxyl ions shifts the opposite way. Because the base 'soaks up' hydrogen ions, the result is a solution with more hydroxyl ions than hydrogen ions. This kind of solution is alkaline."

Dr. Sheldon continued, "Acidity and alkalinity are measured with a logarithmic scale called **pH**. Here's why: a strongly acidic solution can have one hundred million million (100,000,000,000,000) times more hydrogen ions than a strongly basic solution! The flip side, of course, is that a strongly basic solution can have 100,000,000,000,000 times more hydroxide ions than a strongly acidic solution. Moreover, the hydrogen ion and hydroxide ion concentrations in everyday solutions can vary over that entire range.

"In order to deal with these large numbers more easily, scientists use a logarithmic scale, the pH scale. Each one-unit change in the pH scale corresponds to a tenfold change in hydrogen ion concentration. The pH scale ranges from 0 to 14. It's a lot easier to use a logarithmic scale instead of always having to write down all those zeros! By the way, notice how one hundred million million is a one with fourteen zeros after it?"

Grace elaborated, "It all has to do with hydrogen ions (abbreviated with the chemical symbol H+). In water (H_2O), a small number of the molecules dissociate (split up). Some of the water molecules lose a hydrogen and become hydroxyl ions (OH–). The 'lost' hydrogen ions join up with water molecules to form hydronium ions (H_3O+). For simplicity, hydronium ions are referred to as hydrogen ions H+. In pure water, there are an equal number of hydrogen ions and hydroxyl ions. The solution is neither acidic or basic."

Dr. Sheldon explained further, "Pure water has a neutral pH of 7. pH values lower than 7 are acidic, and pH values higher than 7 are alkaline (basic). Examples of acidic substances (those below 7 on the pH scale) include urine/ milk (pH 6), black coffee/bananas (pH 5), tomato juice/acid rain (pH 4), sodas/

orange juice (pH 3), lemon juice/vinegar (pH 2), sulfuric acid (pH 1) and the strongest acid – battery acid (pH 0). Examples of alkaline substances (those above 7 on the pH scale) include sea water/eggs (pH 8), baking soda (pH 9), milk of magnesia (pH 10)."

WRAP-UP

Will looked at his watch and out of respect for Dr. Sheldon's time said, "We realize that we had a one-hour appointment and that we are nearing our time. We want to thank you for giving us a better understanding of biochemistry and how it relates to the human body." Dr. Sheldon smiled and concluded, "My pleasure. The things we talked about today should give you enough background to make it through the first level of Anatomy & Physiology at the college level. When you get ready to take the second half of Anatomy & Physiology, come back and we'll delve into some different topics like solutions, pH, functional groups of organic compounds, and metabolism."

Self-Check – Continued

13. What is an example of a very strong acid? An example of a very strong base?

____

____

14. Salts, as well as acids and bases, are electrolytes. What does that mean?

____

____

15. On the pH scale, substances that are below 7 are considered what? Explain.

____

____

____

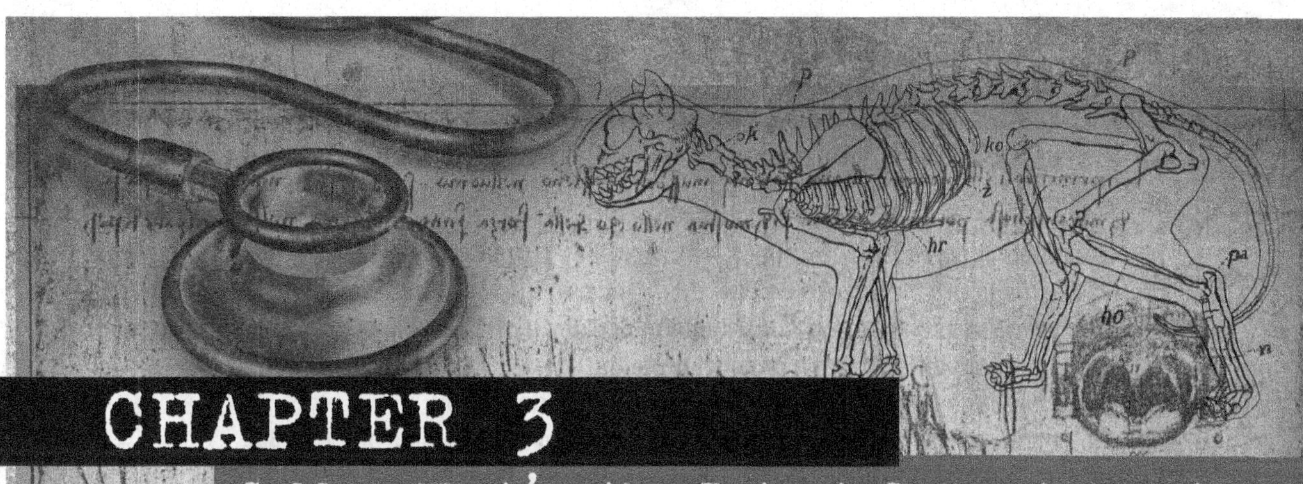

CHAPTER 3

KEY WORDS

The study of Anatomy & Physiology involves many new vocabulary words. It is helpful to gain familiarity with these new key words, just as you would a foreign language.

anaphase	gates	organelles
ATP	gene	peroxisome
cancer	genome	phase (G1)
cell	genomic medicine	phase (G2)
cell (plasma)	glycocalyx	phase (S)
cell cycle	Golgi apparatus	phospholipids
centrioles	Human Genome Project	prophase
centromere	interphase	reticulum (ER)
centrosome	kinetochore	ribosome
channels	lysosome	ribosomal RNA
chromatin	membrane	RNA
chromosomes	messenger RNA	rough ER
cilia	metaphase	second gap
cytokinesis	microvilli	sister chromatids
cytoplasm	mitochondria	smooth ER
cytoskeleton	mitosis	synthesis
daughter cells	mutations	telophase
DNA	nuclear envelope	transcription
DNA polymerase	nucleoli	transfer RNA
endoplasmic	nucleosomes	translation
first gap	nucleotides	
flagella	nucleus	

It is helpful to have an idea of what you need to learn before proceeding. Here is a study guide to assist you:

1. Discuss different types of microscopes and what they permit you to see.
2. Outline the major organelles of a cell including their composition and function.
3. Describe the structure and main constituent molecules of the plasma membrane.
4. Explain how a gate differs from a channel protein.
5. Detail at least three things the glycocalyx of your cells do for you.
6. Describe the three main surface extensions of cells—microvilli, cilia, and flagella.
7. Explain the structure of DNA including its base pairs.
8. Explain the structure of RNA, the three types, and what each type does.
9. Describe genes.
10. Outline the cell cycle and what happens in each stage. Be sure to include the terms aster, centrioles, spindle, centromere, kinetochore, sister chromatids, and cytokinesis.
11. Explain how a mutation happens.
12. Describe what it takes for cancer to develop in a cell.

Grace and Will, on special assignment from their college Anatomy & Physiology professor, have been challenged to visit a local cytology (cell) lab to find out more about the latest developments in the world of cells. They also have several questions that they are curious about when it comes to cells. First, is there a microscope powerful enough to "see" the double-helical strand of a DNA molecule? Secondly, can you actually see atoms under a microscope? And, what's really happening right now in the world of cells and cellular research?

Grace and Will enter the cytology lab and ask to speak with Andy "Buzz" Johnson. Buzz helps run one of the main divisions of the lab and is on duty today to meet with them. Red-headed and shockingly tall at about 6' 8", he greets them briskly, "Hey guys! I understand you'd like to find out more about cells—you're definitely in the right place! Come on back…first we'll check out some of the microscopes we use to view cells, then we'll talk more about cytology."

Perhaps you know somebody who has had to have a biopsy. Waiting to see if the results are cancerous or benign can be very stressful. And while most people meet with their doctors to get their results, it is actually someone working in the cytology lab like a cytotechnologist, pathologist, or medical laboratory technician that is actually examining the cells under a microscope and making the determination.

Buzz continued, "Do you know what kind of microscopes you use at your college?"

Grace piped up, "I think they are compound light microscopes. They have objective lenses that magnify at 4, 10, 40, and 100 times normal vision."

Buzz followed with, "Makes sense. They are the most common type of microscope. We have compound light microscopes here also. Ours are more powerful though and typically magnify up to 1000 times your normal eyesight. Since compound light microscopes use light to illuminate images, anything that you try to view greater than 1000 times normal vision, becomes very blurry. The greater the magnification, the lower the resolving power or clarity. So they certainly have their limits but as you may have already observed in your own lab work at school, they are usually powerful enough to at least let you observe a cell's plasma membrane, nucleus, cytoplasm, and mitochondria. Like this cell, here, at 100 times normal vision."

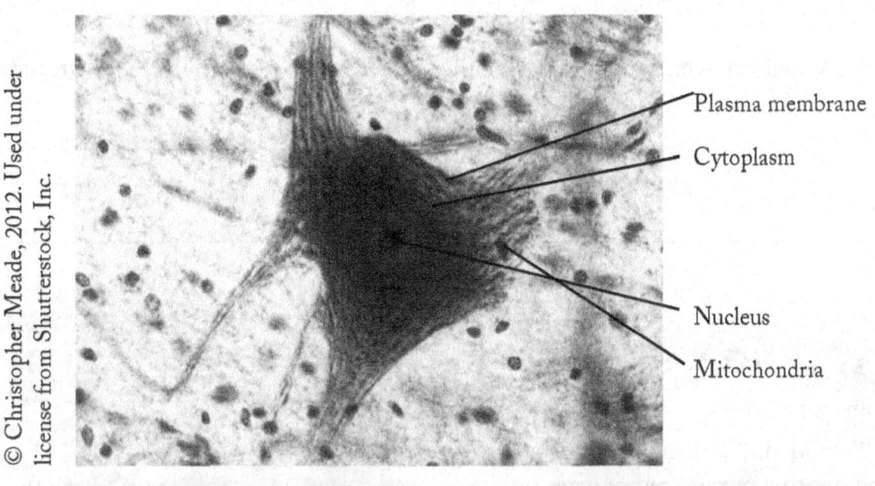

IMAGE 3.1 Cell Under Compound Light Microscope (100x)

Self-Check

1. What do you already know about the nucleus? What does it do? What is it made up of? Maybe something you previously heard or read about it? Write your response here.

____

____

____

2. You may have heard that the genetic material you are made up of is called deoxyribonucleic acid (DNA) and is found tightly coiled in the nucleus of your cells. Have you ever seen a cell under a microscope? Do you remember seeing a nucleus? Describe your experience. For example, what cell parts do you remember actually being able to see with a microscope?

____

____

____

3. DNA is a molecule that has a double-helix structure. Do you think there is a microscope powerful enough to actually see the double-helix structure of a DNA molecule? (Go ahead—take a guess!) _______________

4. What's powerful enough to see the smallest particle of matter—an atom?

____

____

CELL STRUCTURE

Will added, "I brought this diagram that our instructor gave us showing a human cell and all its organelles or structures." Will reached into his pocket and pulled out the cell diagram showing it to Buzz.

Buzz laughed and commented, "I remember those days—my teacher used something just like that! But you will need a microscope with a much higher magnification and resolution to actually see all the other organelles on this diagram."

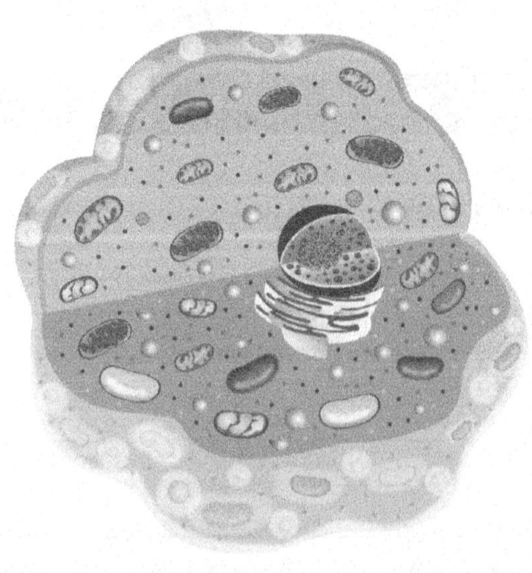

© Matthew Cole, 2012. Used under license from Shutterstock, Inc.

IMAGE 3.2 Animal Cell and its Organelles

Grace added, "Our instructor also gave us this table showing the chemical composition and function of all the organelles." Grace set this piece of paper down for Buzz to view.

TABLE 3.1 Organelles and Their Composition and Function

Organelle	Composition	Function
Plasma (cell) membrane	Lipid bilayer; containing phospholipids, steroids, and proteins	Isolation, protection, sensitivity steroids, and proteins support, control of entrance/exit of materials
Cytoplasm	Fluid inside the cell holding all the organelles (except the nucleus); touches the cell membrane and nuclear envelope.	Includes the cytoskeleton or framework of the cell
Cytoskeleton (microtubules & microfilaments)	Proteins organized in filaments (long straw-like tubes)	Strength and support, movement; overall shape of the cell
Microvilli	Non-moveable membrane extensions containing micro-filaments; almost like mini-fingers	Increase surface area to facilitate absorption of extra-cellular materials
Centrioles	Microtubule subunits	Essential for the movement of chromosomes during division (form the spindle apparatus), organize cytoskeleton
Cilia	Mobile membrane extensions containing microtubules	Movement of materials over cell containing microtubules surface
Ribosomes	RNA and protein structures	Protein synthesis
Rough Endoplasmic Reticulum	Network of membranous channels with attached ribosomes	Modification and packaging of newly synthesized polypeptides
Smooth Endoplasmic Reticulum	Network of membranous channels	Lipid and carbohydrate synthesis without ribosomes attached
Golgi Apparatus	Stacks of flattened membranes	Storage, alteration, and packaging of newly synthesized proteins

Lysosomes	Vesicles containing digestive enzymes	Intracellular removal of damaged enzymes, organelles, &/or pathogens
Peroxisomes	Vesicles containing degradative enzymes	Catabolism of fats and other enzymes' organic compounds
Mitochondria	Double-membraned structures filled with metabolic enzymes	Converts 95% of organic energy with metabolic enzymes into cellular energy (ATP)
Nucleus	Double-membraned structure containing DNA	"Boss" of the cell; storing and processing of genetic material

Buzz proudly walked them into a separate room and announced, "This is a transmission electron microscope (or TEM). With this baby, we can see things a million times smaller than the period at the end of this sentence and hundreds of times better than the light microscope. We can even see ribosomes as well as the actual protein molecules that they assemble!"

© Pan Xunbin, 2012. Used under license from Shutterstock, Inc.

IMAGE 3.3 A Transmission Electron Microscope

Buzz continued, "If you look at the surface of this cell under the electron microscope, it reveals that the cell and many of the organelles are bordered by a membrane. The membrane surrounding the entire cell called the **plasma membrane,** appears as a pair of dark parallel lines and defines the boundary of the cell. It also runs interactions with other cells and controls the flow of materials into and out of the cell."

Grace interjected, "Isn't the plasma membrane made up of phospholipids or something like that?"

Buzz immediately answered, "Yes! Typically about 98% of the molecules in the membrane are lipids. About 75% of those lipids are specifically **phospholipids** arranged in a bi-layer. The rest of the membrane is comprised of cholesterol molecules, glycolipids (phospholipids with short carbohydrate chains of the outside of the plasma membrane that help form the **glycocalyx,** a carbohydrate coating on the cell's surface with multiple functions), and proteins."

CLINICAL APPLICATION – A BETTER UNDERSTANDING OF A CELL'S GLYCOCALYX

The glycocalyx of your cells is chemically unique in everyone except identical twins. It acts like an identification tag that enables the body to distinguish its own healthy cells from transplanted tissues, invading organisms, and diseased cells. Even human blood types and transfusion compatibility are determined by glycolipids. The glycocalyx also helps cells adhere to one another by binding cells together so tissues do not fall apart. Changes in the glycocalyx of cancerous cells enables the immune system to recognize and destroy them.

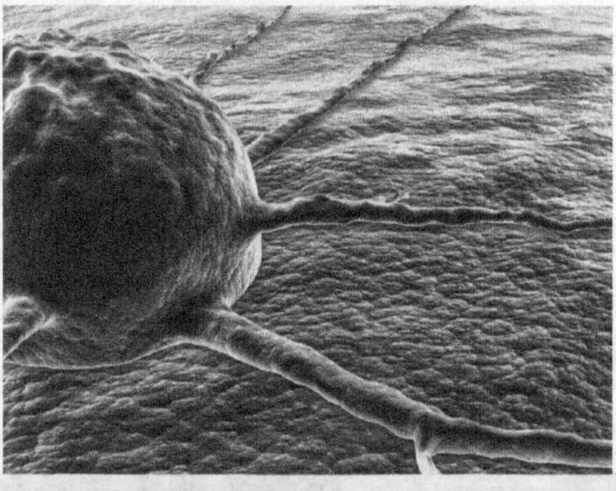

© BioMedical, 2012. Under license from Shutterstock, Inc.

IMAGE 3.4 Cancer Cell at 300x Magnification

Clinical Reflection Questions

1. What is the glycocalyx and what does it do for tissues? Explain in your own words.

____

____

____

2. Explain the uniqueness of the glycocalyx in individuals and in twins.

____

____

____

Buzz added, "Proteins in the plasma membrane are larger than lipids and constitute about 50% of the membrane weight. There are passages called **channels** that allow water and some dissolved substances to move through the membrane. Some channels are always open, whereas others can be thought of as **gates** that open and close under different circumstances."

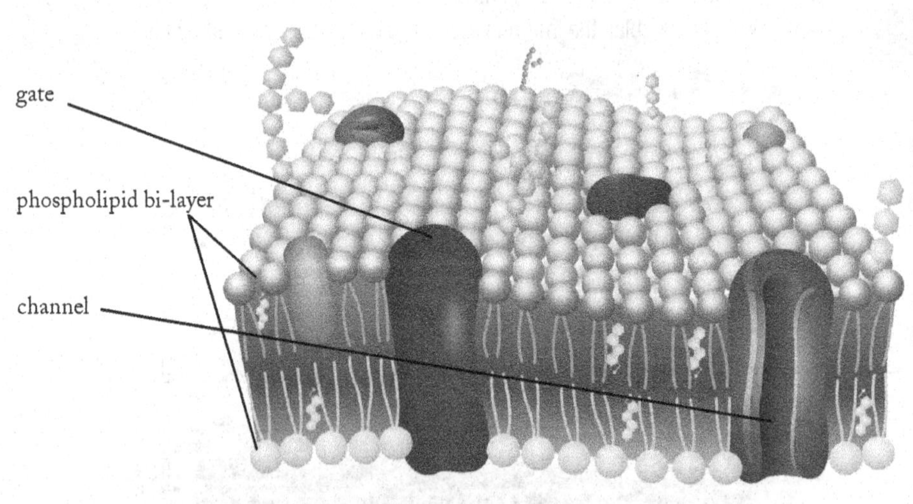

IMAGE 3.5 Cell (plasma) Membrane Showing Phospholipid Bi-Layer

"Oh cool," exclaimed Will as he studied the cell's surface further. "I see some hair-like extensions. What are those?"

Buzz replied, "Those are **cilia**. Some, but not all cilia, move. The lining of the trachea, fallopian tubes and ventricles of the brain have motile cilia that help to move substances like mucus, cerebrospinal fluid, etc. There may be 50 to 200 cilia on the surface of a cell. They usually beat in waves that sweep across the surface of a cell always in the same direction."

Buzz continued, "Some cells have **flagella** which is much longer than cilia and more whip-like such as the tail of a human sperm. It moves in a more undulating or corkscrew fashion to achieve propulsion. Some cell's surface area is increased by extensions as in the case of **microvilli**. They are best developed in cells specialized for absorption such as epithelial cells of the intestines. They provide cells with more than 15 to 40 times as much absorptive surface area as they would if their outer surfaces were flat."

© Jubal Harshaw, 2012. Used under license from Shutterstock, Inc.

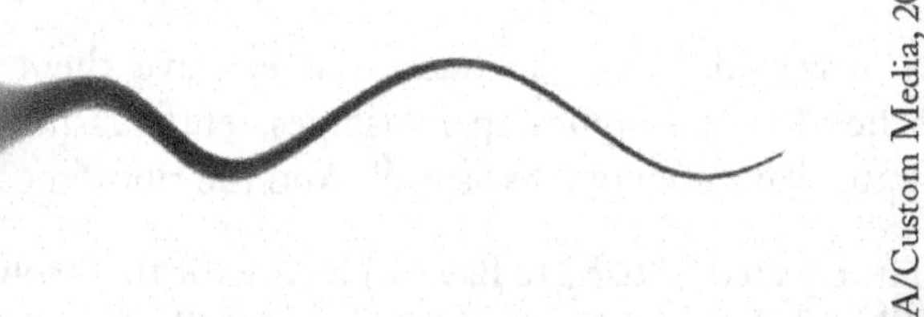

© CLIPAREA/Custom Media, 2012. used under license from Shutterstock, Inc.

Image 3.6 Cilia from Lung Bronchiole under SEM (top) and Sperm with Long Flagella Tails (bottom)

Nucleus

Buzz moved toward an adjacent microscope and commented, "In this cell you can easily see the largest organelle—the **nucleus.** The nucleus is essentially the genetic control center of the cell. It also directs protein synthesis and shelters the DNA. It is usually the only organelle that can be seen under a light microscope, but with this transmission electron microscope (TEM), you can actually see two nuclear membranes surrounding it which together form the **nuclear envelope.** The nuclear envelope is made up of a ring of proteins with nuclear pores interspersed so molecules that serve as the raw materials for DNA and RNA can enter or pass outward. The material inside the nucleus includes **chromatin** as well as one or more dark-stained masses called the **nucleoli,** where ribosomes are produced."

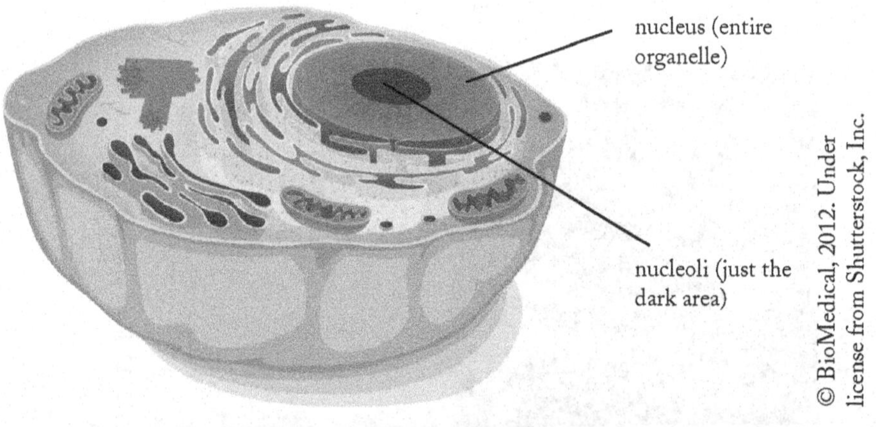

IMAGE 3.7 Nucleus and Nucleoli of an Animal Cell

"Aren't some of the organelles bound by membranes and some are not?" asked Grace.

Buzz answered, "Yes, the nucleus does have them as you have seen; the mitochondria, lysosomes, peroxisomes, endoplasmic reticulum, and Golgi apparatus have membranes as well. And the ribosomes and centrioles do not."

Will interjected, "I used to have an instructor that would compare a cell with a city. She used to state that a city has parts—like an energy plant, a garbage area, etc. and that a cell is similar in the way its parts carry out specific functions."

Buzz commented, "I guess that's a good way to think about things."

Endoplasmic Reticulum

Buzz added, "Well then, if you want to compare a cell to a city, then I guess this next structure would be like a network of super highways being built by highly skilled workers who can read blueprints and complete the detailed construction. The **endoplasmic reticulum** (also known as ER) is a system of interconnected tubules and flattened sacs that serve a variety of functions in the cell. There are two types of ER that differ in both structure and function. One type is called **rough ER** because it has ribosomes attached to it. The other type is called **smooth ER** because it lacks attached ribosomes. Typically, the smooth ER is a tubule network and the rough ER is a series of flattened sacs. The ER is very extensive, extending from the cell membrane through the cytoplasm and forming a continuous connection with the nuclear envelope."

"I think I heard something about the rough ER making phospholipids and proteins for the plasma membrane?" asked Will.

Buzz replied, "Yep—all true! And it also synthesizes the proteins that are either packaged inside other organelles or excreted outside the cell. Rough ER is mostly found in cells in the digestive system and antibody-producing cells."

Grace spoke up, "And I heard that the smooth ER is found in cells that help detoxify the body such as liver and kidney. And, also in the cells of the testes or ovaries that make steroid hormones. Smooth ER helps detoxify alcohol and other drugs."

"Right on!" answered Buzz and then questioned further, "So then, do you think your cells view alcohol as a friend or a foe?"

Grace and Will both looked at each other and laughed saying in unison, "Foe."

Buzz added, "The smooth ER also manufactures the Golgi apparatus, which we will talk about momentarily, by pinching off parts of itself. These bits of membrane add themselves to the Golgi apparatus."

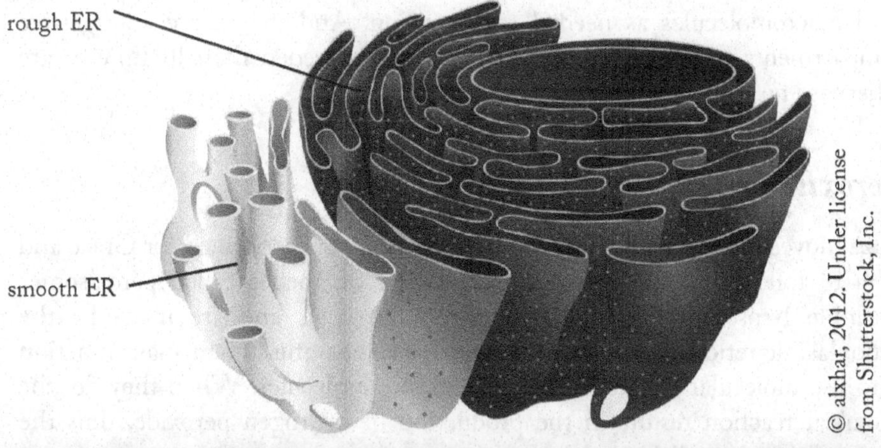

IMAGE 3.8 Endoplasmic Reticulum Showing Rough ER and Smooth ER

Ribosomes

"So we just saw the rough ER with ribosomes on it. But that's not the only place where ribosomes are found in the cell—they're also in the nucleoli, cytoplasm, and nuclear envelope. **Ribosomes** are very small granules of protein and ribonucleic acid (RNA) that assemble amino acids into proteins. In our cellular 'city' so to speak, they are workers who build proteins," explained Buzz.

Golgi Apparatus

Buzz continued, "After leaving the ER, most products are transported to the **Golgi apparatus**. It is comprised of multiple layers of flattened saclike membranes. These sacs sit one on top of the other like a stack of pancakes, and all of the sacs are interconnected. The sacs have the ability to pinch off part of themselves into small compartments or vesicles. Some vesicles become lysosomes (next section).

"When comparing a cell to a city, the Golgi apparatus is analogous to the finishing and packing room in a factory. Once the ribosome finishes manufacturing a protein in the rough ER, the protein needs to be prepared for export."

Lysosomes

Grace chimed in, "Aren't lysosomes able to digest things?"

"And don't they also assist in the death of a cell?" asked Will.

Buzz replied, "Yes and yes. These organelles are actually a package of powerful enzymes within a membrane. These enzymes are first created in the rough ER. They are packaged in a pouch called a vesicle, pinched off and sent to the Golgi apparatus. Lysosomes are responsible for the breakdown and absorption of materials taken in by the cell. White blood cells, for example, use their lysosomes to digest phagocytized bacteria. Lysosomes are also able to digest food macromolecules as needed for nutrition. And they serve as digestive compartments for cellular materials that have exceeded their lifetime or are otherwise no longer useful."

Peroxisomes

Buzz moved toward another electron microscope and motioned for Grace and Will to follow while he commented, "Other organelles called peroxisomes resemble lysosomes but contain different enzymes and are made by the endoplasmic reticulum rather than the Golgi apparatus. Their main function is to use molecular oxygen to oxidize organic molecules. When they do, the chemical reaction results in the production of hydrogen peroxide, thus the prefix of the organelle's name 'perox'."

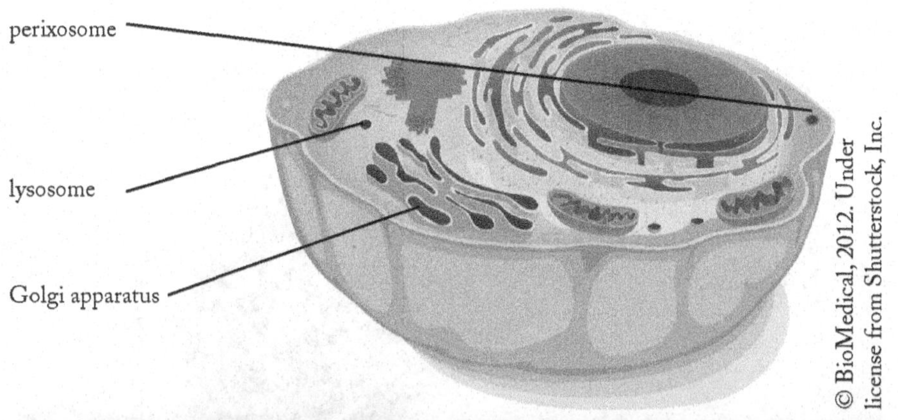

IMAGE 3.9 Lysosome, Peroxisome, and Golgi Apparatus

"I heard they neutralize drugs," exclaimed Will as he chuckled.

Buzz replied, "Right on! Including alcohol! Also free radicals, and a variety of blood-borne pathogens."

Mitochondria

"Now these organelles are the true powerhouses—check 'em out … These are the mitochondria," Buzz said enthusiastically. He continued, "They're interesting because they actively squirm and move—it's like they're really busy and they are! They extract energy from organic compounds and it is transferred to **adensosine triphosphate (ATP)** with the help of enzymes. That's our energy, ya know!"

Grace took a closer look and questioned, "It seems like they are actually different shapes … some look rod-like while others seem like spheres?"

"Yes," answered Buzz, "they do come in a variety of shapes, even thread-like and kidney-shaped."

"And what about what's in them?" asked Will.

Buzz replied, "Well, they are held together by a double-membrane much like the nucleus but the inner membrane has folds called cristae that jut across the organelle like shelves. The space between the cristae contains ribosomes, enzymes, and a small circular DNA molecule called **mitochondrial DNA**."

"So, there is a second place inside a cell that contains DNA besides the nucleus?" commented Grace.

"Yes, and it can be handy in forensics if the nucleus has been destroyed in a crime scene because you could still obtain valuable DNA evidence," answered Buzz.

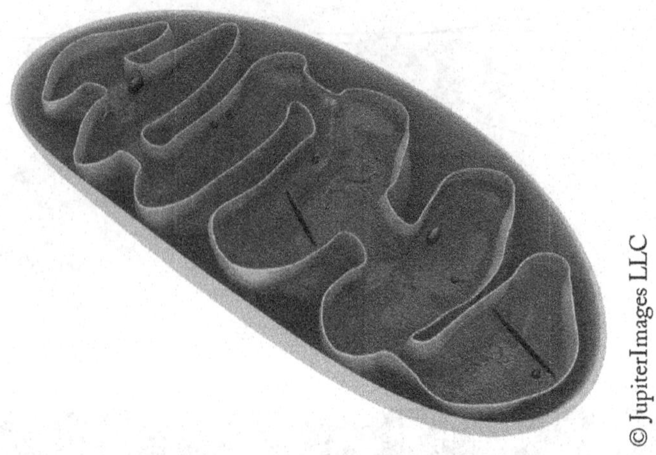

IMAGE 3.10 Mitochondria with Visible Cristae

Cytoskeleton

Buzz continued, "You can probably tell that cells aren't flat. They have a 3-D or three-dimensional shape to them. This comes from their **cytoskeleton** or system of microfilaments and microtubules that branch and tunnel throughout the cell. This system not only helps with physical support but also cellular movement and routing of molecules. The centrioles are part of the cytoskeleton as we will see next."

Centrioles

Buzz added, "If I can find a cell getting ready to grow and divide, I can show you this next particular organelle… ah, here we go. These organelles called **centrioles** are quite curious—they only appear when a cell is getting ready to grow and divide. They're like messengers to give us a clue that a cell is going to undergo cell division. You can see that they look like short cylindrical microtubules—kind of like straws. They usually show up in pairs and the two centrioles always lie perpendicular to each other surrounded by an area of clear cytoplasm called the **centrosome**."

Grace looked and commented, "They look like bundles of sticks."

Buzz replied, "Yes. Can you count the sticks or microtubules?"

Grace peered even closer and answered, "Lets see … um… I see nine bundles of three microtubules each."

Buzz cheered, "That's exactly right! That is their characteristic pattern of arrangement—we always see them exactly like that. During cell division, the centrioles move toward the poles (opposite ends) of the cell. You may also see groups of thread-like microtubules called the **mitotic spindle** (also known as **spindle fibers**) connected to them."

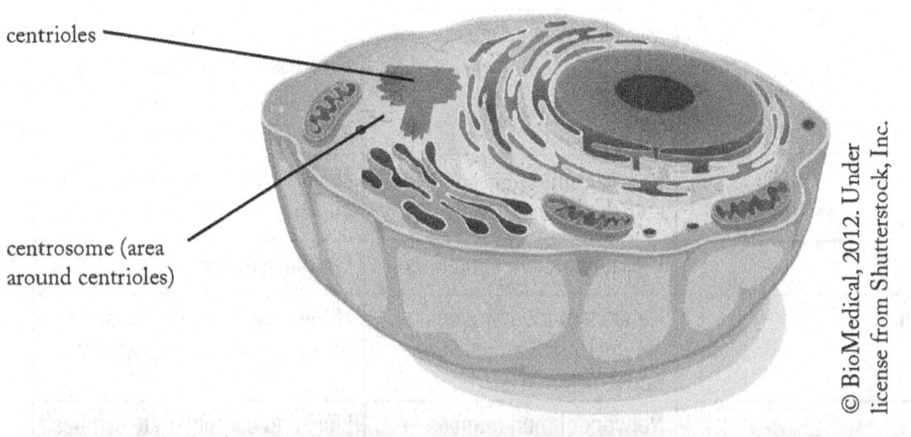

IMAGE 3.11 Centrioles and Centrosome

Self-Check—Continued

5. In your own words, describe what an organelle is.

____

____

6. As you have learned, organelles have specific jobs or functions to carry out. In the following table, practice filling in the organelle name next to its function/description.

Organelle	Composition	Function
a.	Lipid bilayer; containing phospholipids, steroids, and proteins	Isolation, protection, sensitivity steroids, and proteins support, control of entrance/exit of materials
b.	Fluid inside the cell holding all the organelles (except the nucleus); touches the cell membrane and nuclear envelope. *NOTE: Cytosol is an area that is set apart from the rest of the cytoplasm*	Includes the cytoskeleton or framework of the cell
c.	Proteins organized in filaments (long straw-like tubes)	Strength and support, movement; overall shape of the cell
d.	Non-moveable membrane extensions containing microfilaments; almost like mini-fingers	Increase surface area to facilitate absorption of extra-cellular materials

e.	Microtubule subunits	Essential for the movement of chromosomes during division (form the spindle apparatus), organize cytoskeleton
f.	Mobile membrane extensions containing microtubules	Movement of materials over cell containing microtubules surface
g.	RNA and protein structures	Protein synthesis
h.	Network of membranous channels with attached ribosomes	Modification and packaging of newly synthesized polypeptides
i.	Network of membranous channels	Lipid and carbohydrate synthesis without ribosomes attached
j.	Stacks of flattened membranes	Storage, alteration, and packaging of newly synthesized proteins
k.	Vesicles containing digestive enzymes	Intracellular removal of damaged enzymes, organelles, &/or pathogens
l.	Vesicles containing degradative enzymes	Catabolism of fats and other enzymes' organic compounds
m.	Double-membraned structures filled with metabolic enzymes	Converts 95% of organic energy with metabolic enzymes into cellular energy (ATP)
n.	Double-membraned structure containing DNA	"Boss" of the cell; storing and processing of genetic material

Cell Growth and Division

Looking up from the microscope, Buzz questioned, "Speaking of cell growth and division, what do you already know about it?"

Will quipped, "Isn't it known as **mitosis**?"

Buzz quickly answered, "Yes."

Then Grace added, "And isn't it really nuclear division because the DNA of the nucleus, which could also be thought of as the chromosomes, has to divide in such a way that each new cell ends up with the same amount of DNA as the original cell?"

Buzz shook his head in an agreeing manner, then probed further, "Do you know how many new cells you get from the old one?"

IMAGE 3.12 Buzz's Response

Both Grace and Will looked at each other but shrugged their shoulders indicating that they did not know. Buzz held up two fingers and they smiled at his simple answer.

THE STRUCTURE OF DNA

Will was happy to finally be talking about mitosis and DNA and piped up, "I have had a burning question and I'd like to ask it now… can we actually see DNA under the electron microscope?"

Buzz retorted quickly, "Yes, you can! The electron microscope is powerful enough to see it. And there is a lot of DNA packed down into each nucleus. It is tightly coiled and compacted. There are 46 molecules of DNA in human body cells totaling 2 meters (m) in length. Lets take a closer look at DNA."

Buzz continued, "At the molecular level, a DNA molecule resembles a spiral staircase and is often described as a double helix. It is made up of units called **nucleotides** (new-clee-oh-tides). A nucleotide is made up of three things:

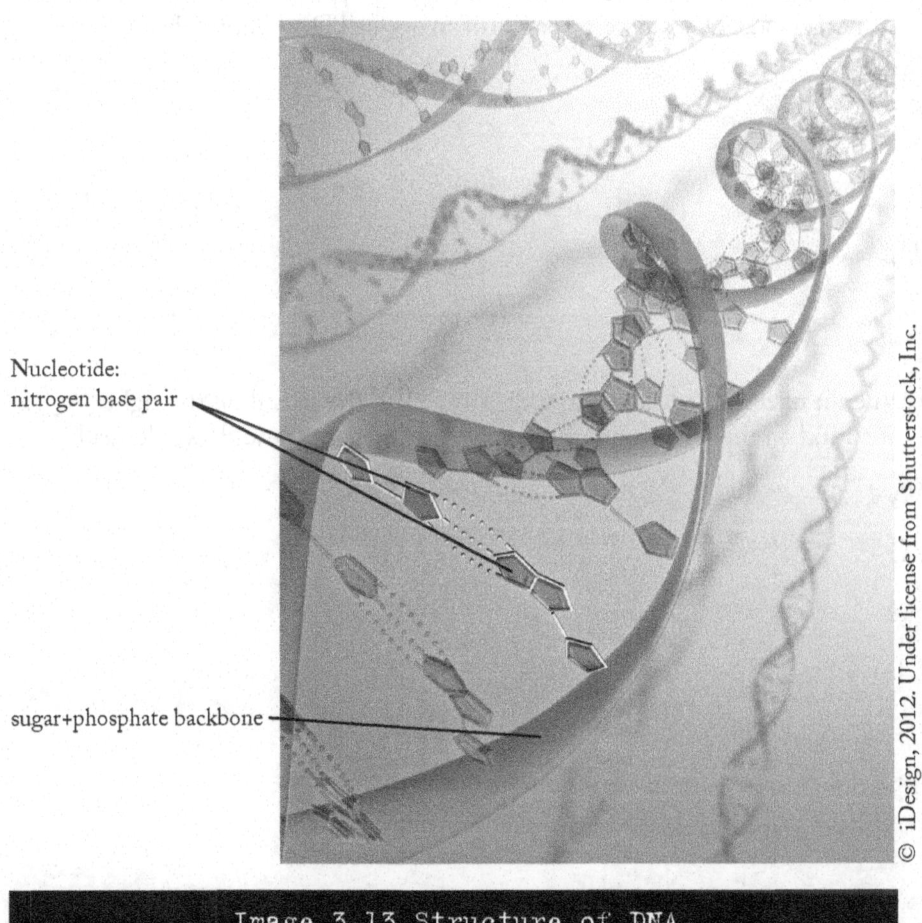

Image 3.13 Structure of DNA

(1) a sugar (an example of the sugar found in DNA is deoxyribose)

(2) a phosphate group

(3) and a nitrogenous base (examples of the four nitrogenous bases found in DNA are adenine, represented by the capital letter "A"; guanine, "G"; cytosine, "C"; and thymine, "T")

"Each sidepiece of the staircase is a backbone composed of phosphate groups alternating with the sugar deoxyribose. Further, the nitrogenous bases face the inside of the helix and hold the two sidepieces together with hydrogen bonds. The step-like connections or 'steps' are pairs of nitrogenous bases. The base pairs always group together in a predictable pattern of adenine (A) with thymine (T) and guanine (G) with cytosine (C)."

A-T and C-G

"The fact that A always binds with T and C always binds with G is called **complimentary base pairing**. Therefore, we can predict base sequences. During the initial start of mitosis, the DNA molecule must make an exact replica of itself—think of it as needing to copy itself on a copier machine. In order to accomplish this task, it unzips itself into two strands. Each strand serves as a template upon which new base pairs are formed. An enzyme called **DNA polymerase** "reads" an intact DNA strand as a template and uses it to synthesize the new strand. For example, a portion of an original DNA molecule might read:"

A-T

C-G

G-C

T-A

T-A

G-C

When it unzips, the leading strand (or strand on the left) would then read:	And the strand on the right (or lagging strand) would read:
A -	- T
C -	- G
G -	- C
T -	- A
T -	- A
G -	- C

Will questioned, "I once heard someone say that each of us has enough DNA to reach from Earth to the Sun and back. Have you ever heard that before?"

Buzz replied immediately, "I have. Mathematically we can substantiate an even greater distance than that! Humans have been estimated to have approximately 50 trillion cells, which works out to be about 100 trillion meters of DNA. Considering the fact that the Sun is 150 billion meters away, we actually have enough DNA to go from the Sun and back **300** times."

"Wow, that's amazing!!" Grace shrieked.

But Will was perplexed and questioned further, "How can that be? That actually seems impossible."

Buzz continued, "The answer lies in thinking about the fact that certain proteins compact chromosomal DNA known as **chromatin** into the microscopic space of a nucleus. Under an electron microscope chromatin appears as tangled, folded spaghetti inside the nucleus. The chromatin is actually made up of units called **nucleosomes**, which contain nine histone proteins and about 166 base pairs of DNA. The histone proteins, or **histones** as they are commonly called, assist in folding the chromatin thus helping it to be more compact. This way a lot of chromatin can fit into the microscopic nucleus, much like coiling up a garden hose will make it easier to store in a small space."

Will asked, "What are chromosomes then?"

Buzz replied, "When a cell is preparing to divide, it makes an exact copy of all of its DNA. The chromatin also begins to condense and becomes more visible. With an electron microscope, it is quite clear that the chromatin it is actually 46 long filaments or **chromosomes.** Each chromosome consists of two genetically-identical strands known as **sister chromatids** (only after replication). They are held together at a pinched spot called the **centromere**. On each side of the centromere lies a protein plaque known as the **kinetochore** (kih-NEE-to-core)."

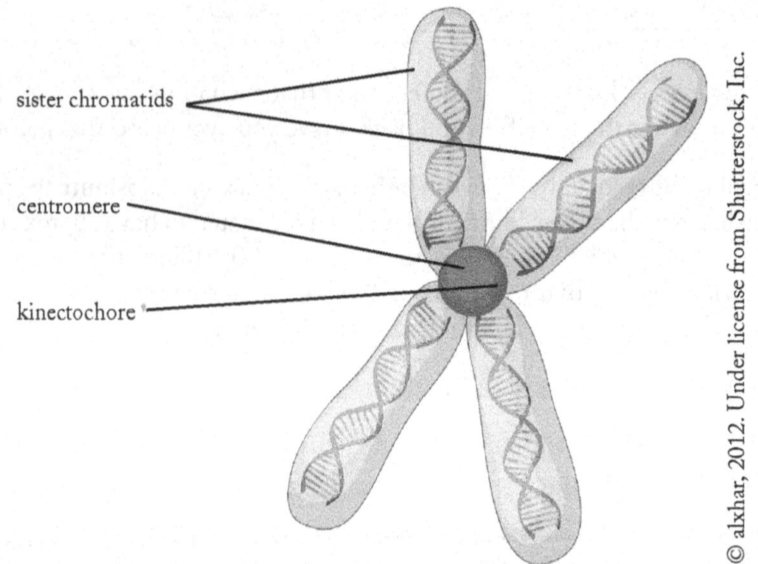

IMAGE 3.14 Chromosome

Self-Check—Continued

7. You try it! What would be the base sequence of the DNA strand across from TTAGCG?

8. Try another! How about GCCAAT?

9. Lets review…In your own words, what does DNA stand for?

10. What is mitosis? How many cells does it result in?

____

____

11. Is DNA the same as chromosomes? Explain your answer.

____

____

12. How many chromosomes do we humans have in our body cells?

____

____

13. Buzz mentioned that each cell has about 2 meters (m) of DNA coiled tightly in the nucleus. If a meter (m) equals 3 feet, how many feet would that equate to?

____

____

14. There is one specific type of cell in our bodies that purposefully does not have 46 chromosomes. Can you think of what that might be? If you can, name it. *NOTE: We will talk about this later in level two Anatomy & Physiology (Biol 2302).*

____

____

GENES

Grace questioned, "So, I'm just curious… what are genes exactly? Don't they have some kind of connection to DNA somehow?"

Buzz answered, "The essential function of DNA is to be a set of instructions, much like reading a set of blueprints, so that proteins can be made. These instructions can also be thought of as a **gene** or a segment of DNA that serves as a recipe on how to build a protein molecule such as ribonucleic acid (RNA). We will talk more about RNA shortly."

Grace replied, "Makes sense for now."

Buzz added, "Every one of us has approximately 20,000-25,000 genes. All of the genes from one person is a **genome**. You may have heard of the Human Genome Project? The goal of this enormous international study conducted between 1990 and 2003 was to accurately map all of the genes in the human body, which has pretty much been accomplished. Well, at least 99% has been mapped. And, we learned some important things we didn't know including the idea that genes can create millions of different proteins. We used to believe that one gene created just one protein but that idea is dead now. We also discovered that one gene averages about 3,000 bases in length, although they can be as long as 2.4 million bases. Finally, we discovered the chromosomal locations of over 1,400 disease-producing mutations. Prior to the project, we knew fewer than 100."

"Wow!" exclaimed both Grace and Will at the same time.

Buzz added, "Yeah, it is pretty incredible."

RNA

Buzz thought for a moment and then commented, "It would be easy to just concentrate on DNA and the nucleus but we have also been talking a lot about genes, proteins and nucleic acids. And when I think of proteins, I think of the small units that they are made up of called **amino acids**. And, I also think about one specific organelle that we talked about earlier in our visit. An organelle that is responsible for making proteins. Can you name it?"

Grace spouted, "Do you mean ribosomes?"

Buzz was tickled that she remembered and belted out, "Right-t!" as he suppressed his laugh. He continued, "Ribosomes have to get their orders from the nucleus because, after all, it is the 'boss' of the cell. So, there is a special macromolecule that is a close cousin to DNA called ribonucleic acid or **RNA.** RNA differs from DNA in several ways including it is a much smaller molecule than DNA, its sugar is ribose rather than deoxyribose, RNA is a single chain rather than a double-helix, and RNA does not contain the nitrogen base thymine but uracil (U) which takes its place."

Buzz continued, "**RNA** helps carry this information from the genes to where proteins are assembled on ribosomes out in the cytoplasm. This is done by **messenger RNA** (mRNA) which acts like a relay team. First, through the process of **transcription**, the sequence of base pairs is transcribed from DNA by an enzyme known as RNA polymerase. Next, the mRNA moves from the nucleus to the ribosomes in the cytoplasm to form proteins. The mRNA actually translates the sequence of base pairs into a sequence of amino acids to form proteins. This process is called **translation**."

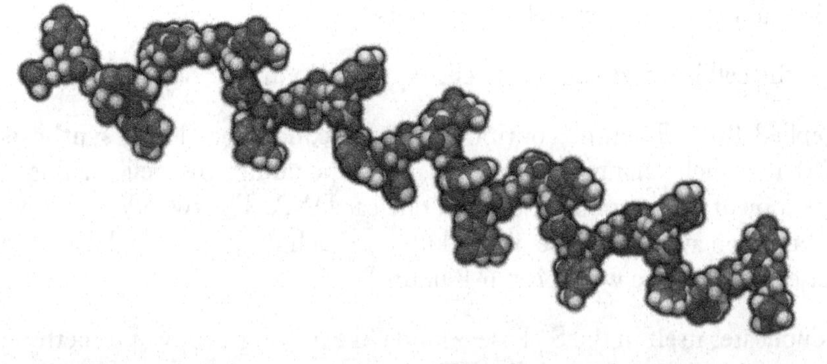

IMAGE 3.15 Structure of mRNA

Grace interjected, "Aren't there different types of RNA?"

Buzz answered, "Very good. Yes, there are actually three different types." Then he continued, "Besides mRNA, there is also **ribosomal RNA** (rRNA) which is RNA that is a structural component of a ribosome and **transfer RNA** (tRNA) which is involved in protein synthesis in that it brings the correct amino acid to the ribosomes."

DNA REPLICATION AND THE CELL CYCLE

Buzz perked up and exclaimed, "There's one more thing I want to share with you before you leave today. Remember earlier when we mentioned mitosis?"

Grace and Will both nodded their heads and said in unison, "Yes."

Buzz continued, "Well, most cells periodically divide into two daughter cells. In order to accomplish this, a cell goes through a series of divisions called the **cell cycle**. Not every cell has the same length of cell cycle—some cells, like skin and intestinal cells, divide and grow quickly while others are very slow, like cartilage cells. Still other types of cells may not divide at all, like most types of nervous tissue."

Grace was puzzled and questioned, "So some types of cells don't repair themselves?"

"That's exactly what I am saying. Take cardiac cells for example. Let's say someone has a massive heart attack and the cells in a large area of the heart die off; this could result in permanent and irreversible damage to the heart," explained Buzz.

Buzz continued, "Mitosis is a phase of the cell cycle. But proceeding mitosis, a few other phases must happen. In the first phase known as **first gap phase (G1)**, a cell is pretty much just being itself—making proteins, growing, and carrying out basic tasks. Lets say the total length of a cell's cycle is 24 hours… then G1 would last approximately 8-10 hours."

"Okay, so the cell is just doing what cells typically do in G1?" asked Will.

"Yes," replied Buzz. Then he continued, "The second phase is the **synthesis phase (S)** in which changes begin to happen including: the cell making a duplicate copy of its centrioles and all its nuclear DNA. The two identical sets of DNA are then available to be divided up into each daughter cell during the next phase. The S phase would last 6-8 hours."

"DNA duplicates itself in the S phase—kinda like making a copy of something on a copy machine?" questioned Grace further.

"Yes," replied Buzz again. Then he added, "And the third phase is known as the **second gap phase (G2)** in which the cell finishes replicating its centrioles and makes enzymes that control cell division. In this phase, soundness of the DNA replication is checked and errors are repaired, if detected. G2 is fairly brief and would last about 4-6 hours. Phases G1, S, and G2 are collectively known as **Interphase**. Finally, mitosis (M) begins and the cell actually replicates its nucleus and then pinches in to form two new daughter cells." Buzz took out a pen and paper and began drawing. He commented, "You can think about the phases like a pie that is divided up into slices."

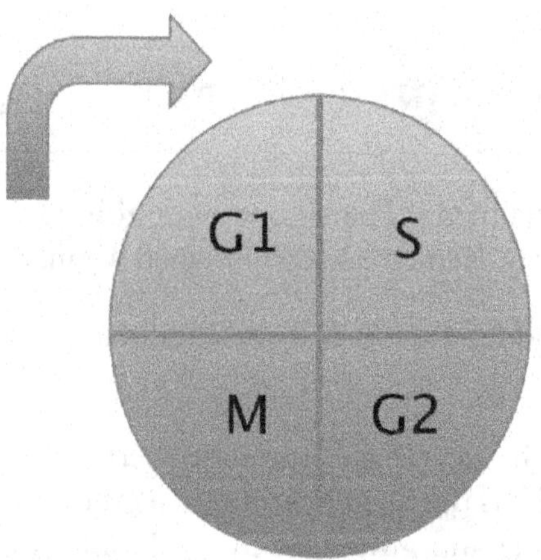

IMAGE 3.16 Cell Cycle

Will said, "I think I know the stages of mitosis. Is it prophase, metaphase, anaphase, and telophase?"

"That's right!" exclaimed Buzz. He continued to draw and added, "Now lets take a closer look at each stage of mitosis, shall we?"

Interphase (not part of mitosis)

Centrioles appear. An aster appears around them. The aster is the beginning of what will eventually elongate into microtubule spindle fibers. The nucleus is intact with chromatin visible.

Prophase (beginning of mitosis)

Nucleus is breaking down. Chromatin has condensed into visible chromosomes held together by centromeres. Centrioles are moving apart. Spindle is developing.

Metaphase

Chromosomes are lined up at the center of the cell. Spindle is across the entire cell and is actually attached to each chromosome pair via the centromeres.

Anaphase

An enzyme called separase triggers chromosome pairs to begin pulling apart resulting in single strands of genetic material called sister chromatids. 50% of DNA is moving to the north pole of cell and 50% toward the south pole.

Telophase

The cell is beginning to pinch in the middle (to form two new cells)—this is called cytokinesis. The spindle fibers will vanish. A nucleus will begin re-forming around the sister chromatids in each cell.

Two New Daughter Cells

Each new body cell is identical to the original cell, meaning 46 chromosomes.

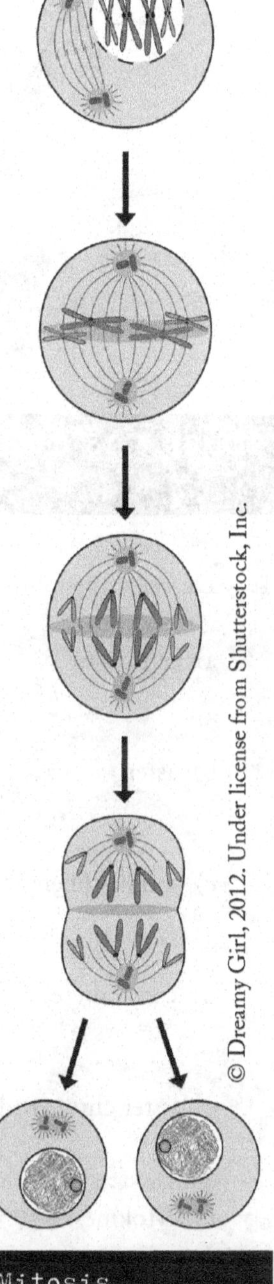

© Dreamy Girl, 2012. Under license from Shutterstock, Inc.

IMAGE 3.17 Cell Cycle with Mitosis

Buzz continued, "Now let's look under the microscope and see the cells in different phases. Which stages can you identify?"

Self-Check—Continued

__________ 15.

__________ 16.

__________ 17.

__________ 18.

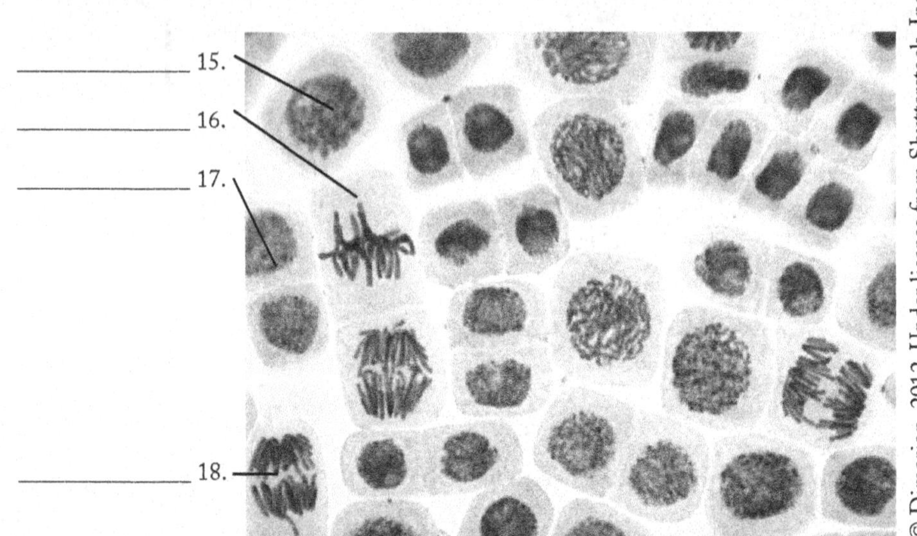

IMAGE 3.18 Check cells in Different Stages of the Cell Cycle

19. Describe the following terms:

    a) centrioles

    b) aster

    c) spindle fibers

    d) centromere

    e) sister chromatids

    f) cytokinesis

20. Draw a cell in the following phases: Interphase (okay to represent with one drawing) and Mitosis (include all four stages of mitosis). Label the appropriate phases with the terms centrioles, aster, spindle fibers, centromere, sister chromatids, and cytokinesis.

CLINICAL APPLICATION: GENOMICS

Many diseases that we face today are hereditary in nature. Take Alzheimers and hemophilia (a clotting disorder), for example, just to name a few. What if there was a genetic test that could determine whether you had the genetic likelihood of developing a certain disease? Would you want to know?

There is a branch of medicine now called *genomic medicine* that uses our knowledge of the human genome to predict, diagnose, and treat diseases. For less than $1,000, you can have your genome scanned for markers indicating disease risk. This could allow for earlier detection and possibly earlier clinical intervention. Knowing about such disease risks and defective genes, expands the potential for future gene-substitution therapies where defective cells may be removed and new genes inserted into them before being reintroduced to your body. The hope being that the new cells would grow and provide whatever was originally lacking. Sound too good to be true?

Like anything, there is a flip side. Once you know your genetic profile, would it strictly be held in confidence between you and your doctor? Or would your insurance company also be privy to the information so they can adjust your premiums to cover catastrophic illnesses? What if they then denied you coverage altogether? What about prospective employers—should they have a right to know about your genome before offering you employment? There are a lot of unanswered ethical questions to be sure. Lets start with your own personal view.

Clinical Reflection Questions

Complete these questions in conjunction with the Clinical Applcations box on Genomics page 77.

1. What type of hereditary diseases (such as cardiovascular, cancer, or Alzheimers) typically run in your family? Would you like to avoid them? Explain your answer.

____

____

____

2. If cost were not an issue, would you want to be screened for these genetic diseases to possibly determine if you have the genetic profile to develop them? Why or why not? Explain your answer.

____

____

____

3. Do you believe an insurance company should have the right to know the outcome of a person's genetic test? What about a prospective employer? Explain your answer.

____

____

____

ERRORS, MUTATIONS, AND CANCER

Gene **mutations** (a change in the normal genes of an organism) have varying effects on health, depending on where they occur and whether they alter the function of essential proteins. The types of mutations include:

Missense mutation

This type of mutation is a change in one DNA base pair that results in the substitution of one amino acid for another in the protein made by a gene.

Nonsense mutation

A nonsense mutation is also a change in one DNA base pair. Instead of substituting one amino acid for another, however, the altered DNA sequence prematurely signals the cell to stop building a protein. This type of mutation results in a shortened protein that may function improperly or not at all.

Insertion

An insertion changes the number of DNA bases in a gene by adding a piece of DNA. As a result, the protein made by the gene may not function properly.

Deletion

A deletion changes the number of DNA bases by removing a piece of DNA. Small deletions may remove one or a few base pairs within a gene, while larger deletions can remove an entire gene or several neighboring genes. The deleted DNA may alter the function of the resulting protein(s).

Duplication

A duplication consists of a piece of DNA that is abnormally copied one or more times. This type of mutation may alter the function of the resulting protein.

Frameshift mutation

This type of mutation occurs when the addition or loss of DNA bases changes a gene's reading frame. A reading frame consists of groups of 3 bases that each code for one amino acid. A frameshift mutation shifts the grouping of these bases and changes the code for amino acids. The resulting protein is usually nonfunctional. Insertions, deletions, and duplications can all be frameshift mutations.

Repeat expansion

Nucleotide repeats are short DNA sequences that are repeated a number of times in a row. For example, a trinucleotide repeat is made up of 3-base-pair sequences, and a tetranucleotide repeat is made up of 4-base-pair sequences. A repeat expansion is a mutation that increases the number of times that the short DNA sequence is repeated. This type of mutation can cause the resulting protein to function improperly."

Buzz continued, "Most mutations are harmless while others drive the cell into steady or uncontrolled cell division, which can lead to tumor development. Tumors that grow but do not spread are benign or non-cancerous. Tumors that spread or metastasize are **cancerous**. Perhaps long-term exposure to the sun's

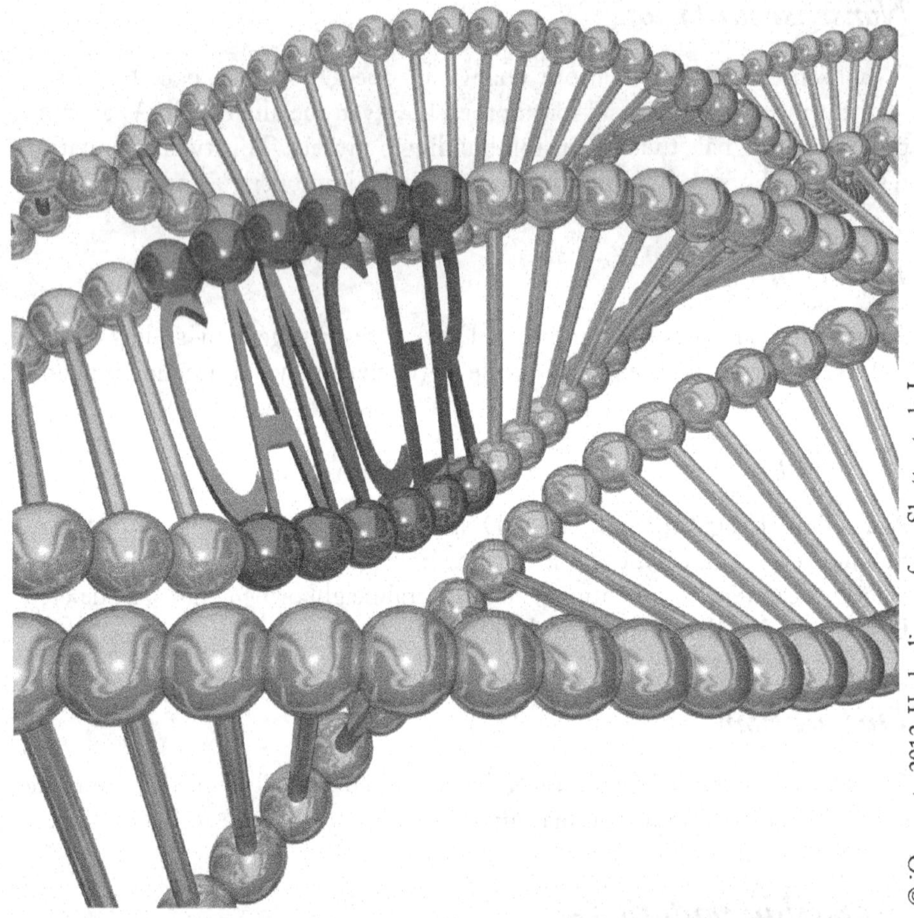

IMAGE 3.19 Cancerous DNA

radiation has caused a mutation in a gene in a skin cell. Or, the cell has just divided and it made a mistake when it copied its DNA."

"So, just one mistake or mutation can lead to cancer?" asked Grace.

"No, not typically. There isn't just one mutation that causes cancer. Usually lots and lots of DNA changes have to occur before a cell begins growing uncontrollably. And once a DNA mutation has happened, more tend to happen so that you end up with lots of DNA changes," replied Buzz.

"I've heard it said that every one of us develops cancer cells in our bodies every day but not everyone gets cancer. Have you heard that?" questioned Will.

"Yes I have and it's a true statement. While DNA polymerase is fast and accurate, it can make mistakes. For example, it might read C but places a T across from it. If nothing were done to correct it, future generations of cells would have thousands of faulty proteins accidentally coded for by DNA that had been miscopied. To prevent these mistakes, thankfully there are multiple ways of correcting replication errors. The DNA polymerase itself double-checks new base pairs and can replace an incorrect or unstable new base pair with a more accurate one. Our bodies actually have a very high degree of replication accuracy resulting in only one mistake for every billion base pairs replicated.

"If lethal mistakes still get through, hopefully your body's immune system is strong enough to detect and destroy the mutated/cancerous cells," commented Buzz.

"So, in summary, depending on how a particular mutation modifies an organism's genetic makeup, it can prove harmless, helpful, or even hurtful" stated Grace.

Self-Check—Continued

21. In your own words, describe cancer formation in cells.

____

____

____

22. What are some ways the cell catches or corrects mistakes? Explain your answer.

____

____

____

WRAP-UP

Grace glanced at the clock on the lab wall and commented, "Buzz, we realize we're at the end of our appointment time. We so appreciate the tour of the lab, information of the different types of microscopes, and getting to see all the organelles."

Will added, "And thank you for helping us understand DNA and cell division!"

"It was great to meet with you both. Keep your interest going in cells!" Buzz concluded as he walked them out.

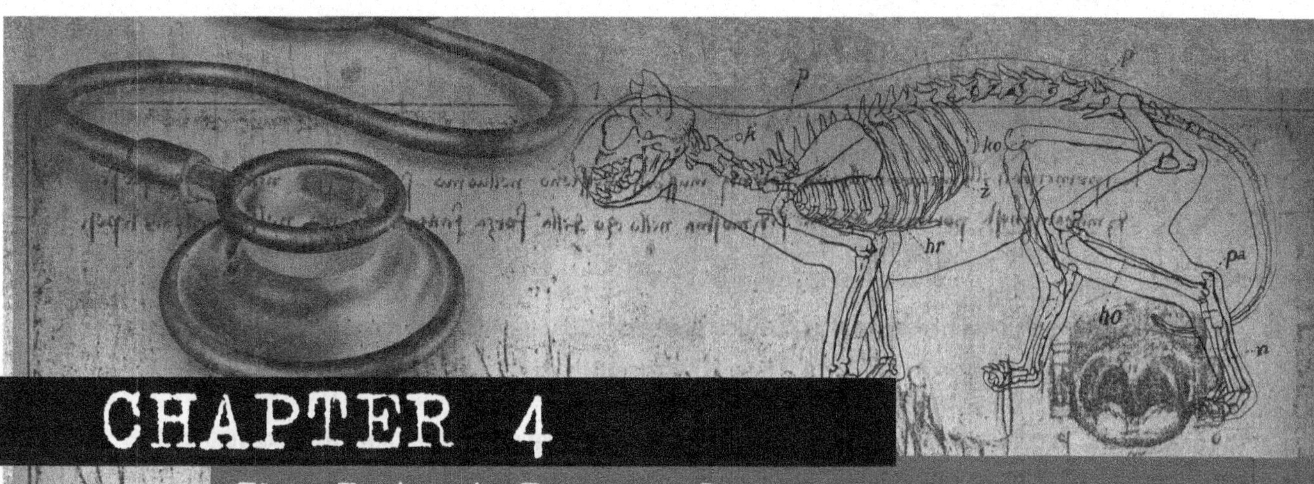

CHAPTER 4

The Latest Tissue Discoveries: Are We Really Re-growing Human Limbs?

KEY WORDS

The study of Anatomy & Physiology involves many new vocabulary words. It is helpful to gain familiarity with these new key words, just as you would a foreign language.

adipose

basement membrane

brown fat

cardiac muscle

chondrocytes

collagen

columnar

compact bone

connective

cuboidal

dense connective

elastic cartilage

endocrine glands

epithelial

erythrocytes

exocrine glands

extracellular matrix

fibroblasts

fibrocartilage

fibrous connective

glial cells

goblet cells

histology

hyaline cartilage

intercalated discs

keratin

leukocytes

loose connective

macrophages

mast cells

muscle

muscle fibers

myocytes

nervous

neurons

osteocytes

platelets

pseudostratified

reticular tissue

simple epithelial

skeletal muscle

smooth muscle

spongy bone

squamous

stratified columnar

stratified cuboidal

stratified epithelial

stratified squamous

striated

tight junctions

tissue

transitional epithelial

It is helpful to have an idea of what you need to learn before proceeding. Here is a study guide to assist you:

1. Explain the field of histology.
2. Describe tissue.
3. Explain the four main classes of issue, their basic function, and give examples of their locations.
4. Describe the matrix.
5. Explain the characteristics of epithelial tissue.
6. List the major types of epithelium, distinguish them from each other and give examples of where each type can be found in the body.
7. Describe the special cells found in fibrous connective tissue.
8. Explain the significance of adipose tissue; include white and brown fat in your answer.
9. Describe the three types of cartilage and give examples of where they are located in the body.
10. What are chondrocytes? Osteocytes?
11. Explain the difference between compact and spongy bone.
12. Explain the three main types of blood cells and what they do.
13. Explain the role of neurons and glial cells in nervous tissue.
14. Describe the three main types of muscle and where they are located in the body.
15. Distinguish between endocrine and exocrine glands.

Grace and Will, on special assignment from their college Anatomy & Physiology professor, have been challenged to visit the McGowen Institute for Regenerative Medicine to talk with Dr. Barr so they can learn more about tissues, including growing new limbs where they were once severed and repairing massive wounds that typically result in amputation.

"Hello, I understand you are here to learn about tissues, including more about what we actually do here to regenerate them?" said Dr. Barr with a friendly smile.

"Hi, yes, we're very curious," replied Will.

"We heard something about a man getting his finger accidentally chopped off and something about some kind of powder that was sprinkled on it and he regrew the entire end of his finger back in about a month. Is that really true?" questioned Grace with intensity.

"Oh yes, it's true! Follow me and we'll head down to the main research area," answered Dr. Barr.

HOW WE STUDY TISSUES

Dr. Barr continued, "This is the histology lab. Maybe you have heard the term histology before? Do you know what it means?"

"Uh, cell study?" replied Will.

"Not exactly," answered Dr. Barr.

"I think tissues and how they form organs," interjected Grace.

"And you'd be right, Grace!" affirmed Dr. Barr. He continued, "Tissues are basically a group of similar cells doing a similar job together. They work together to perform a specific structural or physiological role. There are four main classes of tissue—epithelial, connective, nervous, and muscular. Here's a flyer that I keep handy that may help you remember."

TABLE 4.1 The Four Main Classes of Tissue

Type	Definition	Locations
epithelial	Sheets of closely knit cells that cover organs, form glands, and typically serve to absorb, secrete, or protect	• glandular tissue (examples include sweat glands and salivary glands among others) • linings of cavities (such as mouth, blood vessels, heart and lungs) • germinal epithelial (ovaries and testes) • uterus and Fallopian tubes • epidermis
connective	Holding things together; support, bind, and protect	• ligaments and tendons • bone and cartilage • blood • adipose • spleen
nervous	Contains excitable cells; communication, rapid transmission	• brain and spinal cord • nerves
muscular	Tissue made up of muscle cells	• skeletal, cardiac, and smooth muscle

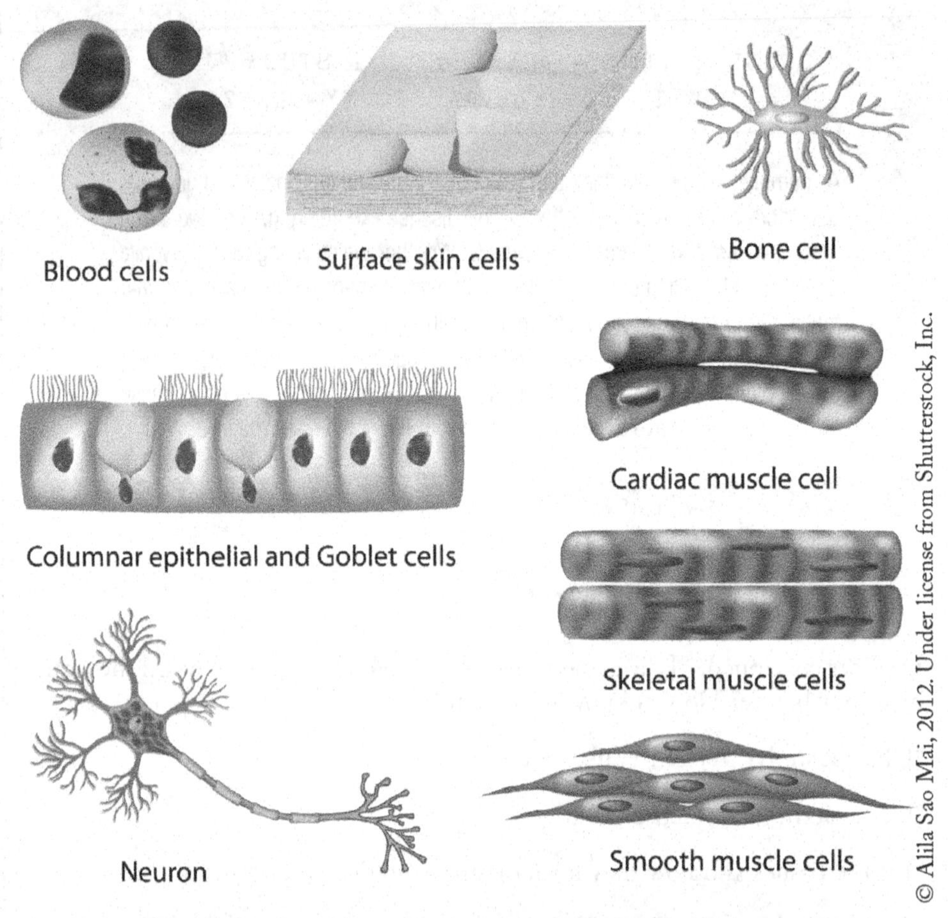

IMAGE 4.1 Examples from the Four Main Tissue Classes

Dr. Barr expounded by adding, "Tissues, like connective and epithelial, have a matrix of extracellular material around them. This **extracellular matrix** is highly regarded in tissue regeneration and is comprised of **collagen** fibers (i.e., collagen is the most abundant protein in the body forming the fibers of many connective tissues including the dermis and tendons) and ground substances including water, gases, minerals, hormones, and other substances. This is the material from which many cells get their oxygen and nutrition, and release their wastes and/or hormones into. Have you heard about how the extracellular matrix from pigs' bladders are helping humans to regrow limbs? Check out the latest research!"

CLINICAL APPLICATION — THE MATRIX OF PIG BLADDERS HELPING TO REGROW HUMAN LIMBS

Have you ever noticed that when a lizard or a salamander looses its tail, it regrows another one? Have you ever wondered why humans can't seem to regrow a severed limb? Well, they couldn't until recently, thanks to a little help from pig bladders.

Bizarre as it sounds, it all started back in 2008, when a researcher was working with extracellular matrix from the bladder of pigs to try and repair strained ligaments in horses, when he got an emergency call from his brother who had been involved in a bad accident and lost the entire end of his finger.

Rather than follow the conventional route of trying surgery and attempting to reconnect the finger and all the associated tissues (bone, tendons, muscle, blood vessels, etc.), he decided to try the extracellular matrix on himself, although no known human trials had ever been done. So, he obtained some of the extracellular matrix powder, also nicknamed "pixie dust," for his brother, sprinkled it on his finger daily, and bandaged it. After just a week, he noticed major changes and the finger actually started to regrow (all of the missing components including muscle, bone, nerves, blood vessels, etc). At the end of just 4 weeks, the tip had completely regrown including the fingernail and it had complete feeling and functionality! Essentially, the powder directed all the tissues to grow fresh tissue instead of forming a scar.

It is thought that the dust kick-starts the body's natural healing process by sending out signals that mobilize the body's own cells into repairing the damaged tissue.

More recently, the extracellular matrix from pig's bladders has been used to regrow human bladders and is set to revolutionize the treatment of larger limbs. In fact, in 2010, it was used effectively on a returning soldier from Afghanistan whose upper thigh had been blown away from explosives. The extracellular matrix was used on him to actually re-grow his entire upper thigh including muscle, nerve and bone tissue.

IMAGE 4.2 Pig

Clinical Reflection Questions

1. What is the source of extracellular matrix that is being used for human tissue regeneration?

____

____

2. What are some of the latest accomplishments in the field of human tissue regeneration?

____

____

____

"Wow, that is truly amazing!" exclaimed Grace.

"That's totally cool. It beats anything I've ever heard," commented Will.

"So, I'd like to tell you a little more about tissues now," added Dr. Barr.

EPITHELIAL TISSUE

Dr. Barr elaborated, "**Epithelial** tissue grows and repairs itself quickly as either a single sheet (simple epithelial) or multiple layers (stratified epithelial). You can think about it like dessert—a lemon bar is a single-layer dessert but strawberry shortcake is stratified!"

Simple Epithelial (one layer sheet)

Stratified Epithelial (multiple layers or sheets)

© Mopic, 2012. Under license from Shutterstock, Inc.

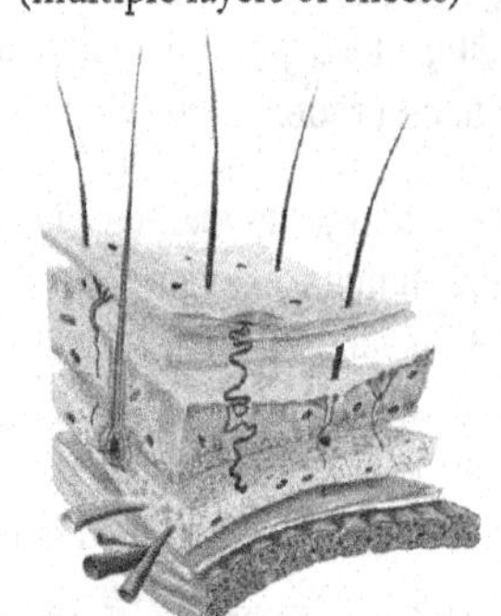

© hkannn, 2012. Under license from Shutterstock, Inc.

© Anna Hoychuk, 2012. Under license from Shutterstock, Inc.

© cdrin, 2012. Under license from Shutterstock, Inc.

IMAGE 4.3 Simple Versus Stratified Epithelium

FUNCTIONS OF EPITHELIAL TISSUE

Dr. Barr continued, "It has many functions including protection (i.e., the epidermis protects against invading bacteria and infection), secretion (it produces mucous, sweat, oil, enzymes, hormones, etc.), excretion (voids waste products such as CO_2 across the air sacs of the lungs), absorption (absorbs nutrients from the lining of the small intestine), filtration (urinary waste is filtered through the kidney), and sensation (nerve endings allow us to feel a touch on our skin)."

SPECIAL CHARACTERISTICS OF EPITHELIUM

Dr. Barr continued, "Epithelial tissue lies on a **basement membrane**. That means that underlying the cells that are the cellular component of the epithelial tissue, there is a layer of acellular ('a-' means not, so 'acellular' means not cellular) material. This basement membrane can be thought of as a sticky layer to keep the epithelial cells attached to whatever underlies them. The bottom edge of the epithelial tissue abuts the basement membrane; this bottom edge is called the basal surface. The edge of the epithelial tissue that faces the lumen (or the outside world) is called the apical surface."

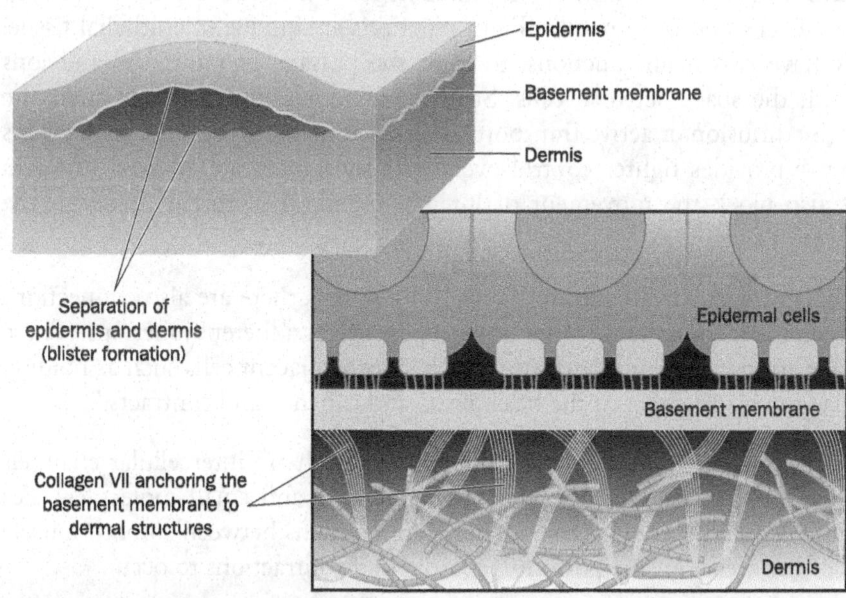

IMAGE 4.4 Relationship of Basement Membrane to Epidermis and Dermis

He continued, "There are also other characteristics of epithelial tissue in general:

"Cells within this tissue readily divide to make more cells. This helps this tissue recover after any sort of abrasions occur.

"This tissue does not have any vasculature. This means that there are no blood vessels within it. This should make sense, since epithelial tissue is likely to get damaged by material moving against it—and you don't want to bleed every time something bangs into your skin or every time you swallow something rough."

SPECIAL CONNECTIONS BETWEEN CELL MEMBRANES

Dr. Barr continued, "The cells within epithelial tissue are firmly attached to each other. As a border-tissue, if the cells weren't adherent to one another, it would be a leaky border."

Grace chuckled and commented, "Yuck!—liquids from inside of us would drip out!"

Dr. Barr added, "So the cells have special junctions with each other called **tight junctions** to help with this. Tight junctions form the closest contact between adjacent cells known in nature. Tight junctions are unique to epithelial tissue. They have two main functions: to limit the passage of molecules and ions through the space between cells. So most materials must actually enter the cells (by diffusion or active transport) in order to pass through the tissue. This pathway provides tighter control over what substances are allowed through; they also block the movement of integral membrane proteins between the surfaces of a cell."

Dr. Barr continued, "In addition to tight junctions, there are also connections between the membranes in other tissue types such as adherens junctions. These provide strong mechanical attachments between adjacent cells such as holding cardiac muscle together as the heart beats and expands and contracts."

Dr. Barr continued, "There are gap junction which are intercellular channels that permit free passage between cells (i.e., ions and small molecules). An example would be as birth approaches, gap junctions between smooth muscle cells of the uterus enable powerful, coordinated contractions to occur."

Dr. Barr concluded, "And finally - desosomes (which are found in simple epithelial and stratified squamous epithelial tissues) are localized patches that hold two cells together (e.g., intermediate filaments of keratin in cytoplasm).

"I heard simple epithelial comes in three basic shapes. Is that true?" asked Grace.

"Absolutely" answered Dr. Barr. He bridged the conversation with, "The first three types of simple epithelial tissue are named for their shapes. It is important to keep in mind that cells like these are three-dimensional rather than simply flat. **Squamous epithelial** is thin and scaly, **cuboidal epithelial** is very square-like, and **columnar epithelial** is tall and thin like columns on the front of a building."

Dr. Barr added, "There is another type of simple epithelial tissue worth mentioning called **pseudostratified columnar**. It deceptively looks like two layers but it is not. All cells are in one sheet but the cells are different heights, like how trees are different heights in a forest but all anchored to the same forest floor. Both columnar and psudostratified epithelial often have wine-glass shaped cells which we call **goblet cells**. Keeping in mind that goblet cells are shaped like a wine glass; they appear to have a narrower stem that reaches down into lower basement layers and an expanded end, like the glass, that is filled with extra secreting vesicles that emit mucous and absorb water."

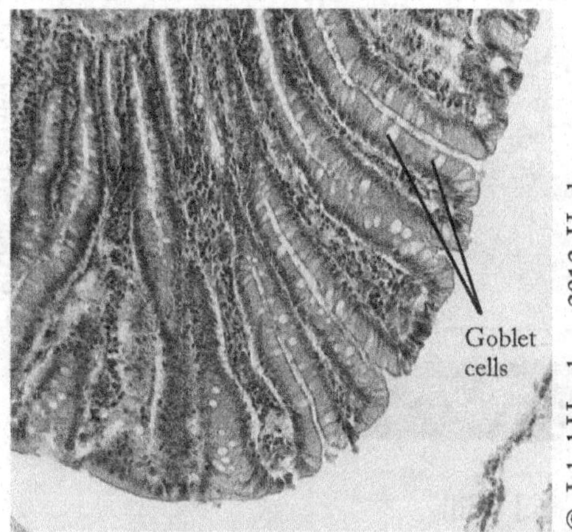

Goblet cells resemble wine glass shape   Wine glass has goblet shape

IMAGE 4.5 Goblet Cells Typically Found in Columnar and Pseudostratified Epithelial

MORE ABOUT STRATIFIED EPITHELIAL TISSUE

Dr. Barr continued, "Besides simple epithelial tissue, you'll often see different types of stratified epithelial in the body. Stratified epithelial tissue can have as little as two layers or twenty or more. Stratified epithelial is also named for its shapes just like simple epithelial. There is **stratified squamous, stratified cuboidal,** and **stratified columnar**. There is also **transitional epithelium**. The most widespread type of stratified epithelial tissue is stratified squamous and it comes in two varieties—keratinized (hard, cornified or horny as in the horns of animals) and non-keratinized (soft). **Keratin** is simply a protein."

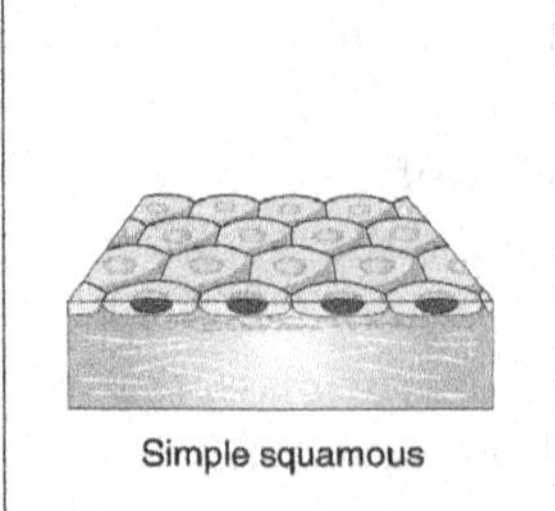

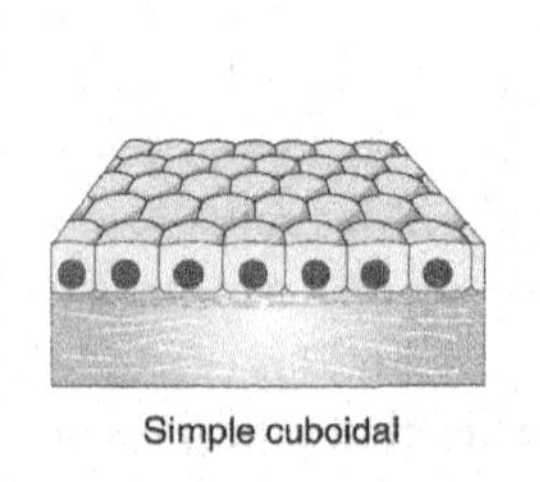

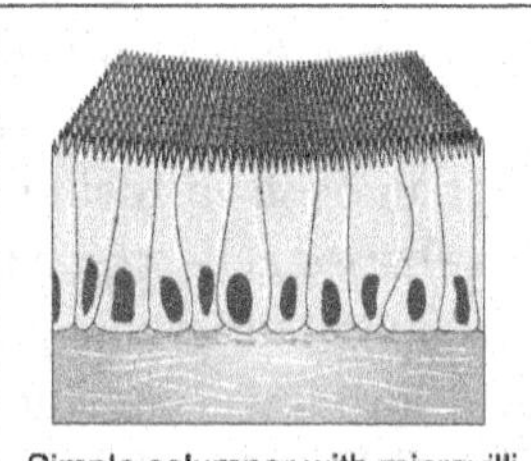

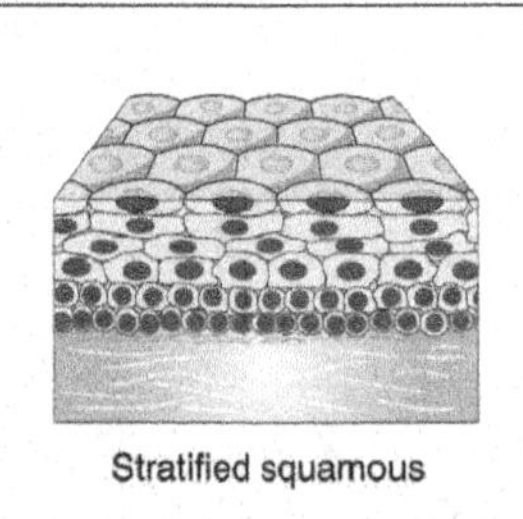

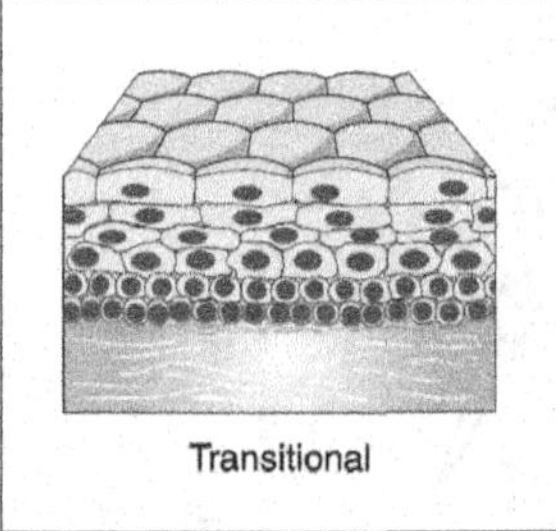

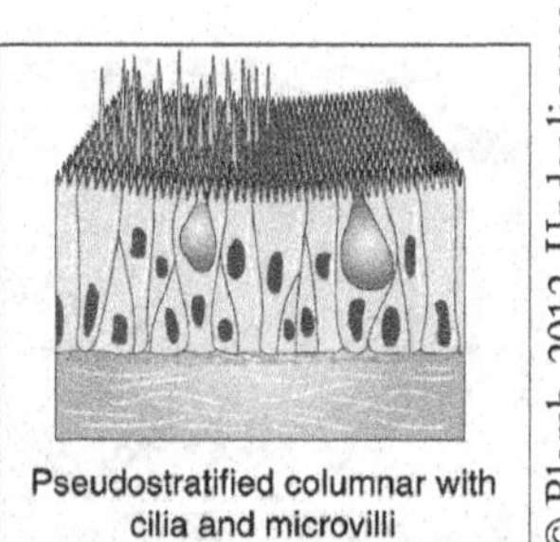

IMAGE 4.6 Examples of Epithelial Tissues of the Body

"Isn't our skin epithelial tissue?" questioned Will.

"Yes, it is," replied Dr. Barr. "In fact, it is stratified squamous epithelial tissue. And it is covered with a layer of compact, dead squamous cells. The cells are packed in keratin and coated with a water-repellant glycolipid which makes the skin's surface dry. But the surface of the skin also exfoliates itself or flakes off. Every time we itch and scratch ourselves, skin cells or epithelials fall off. These epithelials are commonly collected as evidence in crime scenes where DNA can be extracted for circumstantial evidence admissible in court. We can also easily study exfoliated tissue by scraping the inside of your cheek cells with a toothpick, staining it with iodine, and placing it on a microscope slide for a closer look."

"What about the insides of our cheeks? What kind of tissue is that?" questioned Grace further.

"Great question! It is still considered stratified squamous but it is considered non-keratinized (soft), which simply lacks the surface layer of dead cells. We also see non-keratinized stratified squamous making up the tongue, esophagus, and vagina. There is a similar study to that of cheek cells performed on vaginal tissue called a *Pap smear*, where cells are exfoliated from the cervix and examined for signs of cancer. Let's check out some of the different types of stratified epithelial tissues from different parts of the body," replied Dr. Barr.

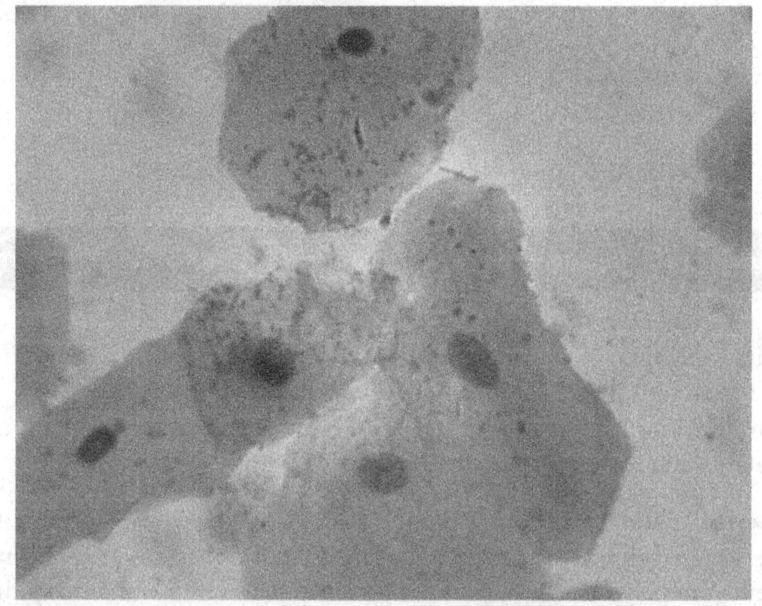

IMAGE 4.7 Cheek Cells

GLANDULAR EPITHELIUM

Dr. Barr added, "Glands are also comprised of special tissue known as glandular epithelial tissue."

"What exactly is a gland?" questioned Grace.

Dr. Barr responded, "A gland is a cell, group of cells, or organ producing a secretion."

"Like the pancreas?" replied Will.

"Yes, that is an example of a gland that is both endocrine and exocrine," Dr. Barr expounded.

"Endo - what? And Exo - what?" Will muttered while looking confused.

IMAGE 4.8 Endocrine and Exocrine Gland Secretions

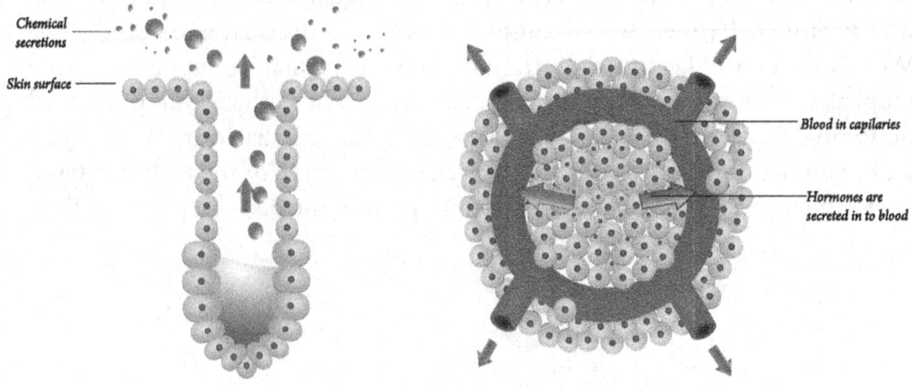

Dr. Barr explained, "**Exocrine glands** have ducts to carry their secretions to specific locations. Examples include the salivary glands whose ducts carry saliva to the mouth, or the pancreas whose duct carries pancreatic fluid to the duodenum (first section of the small intestine).

"**Endocrine glands** are glands of 'internal secretion' whose secretions are usually secreted directly into the blood. Most hormones are secreted in this manner. Examples are follicle stimulating hormone from the anterior pituitary or the hormone thyroxin secreted from the thyroid gland."

Self-Check

1. Explain the difference between simple and stratified epithelial tissue.

   ____

   ____

   ____

2. Describe four of the six functions of epithelial tissue. Give a supporting example for each.

   ____

   ____

   ____

3. Describe the three basic shapes of simple epithelial tissue. Draw an example of each.

Type: Description:	Type: Description:	Type: Description:
Drawing of the Shape:	Drawing of the Shape:	Drawing of the Shape:

4. What are goblet cells? What two types of simple epithelial tissue usually have them?

____

____

____

5. What are the two types of stratified squamous epithelial tissue? Explain each type and give examples of where these types are found in the body.

____

____

____

CONNECTIVE TISSUE

"We've been talking about epithelial tissue for awhile now. I also want to save time to discuss the other tissue types. So, let's talk about connective tissue now. Of all of the types of tissue, connective is the most abundant and widely distributed throughout the body," explained Dr. Barr.

He continued, "Connective tissues do what their name suggests—connect and bind things together like organs. They also function in support (cartilage supporting the ears, nose, etc.), physical protection (the cranium protects the brain), immune defense (white blood cells kill microbes), movement (tendons connect muscle to bone and allow us to contract our limbs), storage (fat is the body's major energy reserve), heat production (i.e., metabolism of brown fat) and transport (blood carries oxygen)."

"From what you have said so far, it sounds like there are many different kinds of connective tissues?" asked Grace.

"Yes, there are," replied Dr. Barr. "They fall into four broad categories: fibrous connective tissue (i.e., dermis of the skin, tendons, ligaments, areolar, and reticular tissue), adipose (fat), supportive (cartilage and bone), and fluid connective tissue (blood)."

FIRST TYPE OF CONNECTIVE TISSUE — FIBROUS CONNECTIVE

Dr. Barr explained, "**Fibrous connective** tissue is the most diverse type. What unites the tissue is several things in common including similar cells like **fibroblasts** (large branching cells that help form the matrix), **macrophages** (large mobile cells that attack foreign invaders and microbes), **leukocytes** (white blood cells), **mast cells** (emits either heparin to inhibit blood clotting or histamine to dilate blood vessels and increase blood flow), and/or **adipose** (clusters of fat cells).

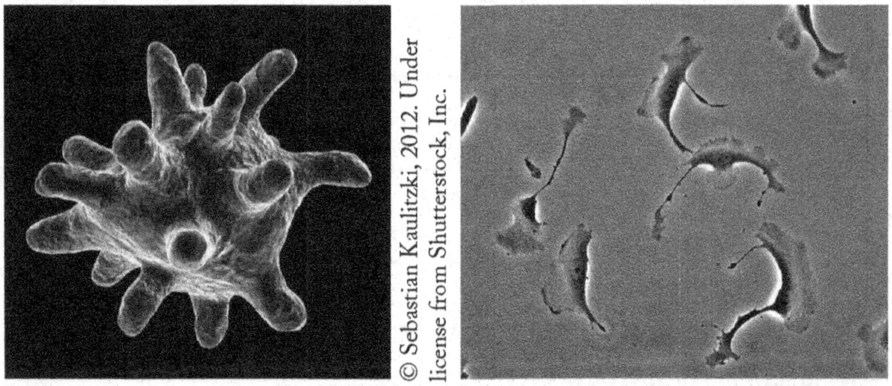

macrophage   fibroblasts

© Sebastian Kaulitzki, 2012. Under license from Shutterstock, Inc.

© T.W., 2012. Under license from Shutterstock, Inc.

IMAGE 4.9 Macrophage and Fibroblasts

"Additionally three types of protein fibers are found in fibrous connective tissues—**collagen** (the body's most abundant protein; it is tough and flexible but resists stretching; examples are tendons, ligaments, and the dermis of the skin), **reticular** (thin collagen fibers coated with glycoprotein; form a sponge-like framework that form the basis for organs such as lymph nodes, spleen, thymus, and bone marrow) and **elastic** (made of a protein called elastin; contributes to stretch and recoil ability).

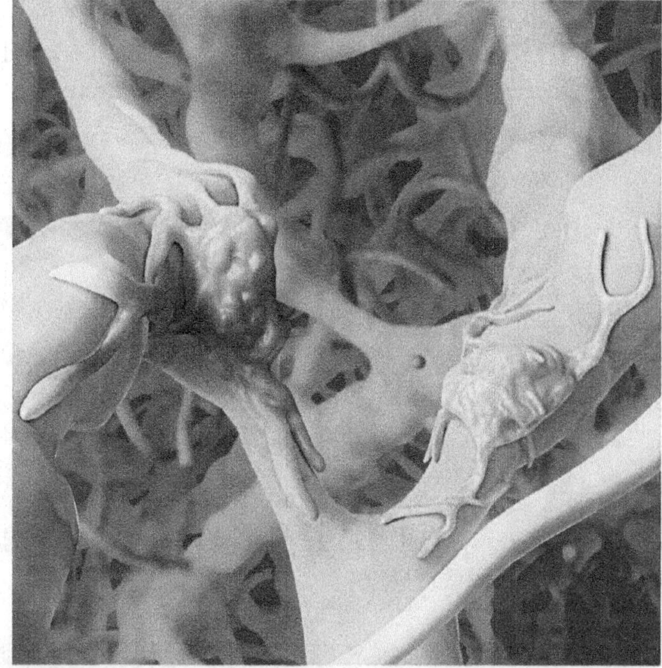

IMAGE 4.10 Human Collagen and Elastin Matrix

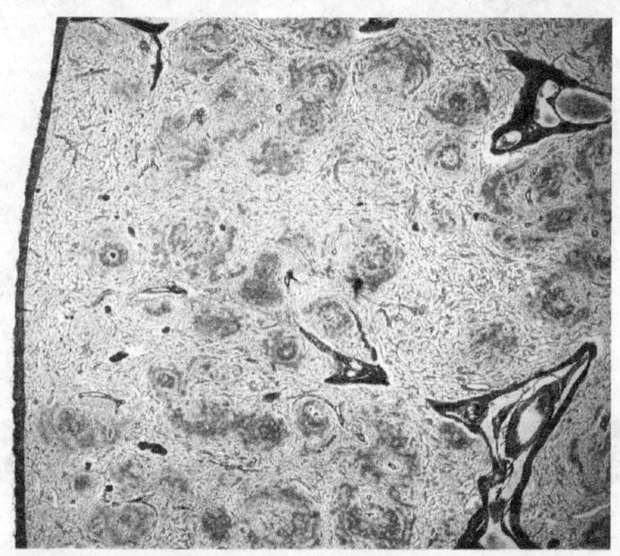

IMAGE 4.11 Red Pulp of Spleen Showing Reticular Fibers 40x

"There are two broad categories of fibrous connective tissues—**loose connective** (examples include areolar and reticular) and **dense connective** (dense regular and dense irregular tissue)." Areolar tissue is found in many locations around the body, such as the skin, around the mucous membranes, around the blood vessels, nerves and organs of the body. Reticular connective tissue is found around the liver, the kidney, the spleen, and lymph nodes, as well as in bone marrow. Dense regular connective tissue is found in ligaments (which link bone to bone at joints) and tendons (connections between bones or cartilage and muscle). Dense irregular connective tissue is found in joint capsules, in the connective tissue that envelops muscles (muscle fascia), and it forms dermis of skin."

areolar tissue

reticular tissue (spleen)

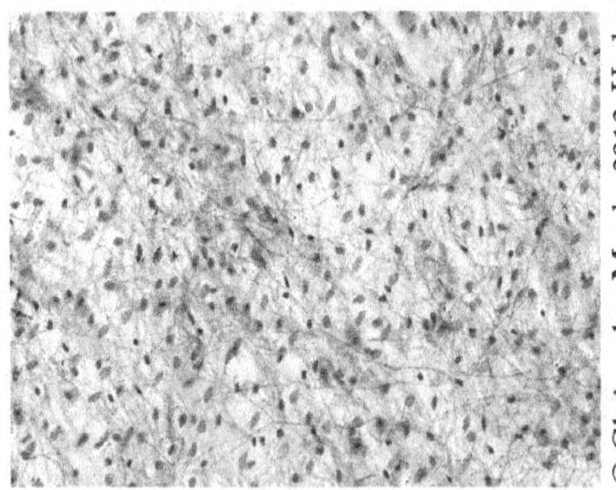

© Christopher Meade, 2012. Under license from Shutterstock, Inc.

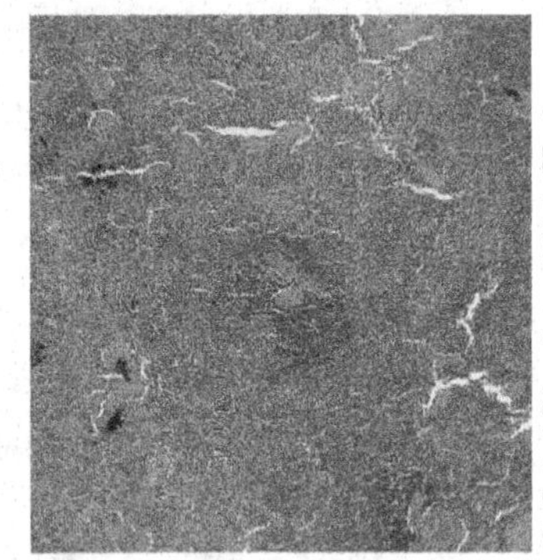

© Jubal Harshaw, 2012. Under license from Shutterstock, Inc.

dense regular tissue (tendon)

dense irregular connective tissue (dermis)

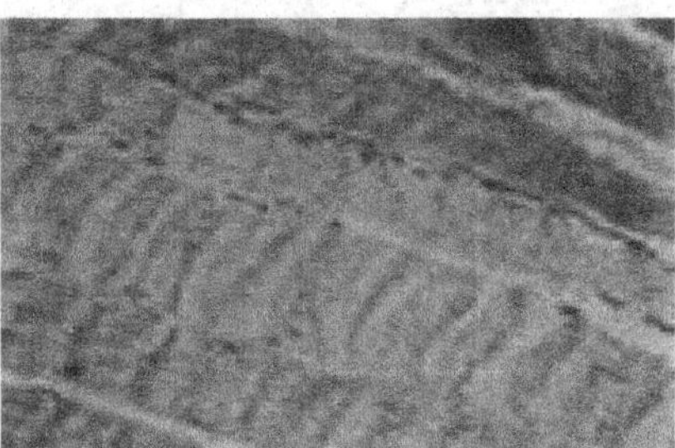

National Institute of Health

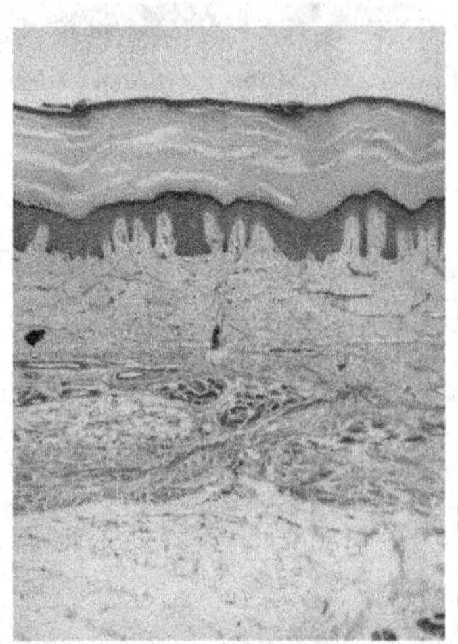

© Jubal Harshaw, 2012. Under license from Shutterstock, Inc.

IMAGE 4.12 Loose Connective and Dense Connective Tissues

SECOND TYPE OF CONNECTIVE TISSUE – ADIPOSE

"There are two types of **adipose** or fat in the human body—white fat and brown fat. Adipose is the body's main fuel reserve. White fat is the more abundant type of adipose and provides thermal insulation, cushions organs like eyeballs, and contributes to body shape especially in females where it pads the breasts and hips," added Dr. Barr.

Grace piped in with, "Is it true that females have more body fat than males?"

"Yes, for women the extra percentage of fat helps meet the caloric need of pregnancy and nursing an infant. Plus having too little fat can interfere with fertility, because body fat plays a significant role in reproduction. Sex hormones are fat soluble and they are stored in the body's fat layers. Women that have too little fat may produce a reduced amount of estrogen which can lead to an abnormal menstrual cycle/ovulation," replied Dr. Barr. Then he continued, "Usually fat cells have a single large, central globule of lipid (triglyceride), a nucleus pushed to one side of the cell, and an extremely thin layer of cytoplasm immediately beneath the plasma membrane. The lipid is in liquid oil form, much like vegetable oil, rather than solid fat at room temperature like shortening. The number of fat cells is usually stable in adults. As we gain weight, we do not gain additional fat cells but rather the cells that we do have become much larger in diameter. As we loose weight they become smaller. So, the number of fat cells we have does not typically increase or decrease—only their size changes."

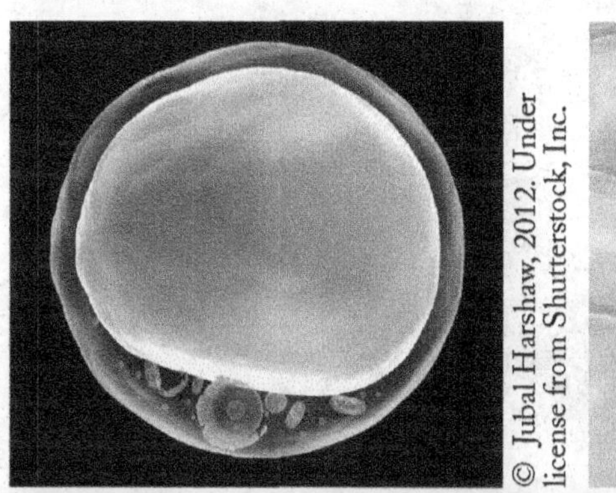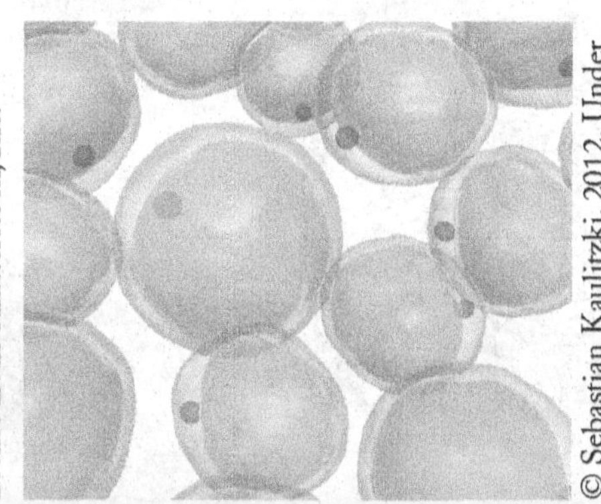

IMAGE 4.13 Adipose Cell (left) and Adipose Tissue (right)

"We also don't want to forget to talk about brown fat. I want to know what that is," interjected Will.

Dr. Barr continued, "Right. **Brown fat** is mostly found in fetuses, infants, and children around the fat pads of the shoulders, upper back, and around the kidneys. It accounts for up to 6% of body weight. It gets its rich color from the abundance of blood vessels that it has, making it appear much darker than white fat. Brown fat is heat and energy generating tissue. It stores lipids a little differently from white fat."

THIRD TYPE OF CONNECTIVE TISSUE – SUPPORTIVE

Dr. Barr continued, "Now that we've discussed epithelial and connective tissues, let's talk about the third main type—supportive. Ever notice the white knobby gristle at the end of a chicken drumstick?"

"Yep, I have!" piped Grace.

"Yeah, my dog loves it," commented Will.

"Ever think about what it is?" questioned Dr. Barr.

"Not really," they both replied.

Dr. Barr explained, "It's cartilage and it pads and cushions the ends of long bones in both humans and other animals. There are actually three different types of cartilage—**hyaline** (the most common type that pads the ends of most bones including the rib cage), **elastic** (comprising the flexible structure of the ear and epiglottis) and **fibrocartilage** (making up the discs in between the vertebrae of the back)."

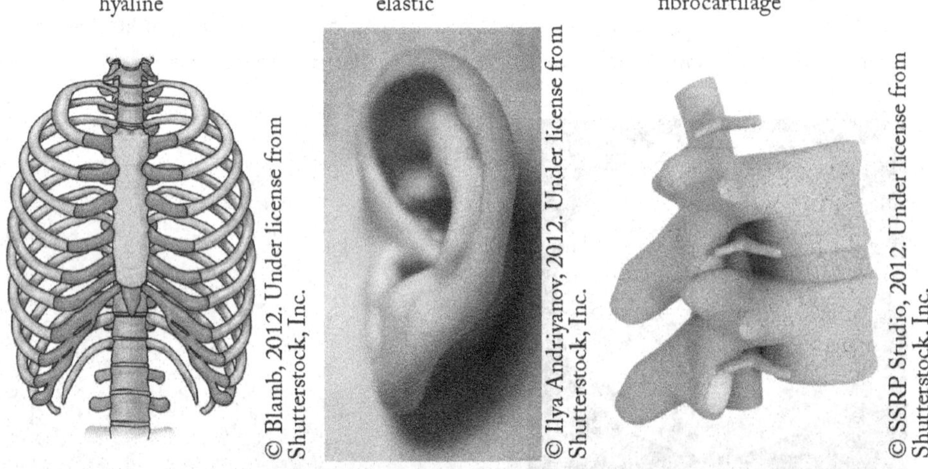

IMAGE 4.14 Locations in the Body and the Three Types of Cartilage

Dr. Barr continued, "Cartilage is made up of cells called **chrondocytes**. The prefix 'chrondo' referring to cartilage, the suffix 'cyte' referring to cell. There aren't a lot of blood vessels running through most cartilage, therefore it does not repair itself very quickly if injured."

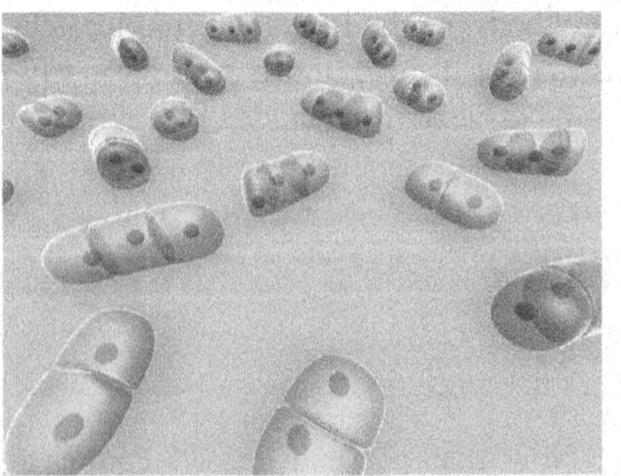

IMAGE 4.15 Chrondocytes

Dr. Barr added, "The other main type of supportive tissue is bone which is hard, calcified connective tissue. Bone is comprised of cells known as **osteocytes**. You already know what the suffix 'cyte' refers to, so can you guess what the prefix 'osteo' means?"

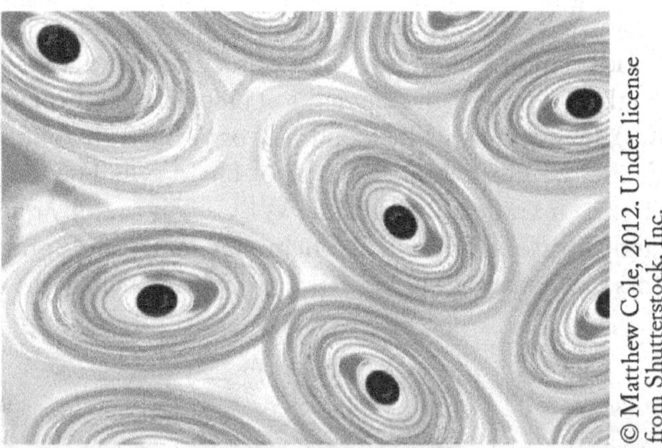

IMAGE 4.16 Illustration of Microscopic Bone Cells

"I think it means bone," replied Will.

"It does mean bone. So, osteocytes are bone cells," confirmed Dr. Barr. "Bone is mostly collagen and a substance called glycosaminoglycan, which enables bone to bend slightly under stress without breaking. The rest of bone consists of calcium and phosphate salts that enables bone to withstand compression under the body's own weight."

Dr. Barr added, "There are really two types of bone—**compact** (the hard outer bone) and **spongy bone** (the softer inner bone tissue where marrow is housed, inside the compact bone)."

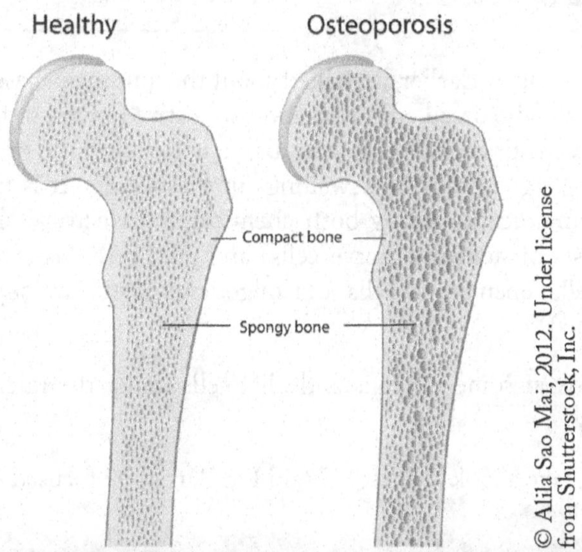

IMAGE 4.17 Compact Bone Versus Spongy Bone in a Healthy Person and a Person with Osteoporosis

FINAL TYPE OF CONNECTIVE TISSUE – FLUID

Dr. Barr continued, "The last type of connective tissue is blood. Most people wouldn't consider blood a tissue but if you think of it, there are solid blood cells suspended in the fluid. The solid elements of blood are **erythrocytes** (red blood cells or RBCs), **leukocytes** (white blood cells or WBCs) and **platelets** (clotting cells). RBCs are the most abundant type of blood cell. They function in the transport of oxygen and carbon dioxide. WBCs play various roles in fighting infections and disease. And platelets help clot injury sites and they secrete growth factors that encourage blood vessel growth and help with maintenance."

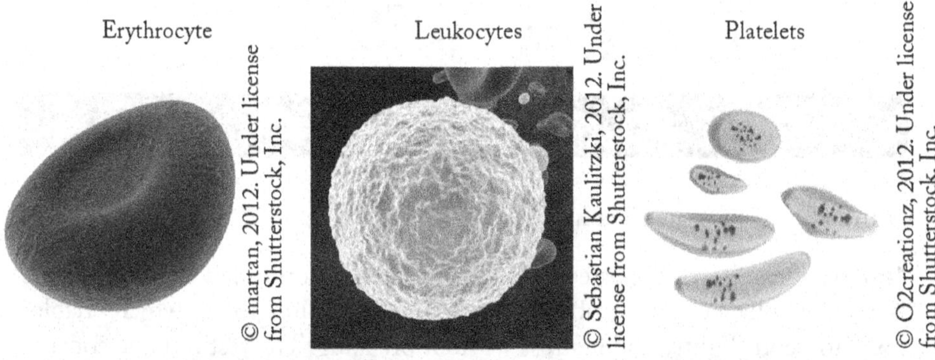

IMAGE 4.18 Three Main Types of Blood Cells

NERVOUS TISSUE

Dr. Barr continued, "Earlier, we talked about the four main classes of tissue and we have already discussed epithelial and connective tissues, so let's spend some time on nervous tissue now. Nervous tissue is found in the brain, spinal cord, nerves, and ganglia (knot-like swellings in the nerves). It is highly branched for fast communication using both chemical and electrical signals. Nervous tissue consists of **neurons** (nerve cells) and **glial cells** (astrocytes, microglia, Schwann cells, ependymal cells and oligodendrocytes are support cells that assist neurons)."

"Can we look at some neurons and glial cells under the microscope?" asked Will eagerly.

"Sure thing, take a look at these…" said Dr. Barr as he focused the microscope.

NOTE: Cells of the nervous system will be handled in much greater depth in Chapter 10.

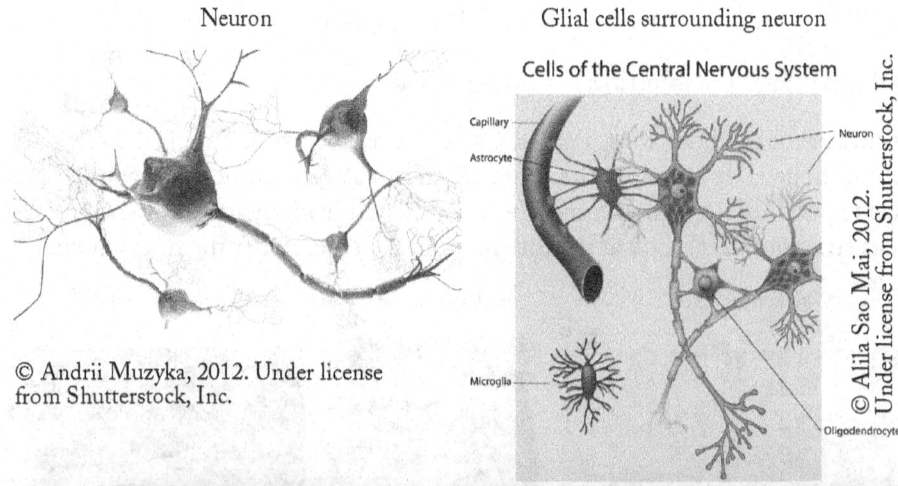

© Andrii Muzyka, 2012. Under license from Shutterstock, Inc.

© Alila Sao Mai, 2012. Under license from Shutterstock, Inc.

IMAGE 4.19 Nervous Tissue

MUSCLE TISSUE

"I think this is the type of tissue I know the most about," said Will as he flexed his bicep and smiled.

"Oh yeah?" questioned Dr. Barr. He questioned Will further, "So what type of muscle is that bicep of yours?"

"Skeletal muscle," quipped Will.

"Yes! And do either of you know what the other two types of muscle are?" Dr. Barr questioned further.

"Heart," shouted Grace.

"And…uh-h-h…" Will attempted to answer but drew a blank.

Grace shrugged her shoulders but gave it a try, "Smooth."

"Right, you guys got them all!" cheered Dr. Barr.

Dr. Barr explained further, "**Skeletal muscle** is considered voluntary because we can consciously control our movements. It is also made up of long threadlike cells called **muscle fibers**. It appears **striated** (banded) under the microscope. These bands are actually overlapping cytoplasmic protein filaments that cause muscles to contract.

"Cardiac muscle is found only in the heart. It is striated too but differs from skeletal muscle in many ways like shorter, branching fibers called **myocytes**. These myocytes are joined end-to-end by junctions called **intercalated discs**. Electrical connections at these junctions enable electrical impulses to travel rapidly from cell to cell so that all the cells contract almost simultaneously. Cardiac muscle is considered involuntary because it is not under conscious control."

Dr. Barr questioned, "Anybody know where smooth muscle is found?"

"I think it lines the intestines," answered Will.

"You'd be right about that! Particularly the small intestine," replied Dr. Barr. He continued, "It also lines the walls of arteries as well as the esophagus, eyes, bladder, and uterus. Smooth muscle is considered involuntary. Smooth muscle lacks striations. Let's look at the different muscle types under the microscope."

skeletal muscle   cardiac muscle   smooth muscle

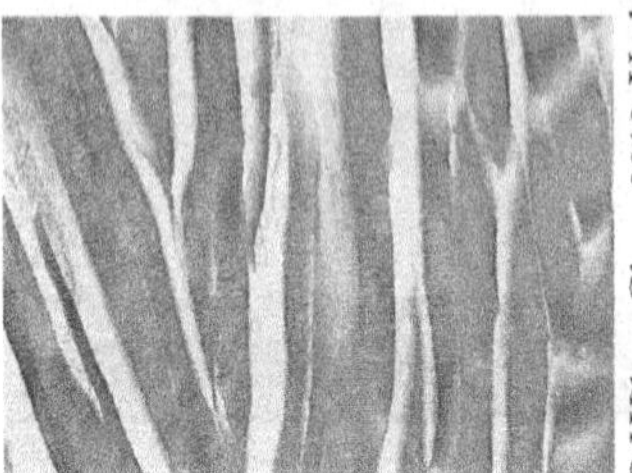
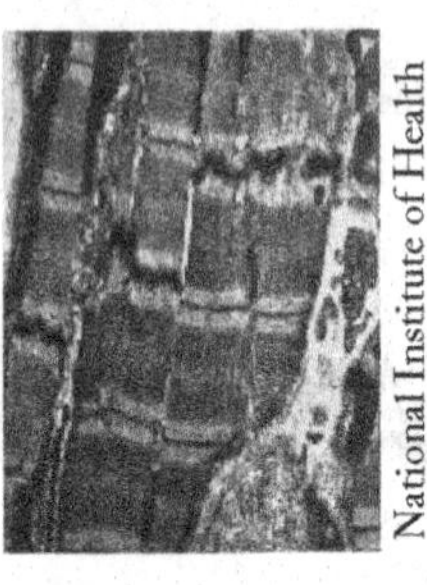
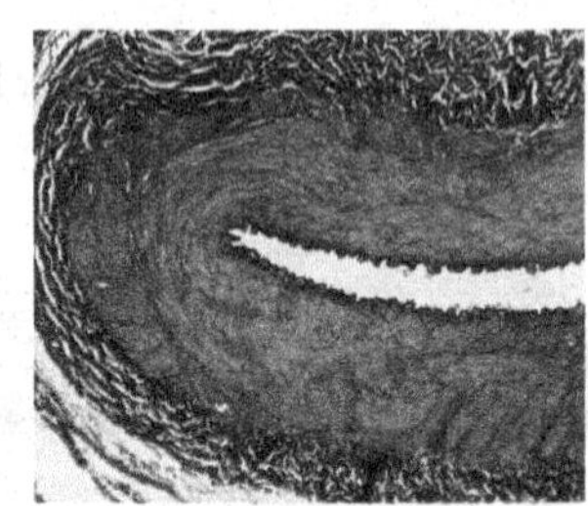

© Wilson Chan, 2012. Under license from Shutterstock, Inc.

National Institute of Health

© Brian Maudsley, 2012. Under license from Shutterstock, Inc.

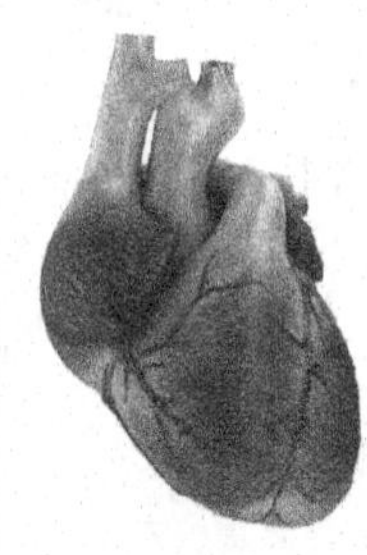
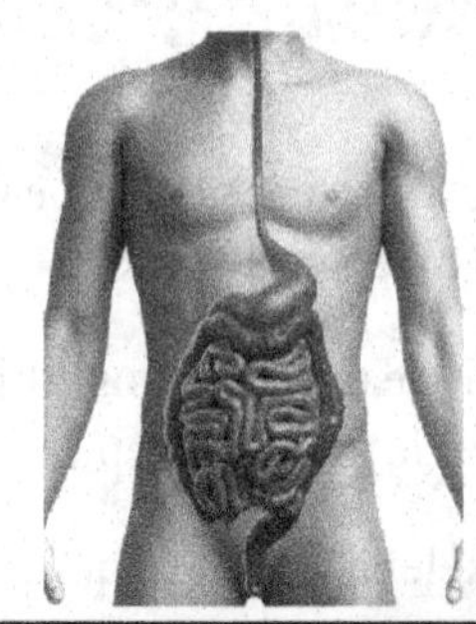

© CLIPAREA l Custom media, 2012. Under license from Shutterstock, Inc.

© Sebastian Kaulitzki, 2012. Under license from Shutterstock, Inc.

© Sebastian Kaulitzki, 2012. Under license from Shutterstock, Inc.

IMAGE 4.20 The Three Types of Muscle Tissue

Self-Check – Continued

6. List the specialized cells of fibrous connective tissue and explain what they do.

7. Explain the difference between white fat and brown fat.

8. Discuss the three main types of protein fibers.

____

____

____

9. Describe the three main types of cartilage and give examples of where they are located in the body.

____

____

____

10. What are chondrocytes? Osteocytes?

____

____

____

11. Explain compact versus spongy bone.

____

____

____

12. What are the three main types of blood cells? What do they do?

____

____

____

13. Explain the role of neurons and glial cells in nervous tissue.

____

____

____

14. List and describe the three main types of muscle cells. Give examples of where they are located in the body.

____

____

____

15. Sketch the three muscle tissue types. Be sure to label any specialized structures that make each type unique.

Type:	Type:	Type:
Drawing of the Shape:	Drawing of the Shape:	Drawing of the Shape:

WRAP-UP

Grace looked at their watch and commented, "Dr. Barr—you've been so gracious to show us so many tissue types—from epithelial to muscle and all in between. It was a like a tour around the human body! Thanks for your help!"

"Yep, I especially enjoyed seeing a macrophage for the first time! And learning about 'pixie dust' from pig's bladders!" added Will.

"Glad I was able to help," concluded Dr. Barr as he walked them out.

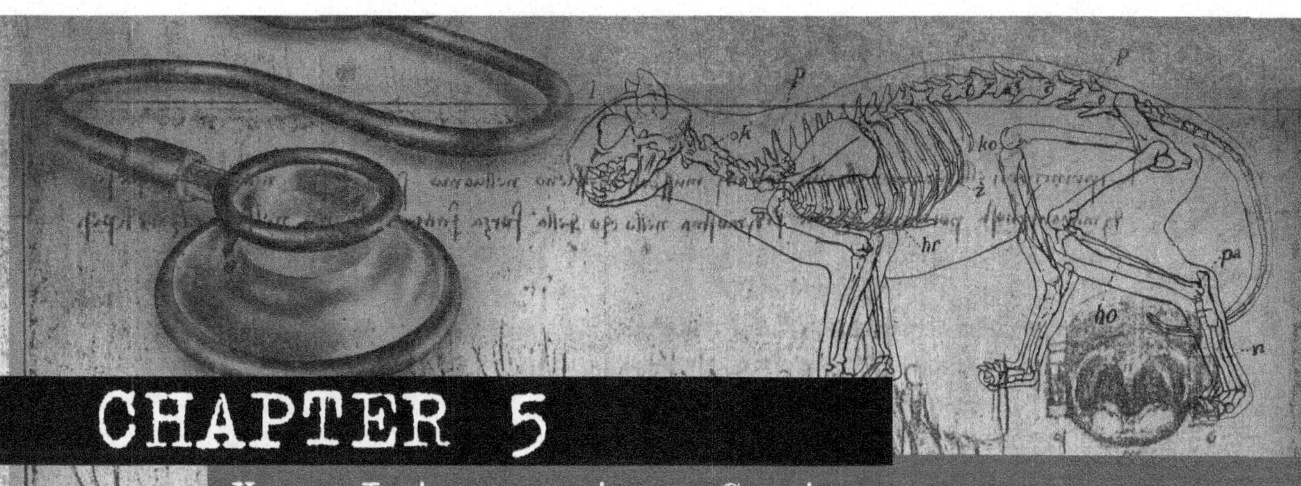

CHAPTER 5

Your Integumentary System: How Deep is Skin Deep?

KEY WORDS

The study of Anatomy & Physiology involves many new vocabulary words. It is helpful to gain familiarity with these new key words, just as you would a foreign language.

acne	hair follicle	metastasis
adipose	hair root	moles
albinism	hair shaft	Pacinian corpuscle
apocrine sweat glands	hemangiomas	pallor
arrector pili muscle	hematoma	papillary layer
basal cell carcinoma	hypodermis	psoriasis
ceruminous glands	integumentary system	reticular layer
cutaneous glands	jaundice	rosacea
cuticle	keratin	sebaceous glands
cyanosis	keratinocytes	sebum
dermatitis	keratinohyaline	spray-on skin
dermis	Langerhans cell	squamous cell carcinoma
eccrine sweat glands	lanugo	stratum basale
eczema	leukocytes	stratum corneum
Ehlers-Danlos syndrome	macrophages	stratum granulosum
epidermis	mammary glands	stratum lucidum
erythema	mast cells	stratum spinosum
fibroblasts	Meissner's corpuscle	striae
first-third degree burns	melanin	sweat glands
follicle	melanocytes	terminal hair
freckles	melanoma	vellus hair
hair bulb	Merkel cell	vitiligo

It is helpful to have an idea of what you need to learn before proceeding. Here is a study guide to assist you:

1. Describe the structure of the epidermis, dermis, and hypodermis. Be sure to include any significant layers or specialized type of cells.
2. Explain skin coloration and variations of normal skin colorations.
3. List and explain some of the common skin markings (freckles, etc.).
4. Distinguish between the three types of skin cancer.
5. Distinguish between the three types of hair.
6. Describe the histology (i.e., layers and significant parts) of hair.
7. Name two types of sweat glands and describe the structure and function of each.
8. Describe sebaceous and ceruminous glands.
9. Distinguish between the three categories of burns.
10. Describe the physiology of the nail and label its anatomical structures.
11. Explain how skin heals.

Grace and Will, on special assignment from their college Anatomy & Physiology professor, have been challenged to visit Dr. Maria Ortiz at the local dermatology office to study the integumentary system, learn more about skin cancers, and check out the latest technology that is saving burn victims.

Grace and Will entered the office and made their way to the front desk. "Hi there. I'm Grace and this is Will. We're with the local community college and we have an appointment with Dr. Ortiz to interview her about the integumentary system."

The receptionist responded, "Just have a seat and we'll let her know you're here."

STUDENT VIEW – WHY SHOULD I STUDY THIS? HOW IS IT RELEVANT TO MY LIFE?

When you hear the word "organs," you may think of the stomach, liver, or heart. However, not all of your organs are internal. Your skin is actually the largest organ averaging about 8 pounds and covering 22 square feet (or 2 square meters). It covers and protects you from foreign invaders. Every day we wear, admire, bathe, and moisturize it. Perhaps we even take it for granted. The skin and the appendages (hair, nails, subcutaneous glands, etc.) make up the **integumentary system**.

After a few minutes, a nurse came out to greet them and take them back to the doctor. "Welcome! I am Dr. Ortiz and I have been expecting you! I'm examining a patient for possible skin cancer. She has given her consent to have students in the room to observe as I conduct the examination, so please follow me!"

"Wow! That's great," commented Grace.

"I'm very interested," exclaimed Will.

"You may not be so enthusiastic once you see some of these cancers firsthand. They're not pretty!" summarized Dr. Ortiz. "In this first waiting room is a patient that I suspect has squamous cell carcinoma. Heard of it before?"

"Um-m… yeah. But I'm not sure what it looks like," answered Will.

"Follow me," Dr. Ortiz added as she entered the room.

"Hello Mrs. Smith, these are the students I mentioned earlier. Thank you for agreeing to allow them to observe today. Now, let's have a look at your shoulder again. Hmmm—it looks like it might be early **squamous cell skin carcinoma**. Although it can **metastasize** or spread, if caught early and surgically removed, the cure rate is in the upper 90 percentile," Dr. Ortiz announced. "I'd like to take a biopsy today so I am going to ask my nurse to prep things. She'll be here momentarily and I'll be back shortly." Dr. Ortiz gestured to Grace and Will to follow her out.

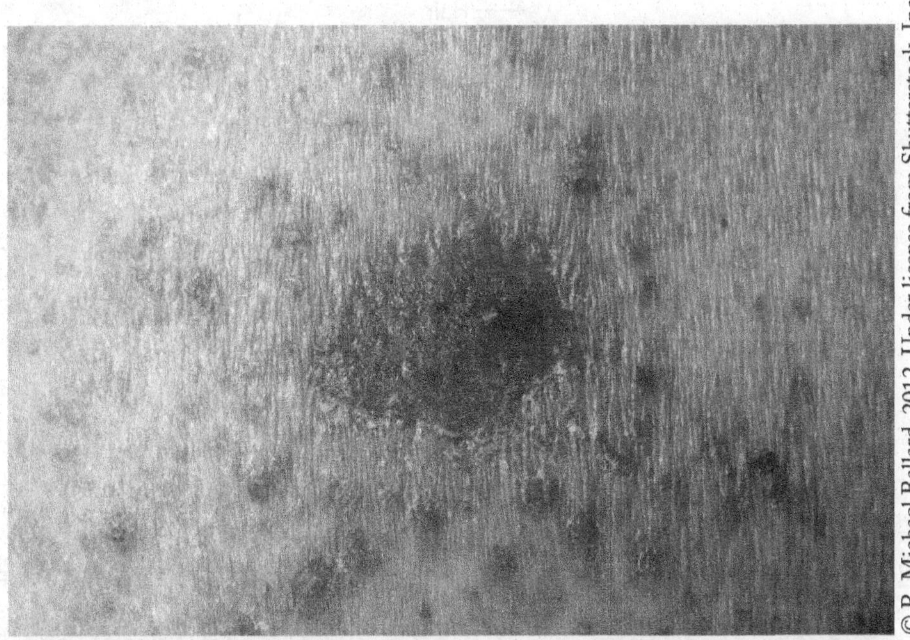

IMAGE 5.1 Mrs Smith's Upper Arm Showing Squamous Cell Skin Cancer

"Wow, that was really raised and red," commented Will.

"Yes, squamous cell carcinoma typically presents itself as raised, reddened, scaly lesions later forming a concave ulcer with raised edges," added Dr. Ortiz.

Grace asked, "Like a fruit loop? You know, the cereal?"

"Yes, I guess that's a good analogy for it," chuckled Dr. Ortiz.

She continued, "It arises out of a layer in the epidermis called the stratum spinosum, which we will look at in a moment. I want you to know that squamous cell carcinoma, if not detected or treated, can metastasize to the lymph nodes and be lethal."

"Yikes, really?" questioned Will.

THE THREE MAIN REGIONS OF THE SKIN

"Yes, it is nothing to play around with," confirmed Dr. Ortiz. "Overall, the skin's functions are protection, temperature regulation, excretion and sensory reception. The skin has three main regions—the **epidermis** or top layer that we can touch, the **dermis**, and the **hypodermis** (subcutaneous tissue). Here is a poster showing the three regions and some of the different structures within each region."

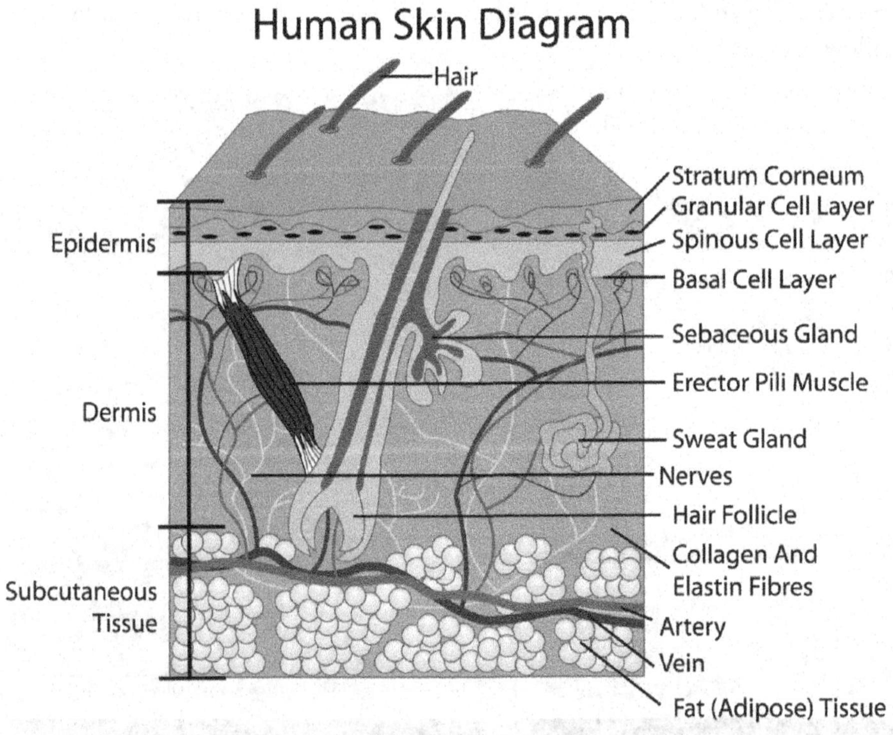

Human Skin Diagram
Hair
Epidermis
Dermis
Subcutaneous Tissue
Stratum Corneum
Granular Cell Layer
Spinous Cell Layer
Basal Cell Layer
Sebaceous Gland
Erector Pili Muscle
Sweat Gland
Nerves
Hair Follicle
Collagen And Elastin Fibres
Artery
Vein
Fat (Adipose) Tissue
© Anita Potter, 2012. Under license from Shutterstock, Inc.

IMAGE 5.2 The Three Regions of the Skin

Self-Check – Answer these questions now before proceeding:

1. Study the human skin diagram carefully. Do you notice any blood vessels in the epidermis? Describe what you see.

   ____

   ____

   ____

2. List the structures of the dermis.

   ____

   ____

   ____

3. What tissue does the hypodermis (subcutaneous tissue) appear to be mainly composed of?

   ____

   ____

   ____

A CLOSER LOOK AT JUST THE EPIDERMIS

Dr. Ortiz continued, "The epidermis has 4-5 layers much like a cake has layers. Each layer is a **stratum**. The oldest layer is **stratum corneum**, which is comprised of dead keratinocytes that flake off. It is the layer we can actually touch or scratch when we itch.

"In the palms of our hands and soles of our feet where our skin is much thicker, we have an extra layer of padding called **stratum lucidum**."

"Doesn't 'lucid' means clear?" questioned Grace.

"Yes it does," answered Dr. Ortiz. Then she continued, "The next layer is **stratum granulosum**, which is also made up of several thin layers of flattened **keratinocytes** (cells made up mostly of keratin). This layer looks very 'grainy' because it is made up of keratinocytes that contain coarse, dark-staining granules of **keratinohyaline** (a blend of keratin and hyaline cartilage) which this layer is named for."

Dr. Ortiz added, "Next, we come to **stratum spinosum,** which is actually several very thin layers of kerantinocytes. This is typically the thickest layer in the epidermis. Special cells called **Langerhans cells** are found in this layer. Langerhans cells are actually immune cells that originated in the marrow of the bones but migrated to epidermis and several other areas of the body including the esophagus, the epithelial lining of the oral cavity, and the vagina. They stand guard against infectious pathogens and toxins. They are also referred to as dendritic cells. These cells have a very spiny look because they have branching pieces projecting off the ends (much like a nerve cell has branches or dendrites) and thus the name 'spinosum.' **Merkel cells** are also found in this layer. Merkels are touch receptors.

"The deepest layer of the epidermis is **stratum basale**. It is also the youngest layer and is mitotically active. This layer also contains pigment cells or **melanocytes** that increase their melanin production in response to sun exposure (thus darkening our skin).

"Here is a flyer I give to my patients to help them understand how their epidermis is constructed. Have a look," said Dr. Ortiz.

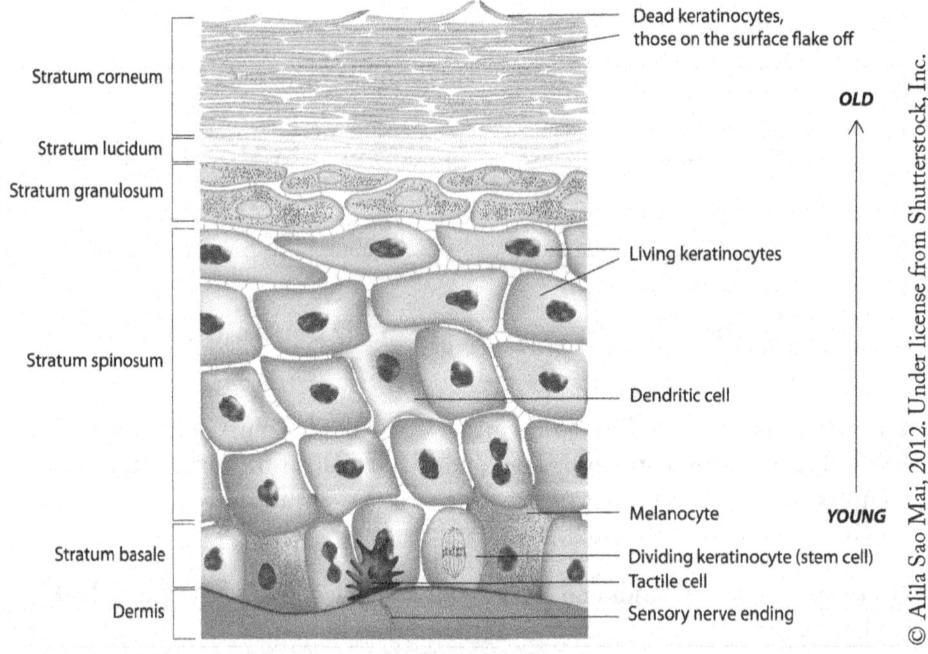

IMAGE 5.3 The Epidermis and its Various Layers

"I'm getting ready to examine my next patient. How about joining me on this visit as well?" asked Dr. Ortiz.

"Sure," answered both Grace and Will at the same time.

"My next patient had melanoma about 2 years ago and has remained cancer free. She found another suspicious area recently, however, and she wants to have it checked," explained Dr. Ortiz.

"For her sake, I hope it is not a reoccurrence," said Grace.

"Okay, here we go. Just follow after me," directed Dr. Ortiz.

"Hello Mrs. Jenkins! Nice to see you again. Sorry that it is under these circumstances though. These are the students that my nurse talked with you about earlier that will be observing today. I understand you saw a new, suspicious area possibly? Can you show me where that area is?" questioned Dr. Ortiz.

The patient pointed to a darkened spot on her left forearm.

Dr. Ortiz continued, "Okay, let me take a closer look. I know things like this can be frightening. There are two different colors apparent. It does look suspicious so I am going to biopsy. I will have my nurse begin prepping for the biopsy and I will be back shortly to do the procedure." Dr. Ortiz silently gestured with her head for Will and Grace to follow her out.

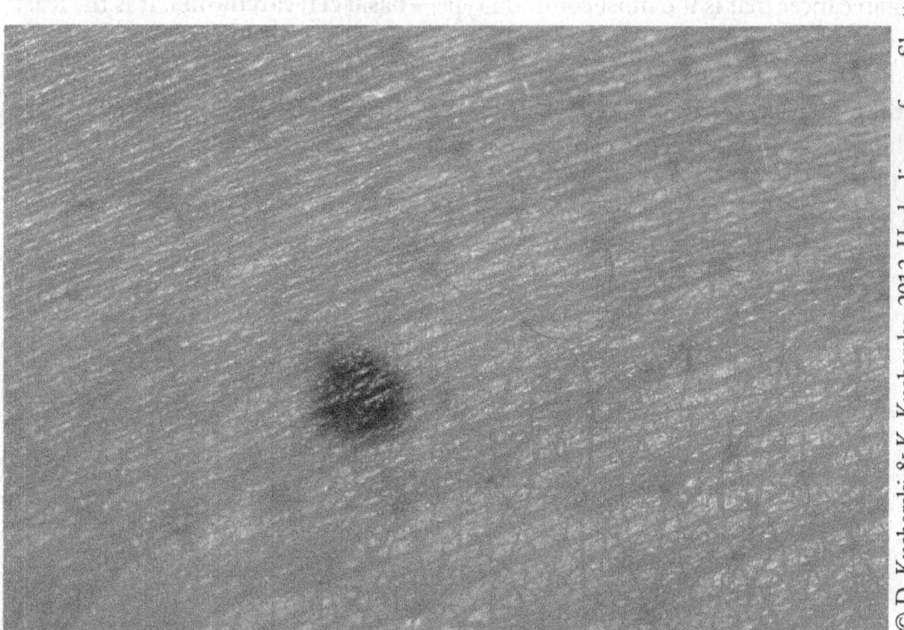

IMAGE 5.4 Melanoma

Out of the patient's ear shot, Dr. Ortiz explained further to Grace and Will, "In addition to being both brown and black, it had a bit of an irregular border which is also suspicious for melanoma. There is a rule called the 'ABCD' rule that makes it easier to remember what to watch for. Here, I'll write it down for you." She handed the piece of paper to Will as follows.

A stands for <u>A</u>symmetry – do the two sides of the pigmented area match?

B stands for the <u>B</u>order – is it irregular or does it exhibit any indentations?

C stands for <u>C</u>olor – is it a combination of black, brown, tan, or sometimes red?

D is for <u>D</u>iameter – is it larger than 6 mm (the size of a pencil eraser)?

Dr. Ortiz continued, "I suspect it is **melanoma** which is actually cancer of the melanocytes. Now it is smaller than the diameter of a pencil eraser so we may have caught it early, which may increase the chances of survival for this patient. It will definitely have to be removed and some chemotherapy will be ordered if the result is positive for carcinoma. Melanoma can also be highly resistant to chemotherapy."

"I've heard having light skin or blonde or red hair can increase your risks for melanoma. Is that true?" asked Will.

"Yes, fair skin, light eyes, light hair—these are all risk factors. So people with these characteristics should take additional precautions in the sun," answered Dr. Ortiz. She continued, "I do want to mention that there is one more type of skin cancer that is the most common type—**basal cell carcinoma**. It is the least deadly and arises from stratum basale in the epidermis but does not usually metastasize and it 99% treatable with surgical removal only. It often arises on the forehead, ears, or chin. When it first appears, it looks like a small, shiny bump. As it enlarges, it can develop a center depressed area and a beaded, 'pearly' edge. Here is a photograph of it."

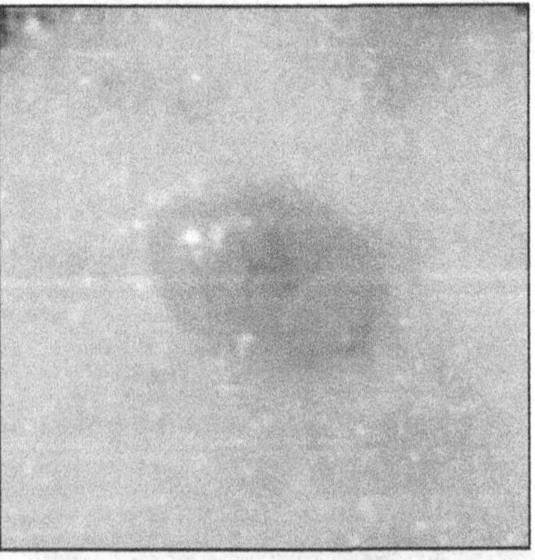

IMAGE 5.5 Basal Cell Carcinoma

Dr. Ortiz checked a clipboard she was carrying and said, "Let's move to the next patient now."

Dr. Ortiz knocked gently on the room door and entered. There sat a young, blonde-haired man in his early twenties. Grace and Will were surprised that he was so young and they secretly wondered to themselves what he was doing at the dermatology clinic—most everyone that they had seen today was middle-aged or beyond. Dr. Ortiz began the conversation, "Hi! I'm Dr. Ortiz. I see this is your first time to the clinic, so, welcome. These are the students my nurse mentioned to you earlier. Thank you for allowing them to observe today. I understand that you have a spot for me to look at so let's take a look."

The young man raised his shirtsleeve to reveal his upper arm. He had a growth there as shown in the photo below.

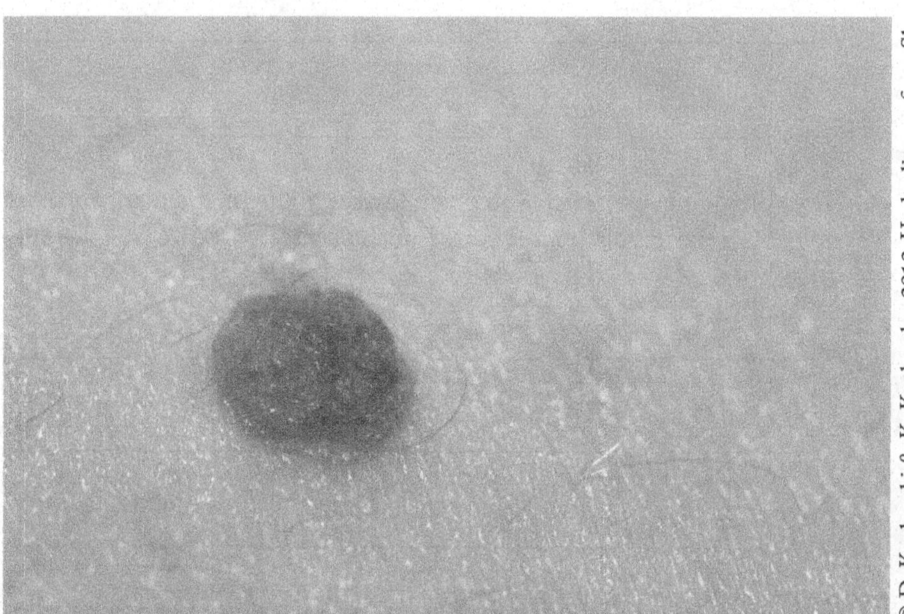

IMAGE 5.6 Can You Identify this Skin Condition?

Self-Check – Continued

4. Based on the photo above, does this look like any type of skin cancer you have seen today while making rounds with Dr. Ortiz? Explain your answer.

____

____

____

5. In order to make a diagnosis on this last patient, what questions would you ask him? Write your questions here.

____

____

____

____

6. If you had to guess, what would you say the young man's mark appears to be? Go ahead, follow your inner hunch or instinct. Write your answer here.

____

____

7. What does the ABCD rule help us with? What do the letters actually stand for?

____

____

____

____

Dr. Ortiz examined the young man further and thanked him for being diligent in his self-care, "It is simply a mole. It does not look suspicious. You'll be okay. Thanks for coming in."

After the young man exited the room, Dr. Ortiz went on to explain to Grace and Will, "Sometimes what may be suspicious to someone is simply a benign skin marking of some type. Although it's good to err on the side of caution, of course! Common skin markings include **freckles, moles,** and **birthmarks**. Freckles and moles are clumps of melanocytes or pigment. Often moles are a little thicker and stand up off the skin or have hair. Birthmarks or **hemangiomas** are patches of skin discolored by benign tumors of the capillaries in the skin. They can range in color from red like strawberry birthmarks to dark purple."

Freckles   Moles   Birthmarks

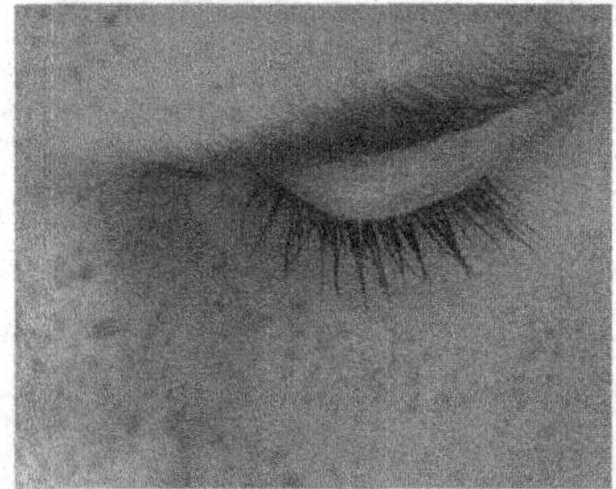 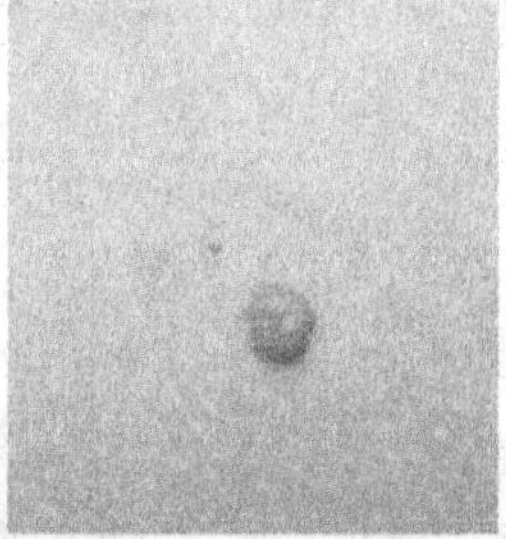 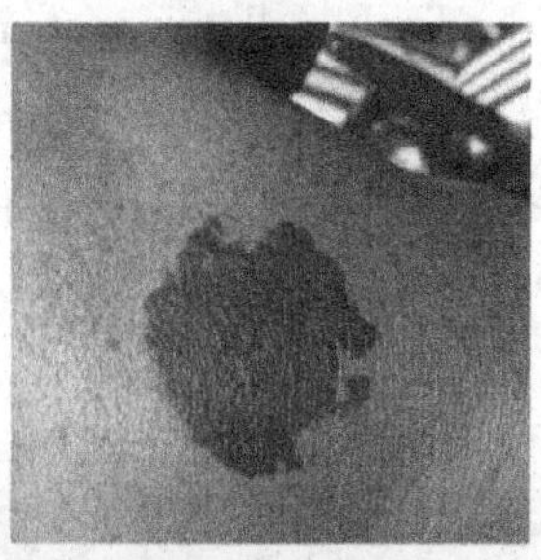

IMAGE 5.7 Common Skin Markings

SKIN COLOR AND VARIATIONS

"So—what makes someone have albinism?" questioned Will further.

"The most significant factor in skin color is **melanin**. Melanin is a brownish-black to reddish-yellow containing pigment. It is produced by melanocytes but accumulates in the keratinocytes of stratum basale and stratum spinosum. In dark skin the melanocytes produce greater quantities of melanin. In albinism, there is little to no melanin in the hair, eyes, and skin. There are other pigments in the skin such as hemoglobin, which is the red pigment of the blood and carotene, a yellowish-orange pigment acquired from eating egg yolks as well as yellow and orange vegetables," Dr. Ortiz answered.

She continued, "Speaking of skin color, some of our patients exhibit different colorations due to their health conditions. Lets say we have a patient with oxygen deficiency due to cardiac arrest. What color do you think his skin might turn?"

"Would it be blue?" asked Grace.

"Right indeed! And that is a condition called **cyanosis**. The prefix 'cyan' actually refers to a shade of blue," replied Dr. Ortiz. She continued, "Here is a chart that shows other diagnostic skin conditions."

CLINICAL SKIN CONDITIONS

Albinism – genetic lack of melanin resulting in white hair, white skin, and pinkish or blue-grey eyes

Cyanosis – blueness of the skin due to oxygen deficiency

Erythema – abnormal redness of the skin due to fever, hot weather, exercise, etc.

Hematoma – also known as a bruise, which is due to a mass of clotted blood showing through the skin due to some type of trauma

Jaundice – yellowing of the skin and whites of the eyes because of high levels of bilirubin in the blood

Pallor – pale coloration from illness, stress, low blood pressure, cold temperatures, or anemia

IMAGE 5.8 Vitiligo

CLINICAL APPLICATION: RARE SKIN DISEASE AFFECTING PIGMENT

Vitiligo is a skin condition where your melanocytes get destroyed causing white patches to appear on your skin. This disease may affect any area of skin. No one knows the exact cause of vitiligo. There is evidence suggesting it may be caused by a combination of auto-immune, genetic, and environmental factors. Only 1-2% of the population is known to have this disorder.

There are also common types of skin disorders that may affect the coloration of the skin such as **dermatitis** (causes a red inflammation from irritation or an allergic reaction), **eczema** (itchy, red, weepy skin lesions caused by skin allergy), **psoriasis** (recurring, reddened scaly plaques caused by an autoimmune response), and **rosacea** (red, patch-like areas typically on the face and neck worsened by hot weather, hot beverages, or spicy food).

Clinical Reflection Questions

1. What cells are destroyed resulting in vitiligo? _______________________

2. What is the suspected cause of vitiligo? _______________________

THE SECOND REGION OF THE SKIN – THE DERMIS

"So the dermis is the region below the epidermis… right?" questioned Will.

Dr. Ortiz replied, "Yes. It's actually a connective tissue layer composed of chiefly collagen but it also has elastic and reticular fibers and specialized cells like **fibroblasts** (they are the most common cells of connective tissue in animals and they make the extracellular matrix and collagen and play a critical role in wound healing), **macrophages** (a type of white blood cell that is a phagocyte or cell that engulfs and then digests cellular debris and pathogens either as stationary or mobile cells), and occasionally **mast cells** (release substances such as heparin and histamine in response to injury or inflammation of bodily tissues) and **leukocytes** (white blood cells).

"And it's a busy place, meaning there are lots of significant structures located in the dermis. While the epidermis does not have blood vessels in it, the dermis does, for example (note the vein and artery in Image 5.12). It also has cutaneous glands (i.e., oil or **sebaceous glands,** milk or **mammary glands,** ear wax or **ceruminous glands**, and **sweat glands**) and nerve endings. Hair follicles and nail roots are also embedded in the dermis.

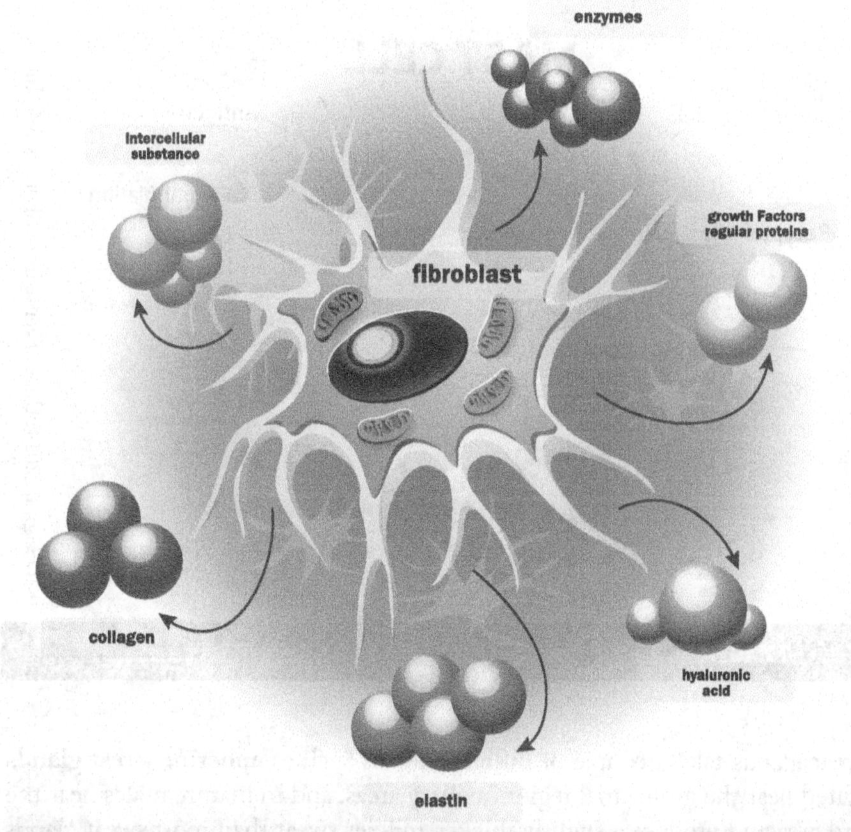

IMAGE 5.9 A Fibroblast is the Most Common Cell in the Connective Tissue of Humans

CELL-MEDIATED IMMUNE RESPONSE

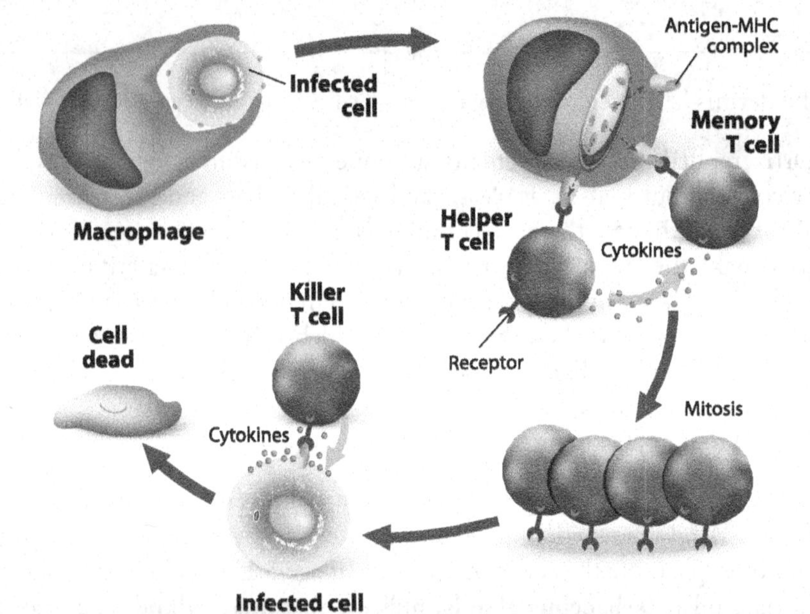

IMAGE 5.10 A Microphage at Work

MAST CELL

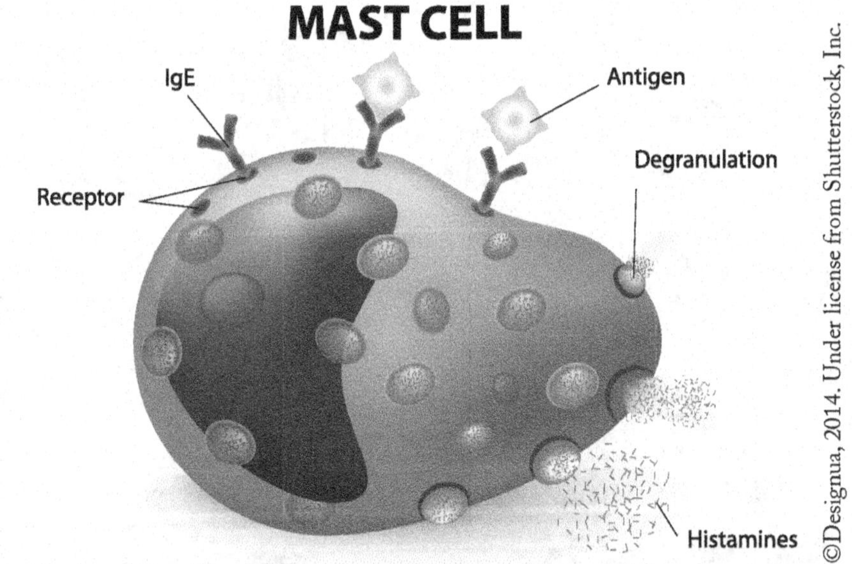

IMAGE 5.11 A Mast Cell Releasing Histamine

"Sweat glands take a couple of different forms such as **apocrine sweat glands** (located near the groin, anal region, axillary areas, and in mature males near the beard area and produce a slightly thicker, milkier sweat than most sweat glands believed to be related to pheromone production) and **eccrine sweat glands** (widely distributed over the entire body especially the palms, soles, and forehead).

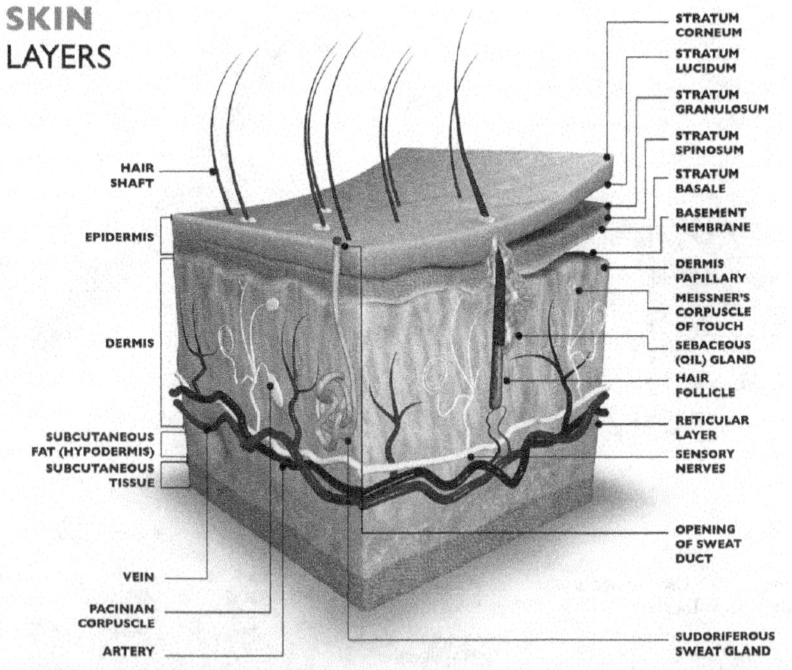

IMAGE 5.12 The Dermis Contains Many Important Structures

"There are two main zones within the dermis. The first or topmost zone is the **papillary layer** which is thin and made up of mainly areolar tissue that is loosely organized to allow for the mobility of leukocytes and other defenses needed for organisms that penetrate through breaks in the epidermis. The papillary dermis contains vascular networks that have two important functions. The first being to support the avascular epidermis with vital nutrients and secondly to provide a network for thermoregulation. The vasculature is organized so that by increasing or decreasing blood flow, heat can either be conserved or dissipated. The papillary dermis also contains the free sensory nerve endings and structures called **Meissner's corpuscles** in sensitive areas like the fingertips. Meissner's corpuscles are mechanoreceptors—they sense low-frequency vibrations (30–50 Hz) that occur when textured objects are moved across the skin.

"The second zone is the **reticular layer** and it is below the papillary layer. It is also much thicker than the papillary layer and is mostly made of dense irregular connective tissue including collagen and elastin fibers. Collagen adds strength and resiliency. Elastin provides for stretch and recoil (see Image 5.15 and the Clinical Application box featuring Ehlers-Danlos syndrome, following the image). Stretching of the skin during pregnancy or excessive weight gain can tear the collagen fibers in this layer and produce stretch marks or **striae**.

"A common sight also in the reticular layer of the dermis are **Pacinian corpuscles** which is a type of touch receptor located in the skin. It is classed as a mechanoreceptor, meaning it is part of the group of sensory receptors that respond to touch and pressure. These type of corpuscles are especially suited to feeling rough surfaces and detecting vibration, and they respond to transient touches rather than sustained

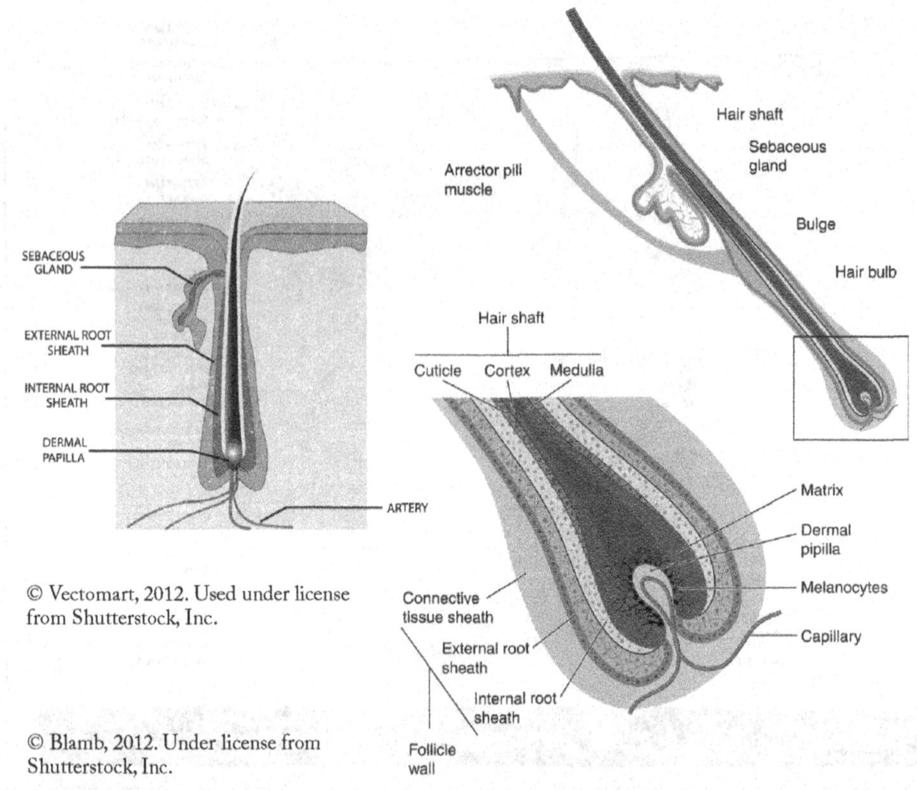

© Vectomart, 2012. Used under license from Shutterstock, Inc.

© Blamb, 2012. Under license from Shutterstock, Inc.

IMAGE 5.13 Hair Bulb with Root (left) and Follicle with Hair Shaft Showing Three Layers (right)

pressure. This is because they are able to quickly adapt to pressure so that it no longer acts as a stimulus. They look like tiny onions. Each corpuscle consists of the end of a sensory nerve fiber with layers of connective tissue wrapped around it. When the skin is touched with enough pressure, the connective tissue capsule is deformed and presses on the nerve ending, triggering an electrical impulse."

APPENDAGES—HAIR AND NAILS

Dr. Ortiz continued, "We mentioned hair follicles and nail roots also originate from the dermal region."

"Don't we have different types of hair on our bodies like fine and coarse?" asked Grace. She continued, "Like, I have to shave the darker hair on my legs but on my sideburn area of my face there is very fine, light blonde hair."

Dr. Ortiz answered, "Yes, good observation. Over the course of our lives we can have several different types of hair. For example, in the last three months of growth, fetuses can be covered in an unpigmented, downy hair called **lanugo**. By the time of birth, most of that hair is replaced with **vellus hair**, which is fine and blonde. Later in life, especially around puberty, coarser, darker hair begins to grow called **terminal hair** (i.e., the hair on our head is terminal hair)."

Dr. Ortiz continued, "Your hair, as well as your skin and nails, is made up of **keratin** (fibrous structural proteins). Hair has three zones: 1) the **bulb** or swelling where the hair originates in the dermis; 2) the **root**, which is enclosed in a follicle. Associated with the follicle are nerve fibers known as hair receptors and muscle fibers known as **arrector pili muscles** that may cause the hair on our arms to stand up in response to fear, cold weather, or other stimuli; and 3) the **shaft** or actual hair that we see sticking out of the skin. Additionally, a piece of hair when looked at under a microscope, has a waxy protective **cuticle** wrapping the outer surface, an outer layer or **cortex**, and an inner layer called a **medulla**."

Will grinned and then added, "Kind of like a cucumber from the grocery store maybe? The peel has wax on it so that would be the cuticle. The peeling is the outer layer so that would be the cortex. The inside would be the medulla."

"Oh Will, you're always making comparisons!" teased Grace.

"But, you're right on," summed up Dr. Ortiz.

Dr. Ortiz added, "Attached to hair follicles are sebaceous glands (small oil-producing glands present in the skin of mammals). They release a fatty substance, **sebum**, into the follicular duct which in turn carries the sebum to the surface of the skin. The glands are distributed over the entire body with the exception of the palms of the hands and the soles of the feet; they are most abundant on the scalp and face."

Will asked, "You mentioned that our sebaceous glands are attached to hair follicles. So, what exactly are hair follicles?"

Dr. Ortiz replied, "Oh yes - A **hair follicle** anchors each hair into the skin. The hair bulb forms the base of the hair follicle. In the hair bulb, living cells divide and grow to build the hair shaft. Blood vessels nourish the cells in the hair root/ bulb, and deliver hormones that modify hair growth and structure at different times of life."

Dr. Ortiz added, "**Acne** is a skin condition that occurs when your hair follicles become plugged with oil and dead skin cells. Acne most commonly appears on your face, neck, chest, back and shoulders."

Will commented, "Yes, I have some acne I deal with now I know more about what causes it. Thanks."

Grace interjected, "So, how does hair grow?"

Dr. Ortiz smiled broadly and commented, "About a half inch a year...". She then continued, "Hair on the scalp grows about .3 to .4 mm/day or about 6 inches per year. Unlike other mammals, human hair growth and shedding is random and not seasonal or cyclical. At any given time, a random number of hairs will be in one of three stages of growth and shedding: anagen, catagen, and telogen."

Anagen

"Anagen is the active phase of the hair. The cells in the root of the hair are dividing rapidly. A new hair is formed and pushes the club hair (a hair that has stopped growing or is no longer in the anagen phase) up the follicle and eventually out.

"During this phase the hair grows about 1 cm every 28 days. Scalp hair stays in this active phase of growth for two to six years.

"Some people have difficulty growing their hair beyond a certain length because they have a short active phase of growth. On the other hand, people with very long hair have a long active phase of growth. The hair on the arms, legs, eyelashes, and eyebrows have a very short active growth phase of about 30 to 45 days, explaining why they are so much shorter than scalp hair."

Catagen

"The catagen phase is a transitional stage and about 3% of all hairs are in this phase at any time. This phase lasts for about two to three weeks. Growth stops and the outer root sheath shrinks and attaches to the root of the hair. This is the formation of what is known as a club hair."

Will asked, "So, what happens when people go bald?"

Dr. Ortiz answered, "The most common kind of baldness is male pattern baldness. Male pattern baldness is related to your genes and male sex hormones. It usually follows a pattern of receding hairline and hair thinning on the crown, and is caused by hormones and genetic predisposition.

"Each strand of hair you have sits in a tiny hole (cavity) in the skin called a follicle. Baldness in general occurs when the hair follicle shrinks over time, resulting in shorter and finer hair. Eventually, the follicle does not grow new hair. The follicles remain alive, which suggests that it is still possible to grow new hair."

Self-Check – Continued

8. What type of gland is in close association with hair follicles? Describe their secretion.

____

____

____

NEW TATTOO TECHNIQUE GIVES HOPE FOR BALD MEN AND CANCER PATIENTS

Ashamed and embarrassed are just two of the emotions that some men experience while losing their hair. But we're going to let you in on a little bit of a secret. There's a new simple, permanent procedure that can give you an A-list look. Hair challenged men have a new alternative starting in 2014.

While they're not getting back actual hair, it might look like it thanks to a new procedure called micropigmentation. Tattoo artists apply small dots to the scalp with a very fine, 3 point micro needle. Ink is applied to the head to make it look like men have hair - almost like a 5 o'clock shadow. It creates the look of a closely shaved head, popular now with Hollywood celebrities like Bruce Willlis, Vin Diesel, Jason Statham and Pitbull.

Many clients are cancer patients, who've lost hair. Eyebrows, and other cosmetic tattooing, that include things like eyeliner, even covering up scars and filling in bald spots can also be done.

The procedure can take two to three treatments. There is a small risk of infection. The tattoo needles do pierce the skin, so it can be uncomfortable. The cost to fill in the scalp can range between $1,600 and $5,000.

IMAGE 5.14 The New Technique Creates a Stubble Illusion

9. Distinguish between the three types of hair.

____

____

____

10. Explain the difference between the hair bulb, root, and shaft.

____

____

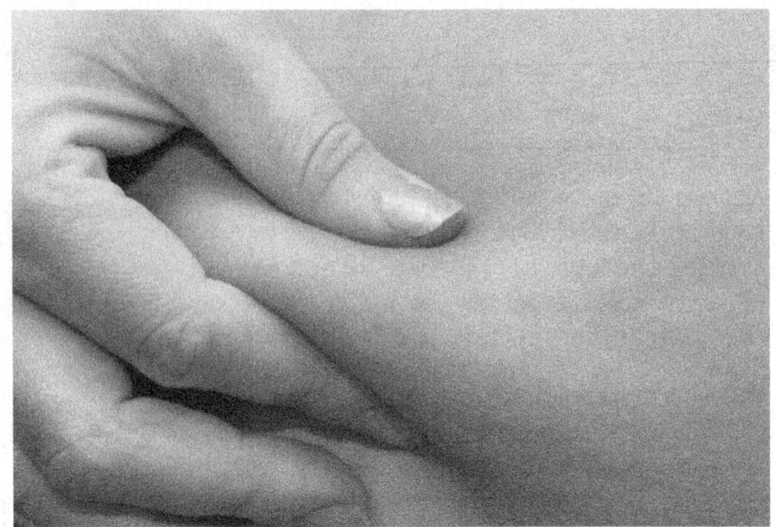

IMAGE 5.15 Stretch-Recoil of Dermis

© Voronin76, 2012. Under license from Shutterstock, Inc.

11. Explain the difference between the hair cuticle, cortex, and medulla.

____

____

____

Grace asked, "What about nails? Do you treat people for nail infections and such?"

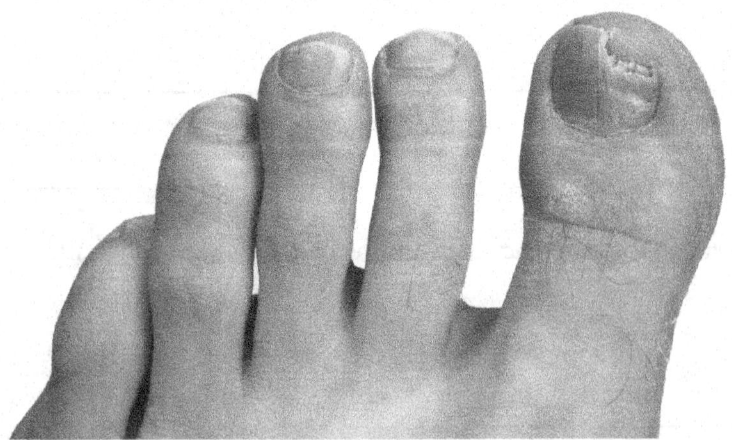

IMAGE 5.16 Nail Fungus

©deepspacedave, 2014. Under license from Shutterstock, Inc.

CLINICAL APPLICATION: EHLERS-DANLOS SYNDROME

If you grab the skin on the back of your hand and pull up, your skin will only stretch so far. And when you let go, it will recoil back into place. This is due to the special stretch-recoil properties in the tissue of the reticular layer of the dermis.

But some individuals suffer from a rare type of disorder caused by a genetic defect in collagen and connective-tissue structure in the reticular layer of the skin's dermis called Ehlers-Danlos syndrome. Although Ehlers-Danlos is collectively known as a group of more than 10 different inherited diseases, one form of the disorder expresses itself as super stretchy skin all over the body. The afflicted person can literally pull their skin away from their body much, much further than normal people. In some case they can take the skin from their neck, for example, and pull it up over their face. There is apparently no pain associated with this behavior or any indication of stretch marks.

There is no specific treatment for this condition. Intelligence is normal. Otherwise, the person has a normal lifespan.

Clinical Reflection Questions

1. How does someone get Ehlers-Danlos syndrome?

   ____

   ____

   ____

2. What causes Ehlers-Danlos? How is the reticular layer of the dermis affected?

   ____

   ____

   ____

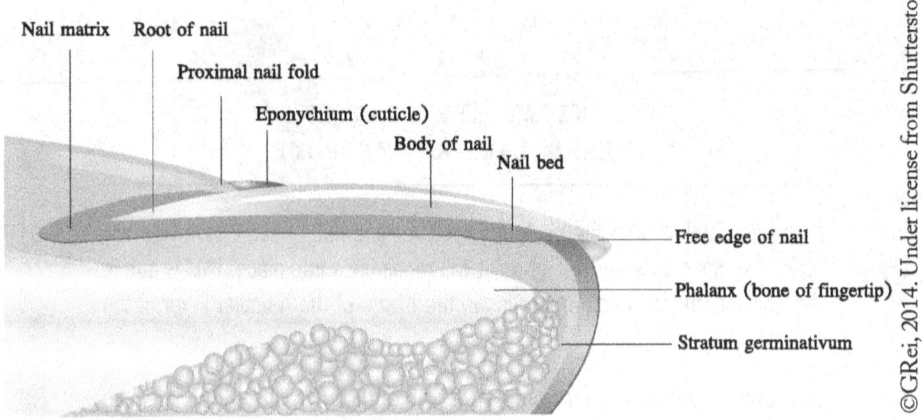

IMAGE 5.17 Nail Anatomy

"Oh yes" replied Dr. Ortiz. She continued, "Treating folks for ingrown nails or nail fungus (a fungal infection that occurs in the nail bed) is very common."

"Did you say nail fungus?" asked Will.

Dr. Ortiz replied, "Yes, nail fungus may begin as a white or yellow spot under the tip of your fingernail or toenail. As the nail fungus spreads deeper into your nail, it may cause your nail to discolor, thicken and develop crumbling edges — an unsightly and potentially painful problem.

"An infection with nail fungus may be difficult to treat, and it may recur."

Dr Ortiz continued, "The fingernail is an important structure made of keratin that has two purposes. The fingernail acts as a protective plate and enhances sensation of the fingertip. The protection function of the fingernail is commonly known, but the sensation function is equally important. The fingertip has many nerve endings in it allowing us to receive volumes of information about objects we touch. The nail acts as a counterforce to the fingertip providing even more sensory input when an object is touched."

She continued, "Nails grow all the time, but their rate of growth slows down with age and poor circulation. Fingernails grow faster than toenails at a rate of 3 mm per month. It takes 6 months for a nail to grow from the root to the free edge. Toenails grow about 1 mm per month and take 12-18 months to be completely replaced."

Dr. Ortiz added, "Anatomically, the structure of the nail is divided into six specific parts - the root, nail bed, nail plate, eponychium (cuticle), perionychium, and hyponychium. Each of these structures has a specific function, and if disrupted can result in an abnormal appearing fingernail. Let's look at the following list that features the different structures, what they do and see an image of nail anatomy in detail (see Image 5. 17)."

Nail Root

The root of the fingernail is also known as the germinal matrix. This portion of the nail is actually beneath the skin behind the fingernail and extends several millimeters into the finger. The fingernail root produces most of the volume of the nail and the nail bed. This portion of the nail does not have any melanocytes, or melanin producing cells. The edge of the germinal matrix is seen as a white, crescent shaped structure called the junula.

Nail Bed

The nail bed is part of the nail matrix called the sterile matrix. It extends from the edge of the germinal matrix, or lunula, to the hyponychium. The nail bed contains the blood vessels, nerves, and melanocytes, or melanin-producing cells. As the nail is produced by the root, it streams down along the nail bed, which adds material to the undersurface of the nail making it thicker. It is important for normal nail growth that the nail bed be smooth. If it is not, the nail may split or develop grooves that can be cosmetically unappealing.

Nail Plate

The nail plate is the actual fingernail, made of translucent keratin. The pink appearance of the nail comes from the blood vessels underneath the nail. The underneath surface of the nail plate has grooves along the length of the nail that help anchor it to the nail bed.

Cuticle

The cuticle of the fingernail is also called the eponychium. The cuticle is situated between the skin of the finger and the nail plate fusing these structures together and providing a waterproof barrier.

Perionychium

The perioncyhium is the skin that overlies the nail plate on its sides. It is also known as the paronychial edge. The perionychium is the site of hangnails, ingrown nails, and an infection of the skin called paronychia.

Hyponychium

The hyponychium is the area between the nail plate and the fingertip. It is the junction between the free edge of the nail and the skin of the fingertip, also providing a waterproof barrier.

THE THIRD REGION OF THE SKIN — THE HYPODERMIS (OR SUBCUTANEOUS TISSUE)

Grace asked, "Isn't there a region under the dermis?"

"Yes," replied Dr. Ortiz. She continued, "It's subcutaneous tissue called the hypodermis. Do you know what the prefix 'hypo' means?"

"I think so… is it under?" Will replied.

"Yes. Under or below. Very good," exclaimed Dr. Ortiz.

Will grinned ear-to-ear. Then he asked, "So what's so special about the hypodermis?"

"Well, its claim to fame is that it is mostly **adipose**," Dr. Ortiz said with a chuckle.

Will continued, "You mean fat?"

"Yes and some areolar tissue too to help bind the skin to the underlying tissues," replied Dr. Ortiz.

CLINICAL APPLICATION: SPRAY-ON SKIN CELLS FOR BURN VICTIMS

There is a new development to help those who have been severely burned. And it is revolutionizing the way we treat patients and helping them recover much, much quicker!

It's called the Skin Gun and it works much like an airbrush gun works when it sprays out thin layers of paint. Only the Skin Gun is loaded with the patient's own solution of stem cells and thin layers are sprayed over the damaged skin tissue.

In many cases, the patient makes a full recovery within days rather than weeks or months by conventional skin graft methods. Additionally, infection rates are low and there is minimal, if any, scarring.

The process was developed by Professor Joerg C. Gerlach and his colleagues at the Department of Surgery at the University of Pittsburg's McGowan Institute for Regenerative Medicine in 2008. If all continues to go well with clinical trials, FDA approval will soon be around the corner and this could be on the market in a few years.

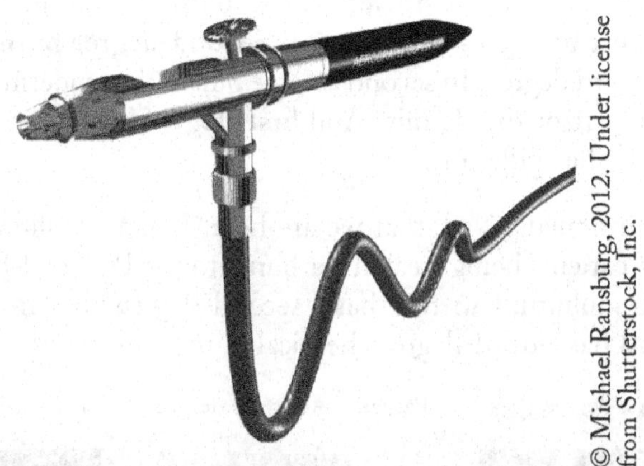

IMAGE 5.18 Skin Gun

Clinical Reflection Questions - refer to Clinical Application box page 130

1. How is spray-on skin applied?

2. What is the advantage over conventional skin grafts?

BURNS

Dr. Ortiz continued, "Follow me… I'm going to take you to the burn unit now at the hospital across the street, so we can better study and observe burn injuries. Are you up to it?"

"Yes," both Grace and Will answered at the same time.

"Okay, let's talk about things along the way. There are three levels of burns. Do you know the difference?" asked Dr. Ortiz.

"I always mix up first and third. Which one is the most serious?" Grace asked.

"That would be third degree," answered Dr. Ortiz. She explained, "With **third-degree burns** the epidermis, all of the dermis, and some deeper tissues like muscle or bone are destroyed. These burns can be life threatening and infections must be aggressively treated. **Second-degree burns** are a little less severe than third degree. In second-degree burns, the epidermis is involved but usually only part of the dermis. And **first-degree burns** are the least serious affecting only the epidermis."

Dr. Ortiz continued, "Now that we are here, I want to allow you to observe some of the patients being treated for burns today. Patient 1 has a first-degree burn from a sunburn. Patient 2 has a second-degree burn from boiling water, and Patient 3 has a third-degree chemical burn from a work-related accident."

Patient 1 – First degree

Patient 2 – Second degree

Patient 3 – Third degree

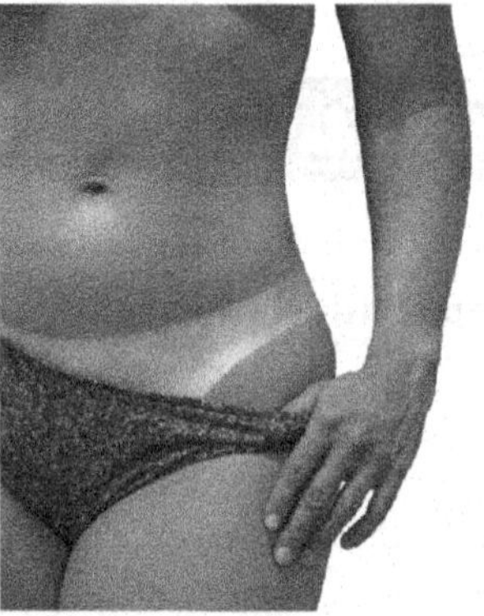

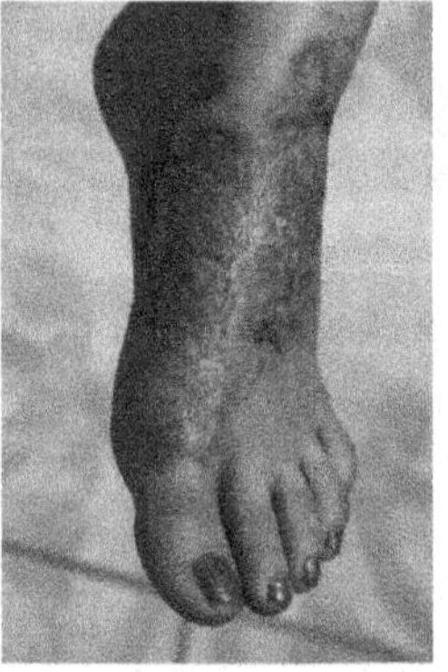

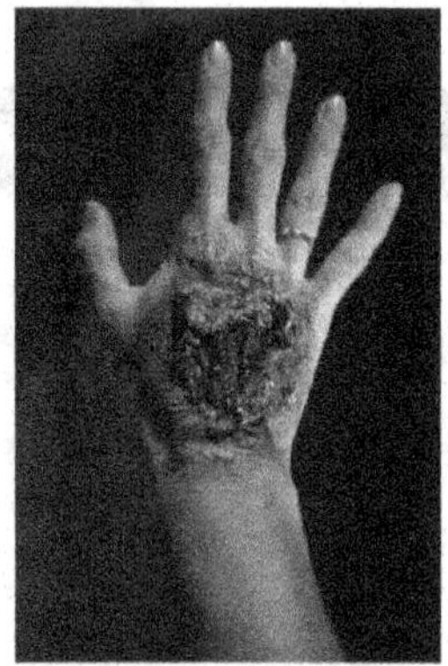

Involves only the epidermis (i.e., bad sunburn)

© Amy Walters, 2012. Under license from Shutterstock, Inc.

Involves the epidermis and part of the dermis (i.e., spilling boiling water from stove onto foot)

© OlegD, 2012. Under license from Shutterstock, Inc.

Extends through the entire dermis and possibly even deeper tissue like muscle (Note: tissue often looks charred)

© Carlos Gutierrez, 2012. Under license from Shutterstock, Inc.

IMAGE 5.19 The Three Types of Burns

Self-Check – Continued

12.   Distinguish between the three types of burns.

____

____

____

Grace asked, "How does our skin heal when we are injured?"

Will added, "Yeah, I'm curious too!"

Dr. Ortiz replied, "When your skin gets cut, your body springs into action to heal the wound. First, the body works to limit blood loss by reducing the amount of blood flowing to the wounded area. Proteins in blood work with the blood platelets already in place and plasma to form a protective covering called a scab. While your skin regenerates underneath the protective layer, the scab protects the wound from outside infection."

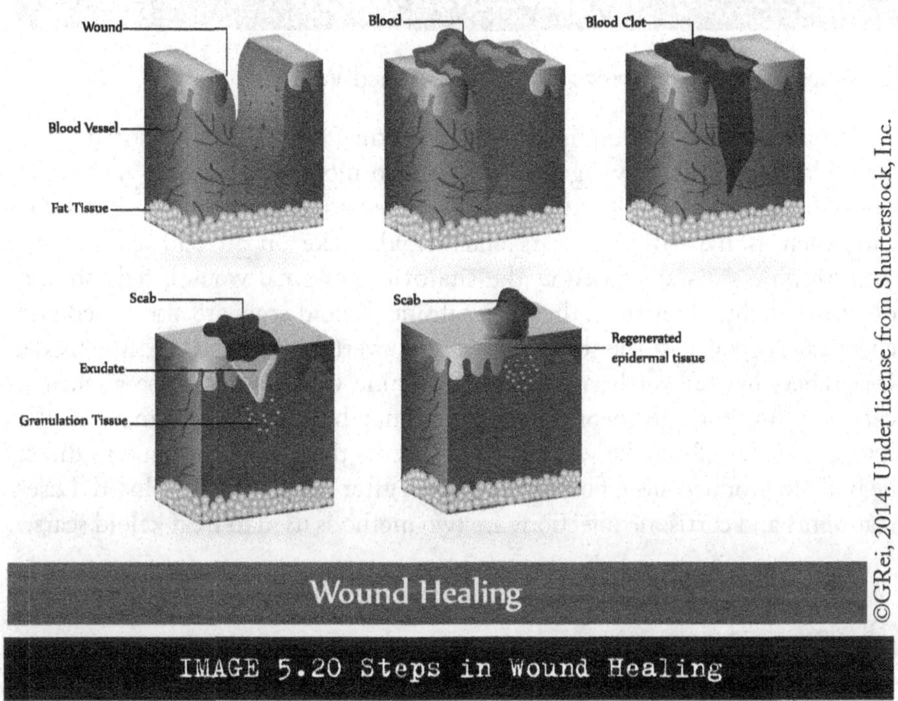

IMAGE 5.20 Steps in Wound Healing

She continued, "The wound is gradually healed as new granular skin tissue begins to generate. Starting at the edges of the wound, the new tissue forms and works its way toward the center until it has covered the entirety of the lesion. Once the wound underneath has sealed itself with another skin layer, the scab will slough off on its own. If the cut or scrape was a shallow one that only affected the outer epidermis layer, then there shouldn't be a scar when your skin heals itself. If the cut went deeper, into the dermis of the skin, then your body moves to create fibrous scar tissue from the granular tissue. In general, the worse the wound, the greater chance that it will result in a scar. Your body needs three to six weeks to bridge a deep cut, producing a protein called collagen at the site of the wound to repair it. Even after the wound is healed, it can take up to two years for a scar to settle into its permanent appearance."

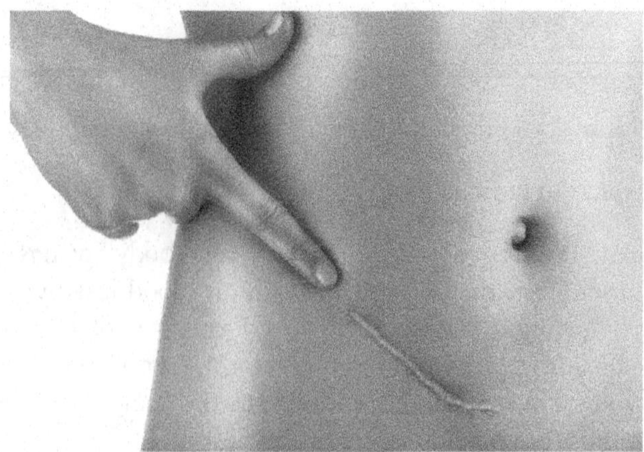

©Artem Furman, 2014. Under license from Shutterstock, Inc.

IMAGE 5.21 Scar

"Does scar tissue have sweat glands?" questioned Will.

Dr. Ortiz replied, "Scar skin tissue isn't like normal skin tissue -- it doesn't have sweat glands or hair growing from it. It's also more vulnerable to ultraviolet rays. Most scars are whitish and lay flat on the surface of your skin. But some scars, such as hypertrophic scars and keloids, take on an odd appearance. Hypertrophic scars are raised at the site of the original wound, reddish and sometimes itchy. Over time, they can subside. Keloid scars are also raised and red, but they grow past the site of the wound, overtaking normal healthy tissue. Researchers haven't yet been able to determine what causes these abnormal scars to form, but one theory is that they may be caused by changes to the signals sent by cells at the wound site. It seems these cells continue to direct the body to produce more fibrous tissue even after the wound has closed. Laser treatments and cortisone injections are two methods used to treat keloid scars."

WRAP-UP

"We've hogged up a lot of your time and we know we need to let you get back to your patients that need you," commented Will.

"Thanks for teaching us so much—I really liked learning about Ehlers-Danlos," said Grace.

"It has been my pleasure to help you better understand the integumentary system," concluded Dr. Ortiz.

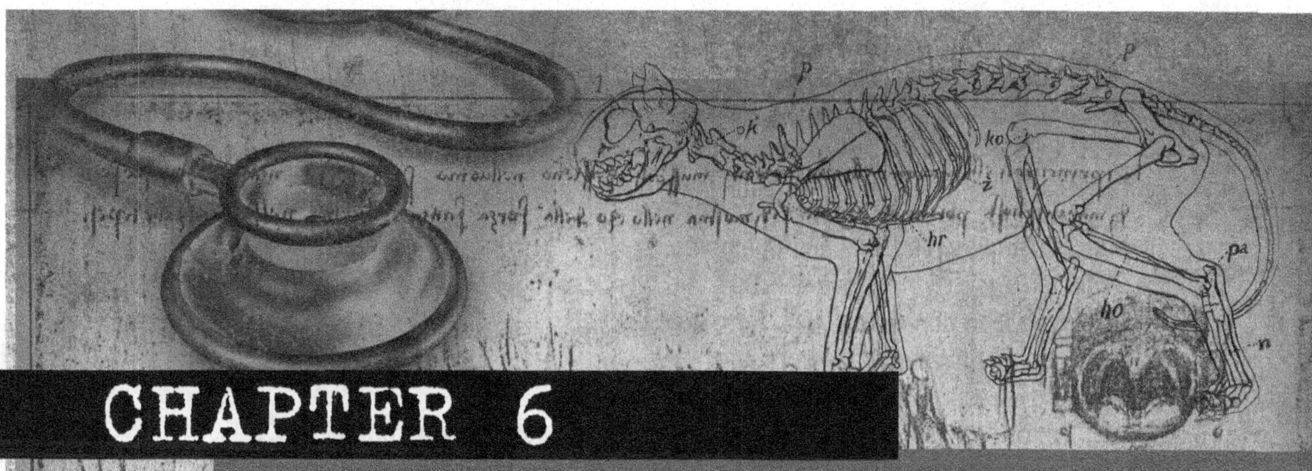

CHAPTER 6

Introduction to Bones: Investigating Bone Tissue and Markings at a Crime Lab

KEY WORDS

The study of Anatomy & Physiology involves many new vocabulary words. It is helpful to gain familiarity with these new key words, just as you would a foreign language.

articular cartilage	long bones
bone marrow	mineralization
canaliculi	osteoblasts
central canals	osteoclasts
chondrocytes	osteocytes
compact bone	osteon (Haversian system)
diaphysis	osteoporosis
diploe	Paget's disease
elastic cartilage	periosteum
endosteum	red bone marrow
epiphyseal line	remodeling
epiphysis	rickets
fibrocartilage	Sharpey's fibers
flat bones	short bones
hyaline cartilage	skeletal system
irregular bones	tendon
lacunae	Volkmann's canals
lamellae	Wolff's Law of Bone
ligament	yellow bone marrow

It is helpful to have an idea of what you need to learn before proceeding. Here is a study guide to assist you:

1. Distinguish between tendons and ligaments.
2. Describe the basic four shapes of bones and give a specific example for each.
3. Distinguish a long bones diaphysis from its epiphysis.
4. Describe the membranes of a long bone.
5. Describe diploe and where is it located.
6. Distinguish osteoblasts from osteoclasts. Explain their role in bone remodeling.
7. Explain an osteocyte and its relationship to lacunae and canaliculi.
8. Differentiate between central canals and Volkmann's canals in compact bone.
9. Explain Wolff's Law of Bone and why you need to know about it.
10. Discuss some of the common bone diseases and disorders.
11. State the functions of bones.

Grace and Will are both extremely interested in bones. Today, they are going to the crime lab to study more about bone cells and tissues, features such as markings, and how bones are used in forensic cases.

INTRO TO BONES

Highly energetic and with a sense of humor, the lead CSI greeted them, "Hello, you guys! I'm Dr. T—that's short for Tommilinski—now you know why folks just say Dr. T!"

"Oh, wow! I never met a real CSI before. I've only seen them on TV on shows like *CSI: Miami* and *Bones*," exclaimed Grace.

"What's it like to be a CSI?" questioned Will.

Dr. T continued, "Why don't you both come on back and I'll show you. I'm working on a case right now involving a homicide scene where bones were found buried in a shallow, unmarked grave near a park. Here's a photo from the actual sight where the bones were found. Have a look… (see Image 6.1).

"I've got the job of figuring out what happened. In order to accomplish that, I'll have to study each bone closely for further analysis. I'll also eventually have to put the body back together from the remaining bones. It doesn't appear that we have the entire skeleton but that's not so unusual in crime scenes," said Dr. T.

© Marques, 2012. Under license from Shutterstock, Inc.

IMAGE 6.1 Found Skeletal Remains

TENDONS AND LIGAMENTS

"In cases like this where the bones may have been in the ground for a couple of years already, there is not much, if any, tendon or ligament evidence—just the bones. Are either of you familiar with what tendons or ligaments do in the human body?" questioned Dr. T.

"Uh-h, don't **tendons** hold skeletal muscle to bone?" Will answered hesitantly.

"And, don't **ligaments** bind bone directly to bone?" winced Grace.

"Yes. And what about cartilage?" Dr. T questioned further.

"It pads the ends of the bones and helps hang the skeleton together I think. But I don't see any cartilage present either," said Will.

Dr. T questioned Will and Grace further, "Do you know there are different kinds of cartilage?"

"Not really" answered Will.

Dr. T continued, "Well - there's **articular cartilage** (the smooth, white type that pads the articular surfaces (ends) of most bones and is typically made up of hyaline cartilage). Here is an example of articular cartilage in the knee (see Image 6.2)."

Anatomy of the Human Knee Joint

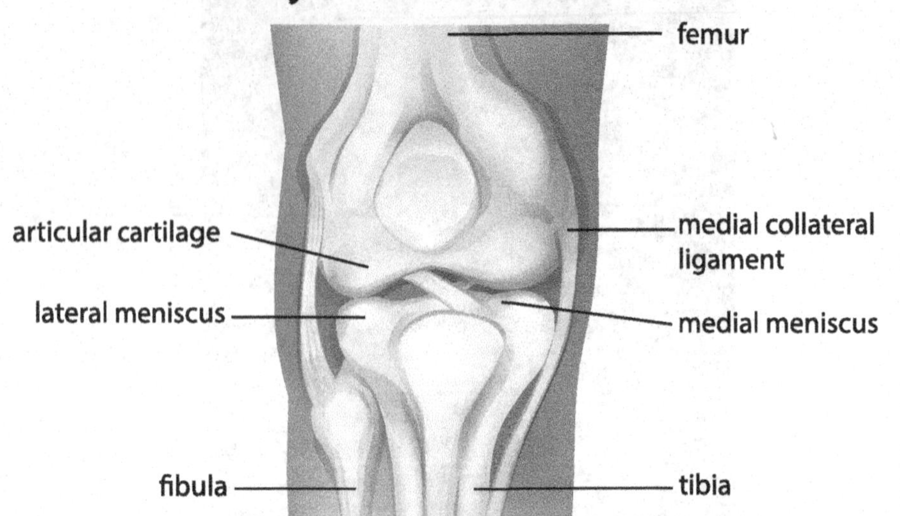

IMAGE 6.2 An Example of Articular Cartilage in the Knee

Grace asked, "Isn't **hyaline cartilage** the most common type in the body?"

Dr. T replied, "Yes, very common - found on the ends of most bones, especially, long bones. It also pads the ribs (rib cartilage is also known as costal cartilage). It is also found in the nose, larynx, and trachea. It is comprised of mostly collagen fibers." He continued, "The second type of cartilage found in the human body is elastic cartilage. Do you know about it?"

Will and Grace both shook their heads to indicate no.

Dr. Barr replied, "**Elastic cartilage** is as the name states - very 'elastic as is made up of mostly elastic fibers'. It is found in the external ear, epiglottis and larynx." He continued, "And then there is **fibrocartilage** which is the strongest type of cartilage because it has alternative layers of both hyaline cartilage matrix and collagen fibers. It is found in the vertebral discs, joint capsules, ligaments."

"Special cells known as **chondrocytes** are found in cartilage," said Will.

"Yeah true. All together, the bones, cartilages, and ligaments actually make up the **skeletal system,**" confirmed Dr. T. Then he continued, "We need to look at the gross anatomy first and do a general cataloging of what bones we have here. I'll start with the victim's skull."

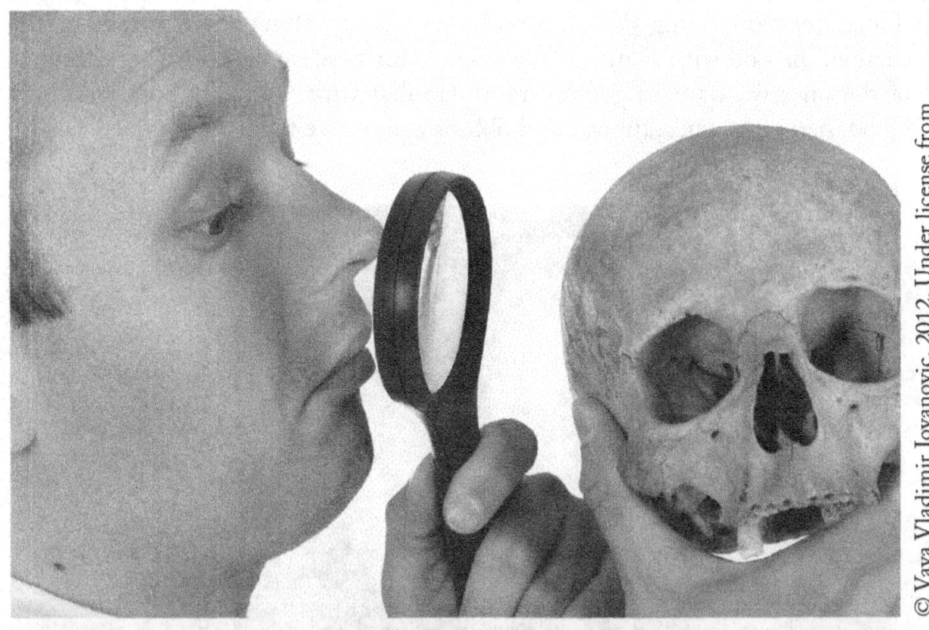

IMAGE 6.3 Dr. T Examining Victim's Skull

STUDENT VIEW – WHY SHOULD I STUDY THIS? HOW IS IT RELEVANT TO MY LIFE?

Perhaps you have broken a bone? Or know someone who suffers from osteoporosis? By studying bone tissue and how bones repair and remodel, you can better understand how you heal from a fracture or deal with a chronic debilitating disease in a loved one.

BONE AND OSSEOUS TISSUE

"Do you know much about bone as a tissue?" asked Dr. T.

"Not much," answered Will.

"I know it is made up of connective tissue hardened by deposits of calcium, phosphate, and other minerals," answered Grace.

"Ah-h! We're off to a good start," exclaimed Dr. T. He continued, "The hardening process is called **mineralization**."

"Is bone the hardest substance in the body?" asked Will.

"Good question, young Will. It is not! That distinguishing honor goes to tooth enamel," he said with a smirk. "There are many tissues making up bone. Some of the ones we want to get you more familiar with today are bone marrow, blood, nervous tissue, adipose, and fibrous connective tissue," continued Dr. T.

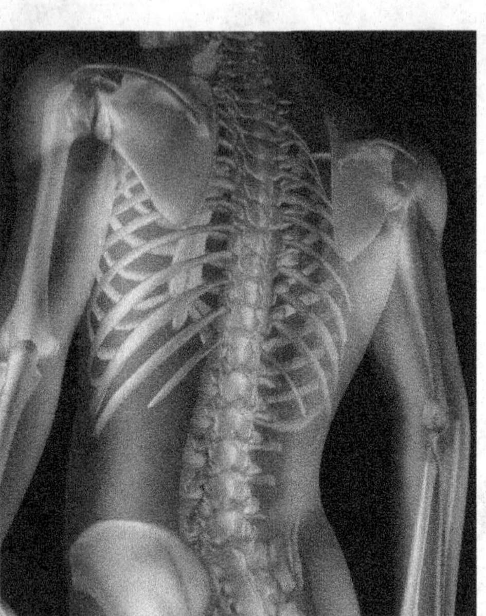

FIGURE 6.4 Axial Skeleton is the Central Axis of the Body — Skull, Vertebral Column and Ribs

"First let's get these bones classified and grouped. Bones are first classified according to bones of the **axial skeleton** such as the skull, vertebrae, and ribs. The axial skeleton is made up of 80 different bones. The **appendicular skeleton** is made up of 126 bones comprising the limbs, shoulders and hips—all areas of movement and motility. The word "appendicular" refers to appendages (things on the peripheral like arms and legs).

"Bones are classified next by shape. There are **flat bones** like the bones of the skull. The sternum is also a flat bone. And, **long bones** that are longer than they are wide like the upper arm bone—the humerus, and the upper thigh bone—the femur. There are **short bones** which are little and squarish in shape like wrist and ankle bones. And finally, there are also **irregular bones**—ones that defy description because they are odd shapes. A good example is the pelvis and individual vertebrae in the backbone. Let's view some examples…" expounded Dr. T.

TABLE 6.1 Examples of Bone Shapes

Bone Shape and Description	Bone Example
Long bones—longer than they are wide, such as the humerus (upper arm) and femur (thigh bone)	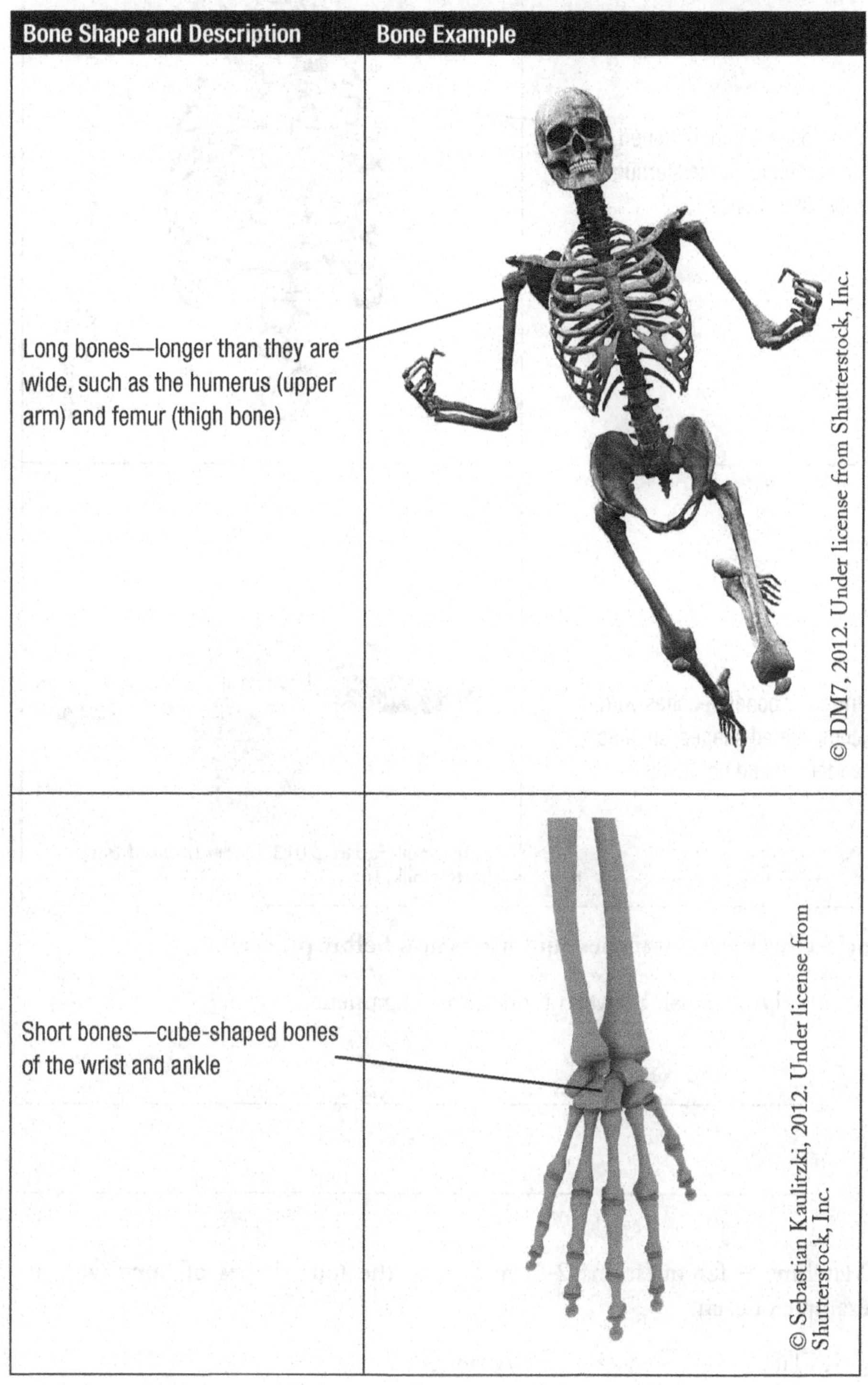 © DM7, 2012. Under license from Shutterstock, Inc.
Short bones—cube-shaped bones of the wrist and ankle	© Sebastian Kaulitzki, 2012. Under license from Shutterstock, Inc.

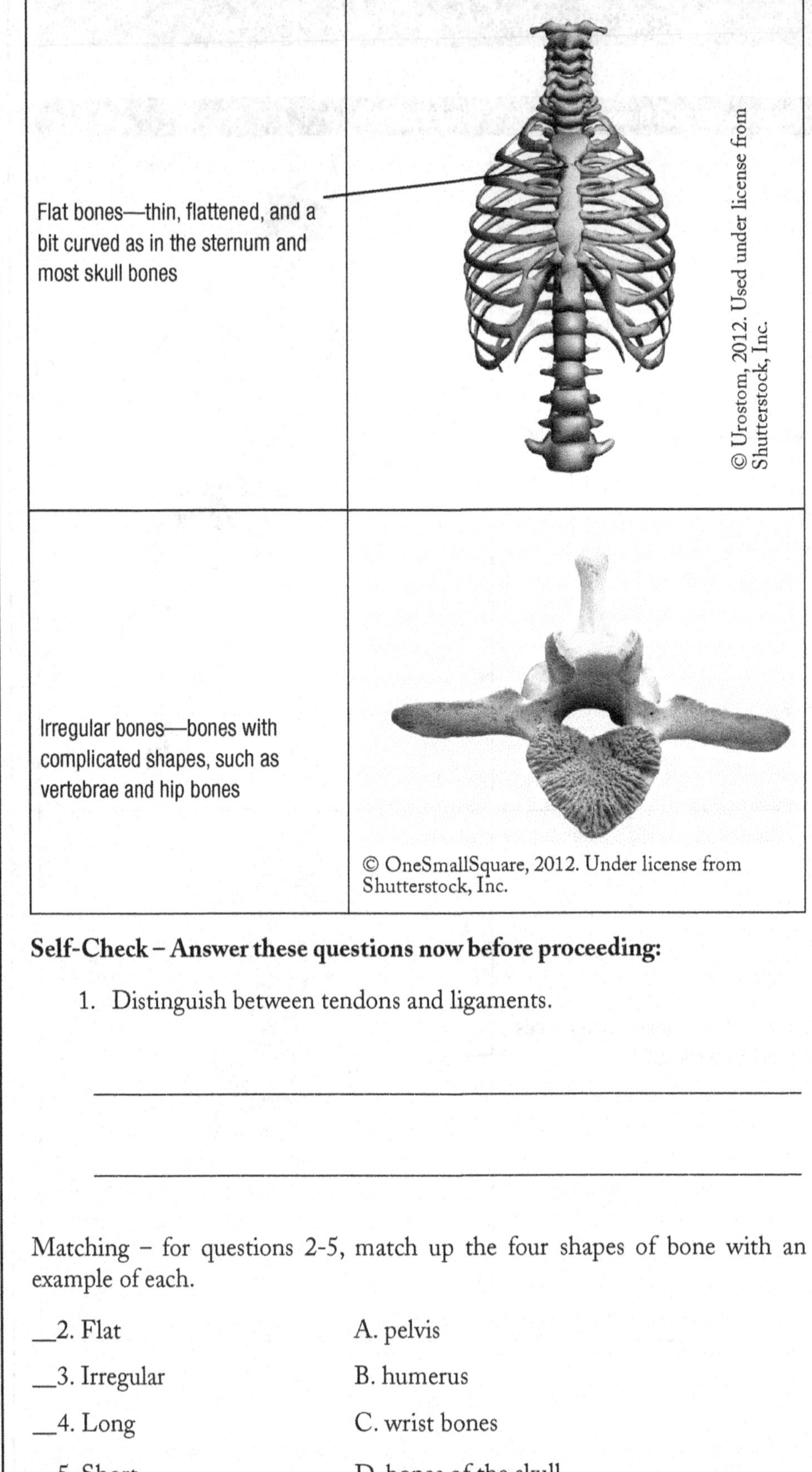

Self-Check – Answer these questions now before proceeding:

1. Distinguish between tendons and ligaments.

____

____

Matching – for questions 2-5, match up the four shapes of bone with an example of each.

___2. Flat   A. pelvis

___3. Irregular   B. humerus

___4. Long   C. wrist bones

___5. Short   D. bones of the skull

BONE MARKINGS

Dr. T explained, "Bones are not smooth. They have anatomical features such as ridges, bumps, depressions, canals, slits, etc. These features are known as **bone markings**. Bone markings are great places for muscles to attach to or for nerves and blood vessels to run through. Sometimes, knowing the bone markings can help us identify a bone if the bone is disarticulated (removed from the rest of the skeleton). Let's examine some of the bone markings on the victim's bones…"

TABLE 6.2 General Bone Markings

Bone Marking	Description and Location
Facet—typically seen on a vertebrae but can be where almost any two bones touch or articulate; a smooth, flat articular (adjoining; abutting) surface.	© zimmytws, 2012. Under license from Shutterstock, Inc.
Foramen—typically seen on the skull; a round or oval opening through a bone.	© Convit, 2012. Under license from Shutterstock, Inc.

Trochanter—only seen on the upper leg bone; a blunt, irregularly shaped process.

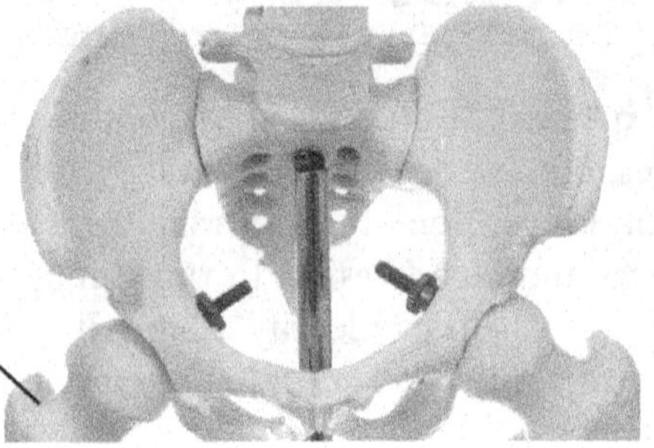

© val lawless, 2012. Under license from Shutterstock, Inc.

Sinus—air-filled cavity lined with a mucous membrane within a bone (as seen in the skull).

© val lawless, 2012. Under license from Shutterstock, Inc.

Crest—typically seen on the pelvis; a narrow ridge of bone that is usually prominent.

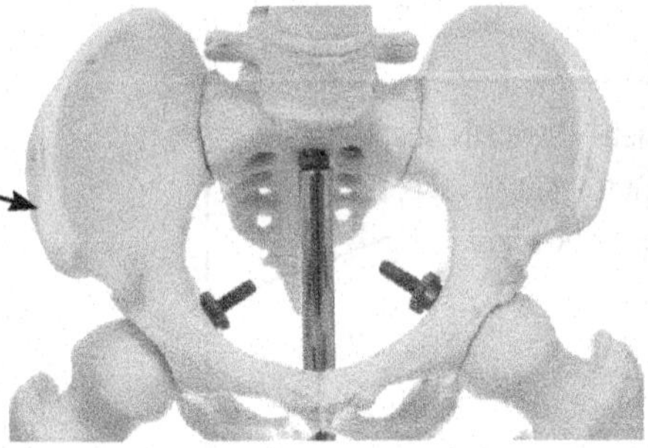

© val lawless, 2012. Under license from Shutterstock, Inc.

Head—an example is the upper leg bone; a bony expansion carried on a narrow neck.

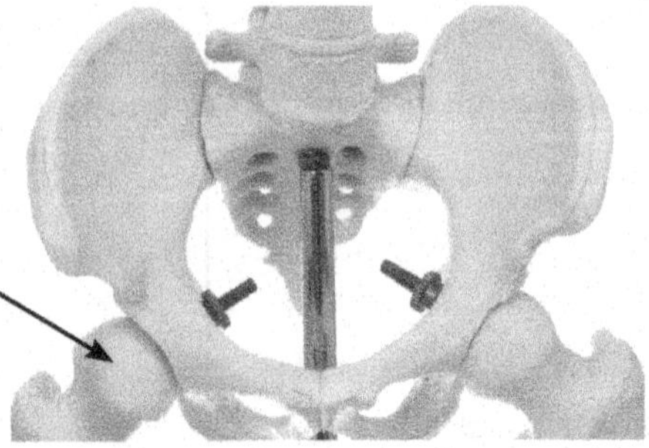

© val lawless, 2012. Under license from Shutterstock, Inc.

Ramus—seen on the lower jaw; an arm-like bar of bone.	© leonello calvetti, 2012. Under license from Shutterstock, Inc.
Tubercle—seen on posterior sacrum and femur; small rounded projection or process.	©Puwadol Jaturawutthichai, 2014. Under license from Shutterstock, Inc.
Tuberosity—seen on the lower arm bone in line with the thumb; large rounded projection; may be rough.	© Henri et George, 2012. Under license from Shutterstock, Inc.

Fossa—pocket; a shallow basin-like depression or pocket in the bone.	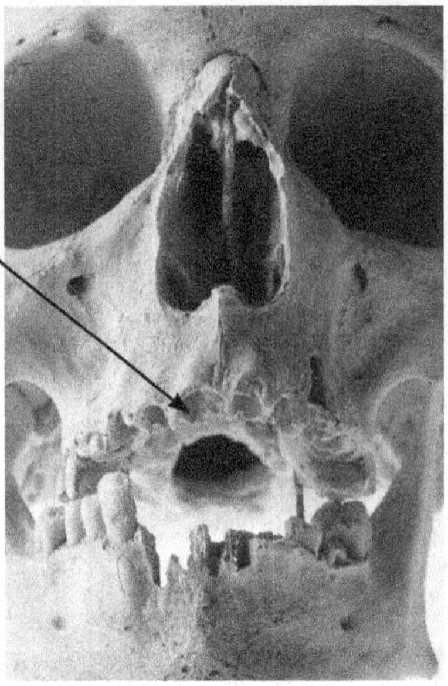 © Fotografiche, 2012. Under license from Shutterstock, Inc.
Fissure—seen in the eye orbits; a narrow slit-like opening.	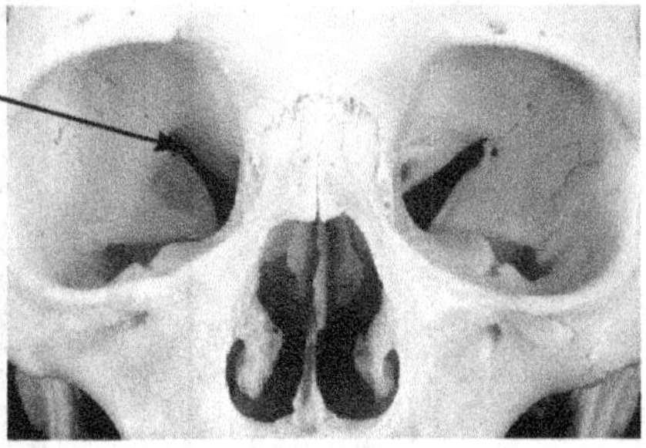 © Anthony Ricci, 2012. Under license from Shutterstock, Inc.
Meatus—as seen in the ear canal; a canal-like passageway.	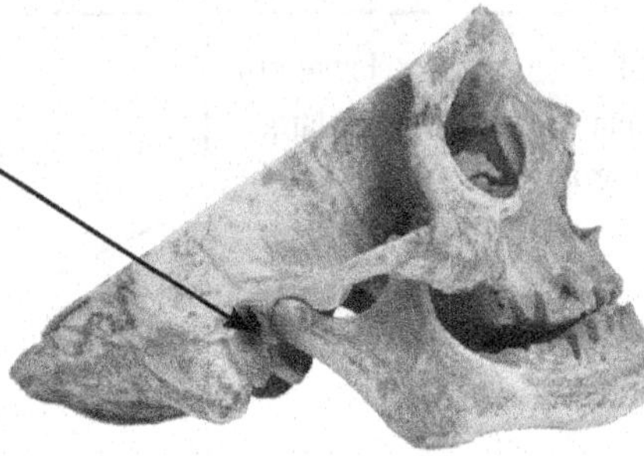 © Kacso Sandor, 2012. Under license from Shutterstock, Inc.

Epicondyle—seen on the upper femur bone; raised area on or above a condyle.

© JupiterImages LLC

Line—seen on the pelvis; a narrow ridge of bone less prominent than a crest.

© val lawless, 2012. Under license from Shutterstock, Inc.

Groove—seen on the costal area of the ribs; a furrow.

© 3DDock, 2012. Used under license from Shutterstock, Inc.

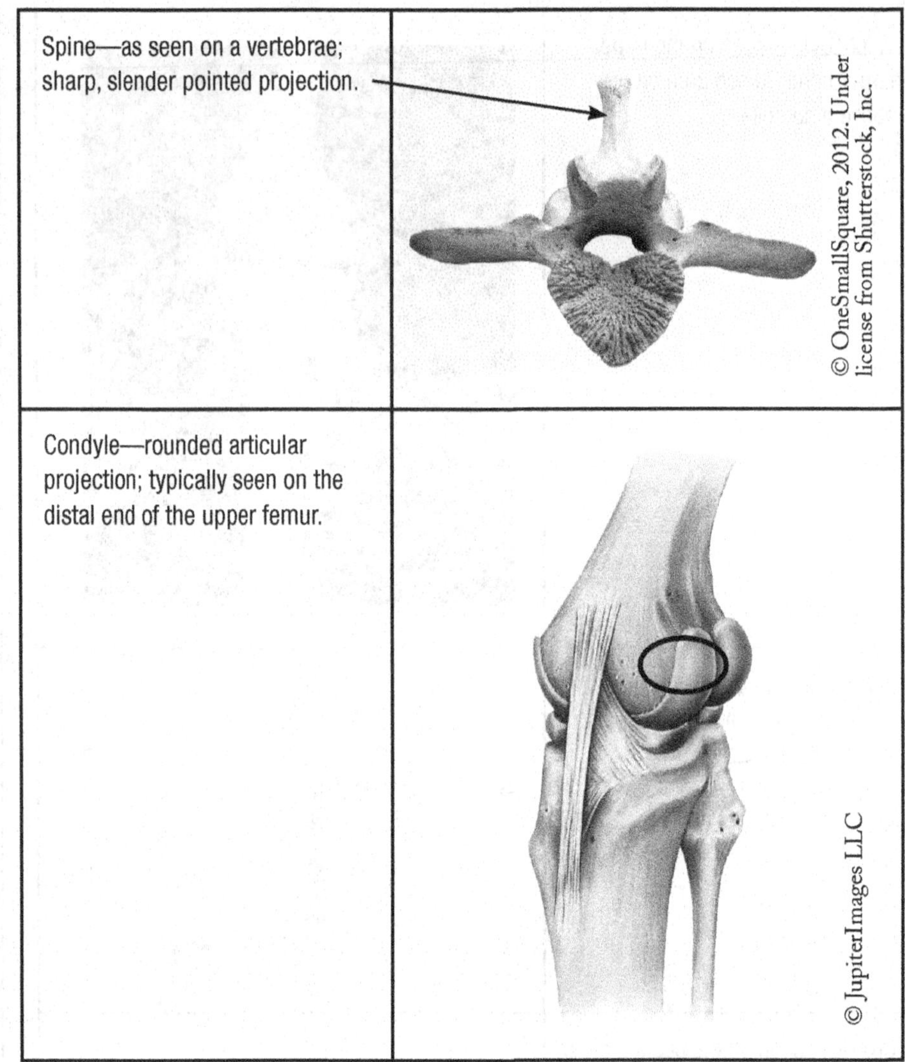

Self-Check – Continued

6. Your turn! Match up the bone marking with its general description:

Bone Marking	Answer Choices
facet answer: ___	A. An example is the femur – a bony expansion carried on a narrow neck
foramen answer: ___	B. Air-filled cavity lined with a mucous membrane within a bone (as seen in the skull)

trochanter answer: ___	C. Seen on femur – small rounded projection or process
process answer: ___	D. As seen on a vertebrae – sharp, slender pointed projection
sinus answer: ___	E. Seen on the mandible – an arm-like bar of bone
crest answer: ___	F. Seen on the pelvis – a narrow ridge of bone less prominent than a crest
head answer: ___	G. Only seen on the femur – a blunt, irregularly shaped process
ramus answer: ___	H. As seen in the ear canal – a canal-like passageway
tubercle answer: ___	I. Rounded articular projection (typically seen on the femur)
tuberosity answer: ___	J. Seen in the eye orbits – a narrow slit-like opening
fossa answer: ___	K. Typically seen on the mandible – a round or oval opening through a bone
fissure answer: ___	L. Seen on the femur – raised area on or above a condyle
meatus answer: ___	M. shallow basin-like depression in a bone; like a pocket

epicondyle answer: ___	N. Seen on the costal area of the ribs – a furrow
line answer: ___	O. Typically seen on the iliac – a narrow ridge of bone that is usually prominent
groove answer: ___	P. Seen on the radius – large rounded projection; may be rough
spine answer: ___	Q. Typically seen on a vertebrae – a smooth, flat articular surface
condyle answer: ___	R. Any bony prominence

GENERAL ANATOMY OF BONES

"This victim's bones are dried out so there is not much marrow. But in living bones, there is a lot of **bone marrow** (the soft bone that makes blood), especially in the upper arms, upper legs, and pelvis. I'm warming up some steak to have for lunch. Inside the steak bones is marrow. Have you seen something like this before?" asked Dr. T.

© Chiyacat, 2012. Under license from Shutterstock, Inc.

IMAGE 6.5 Bone Marrow

"Absolutely," responded Will.

Dr. T continued, "There are two types of bone marrow - red and yellow. The difference between red and yellow bone marrow is the function that they carry out. **Red bone marrow** produces blood cells, is active and undergoes cell division. **Yellow bone marrow** is the producer of blood cells during an emergency, is generally inactive and contains fat.

"Okay, now let's have a look at basic bone anatomy," stated Dr. T.

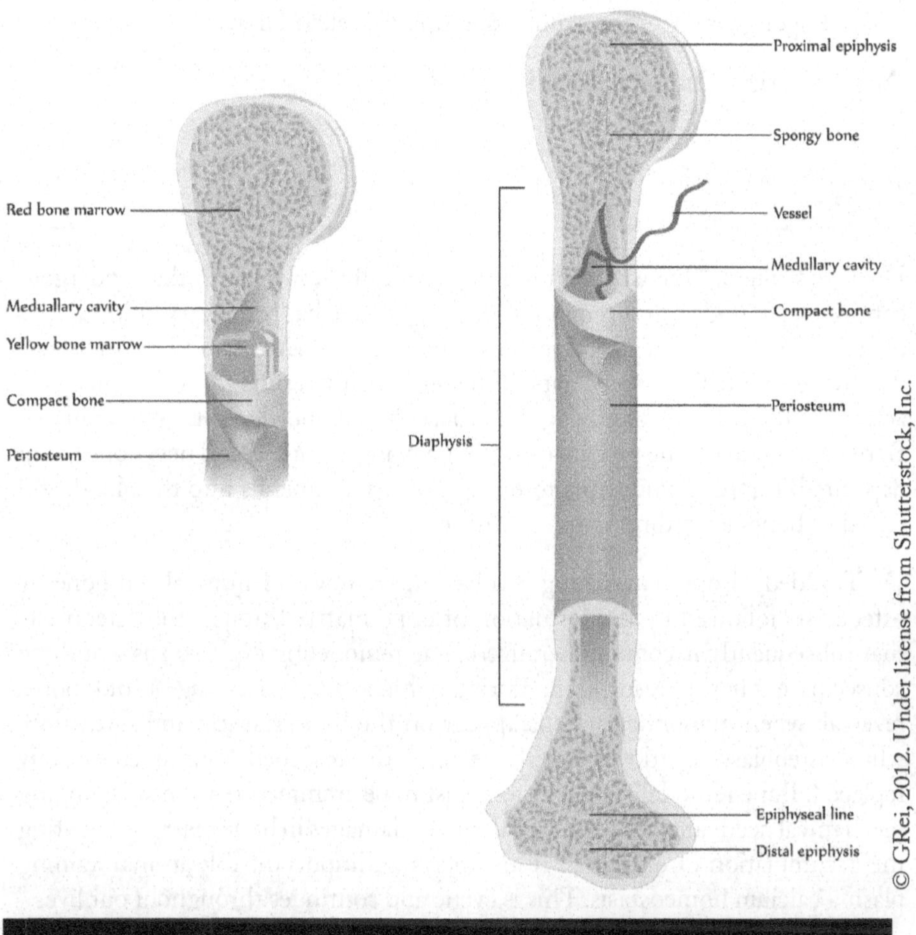

IMAGE 6.6 Bone Anatomy

Dr. T continued, "The bone marrow is wrapped by hard outer **compact bone** made up of a few different membranes—the outermost membrane or **periosteum** and the inner membrane or **endosteum** (the tissue lining the medullary cavity of a bone). The periosteum is the thick fibrous membrane covering the entire surface of a bone except its articular cartilage and serving as an attachment for muscles and tendons. The shaft of a long bone is called the **diaphysis**. The ends are called the **epiphysis**. There is often a visible **epiphyseal line** (an actual line that looks engraved) separating the shaft from the end of the bone. **Sharpey's fibers** are continuous collagen fibers that help anchor the periosteum to the tendons that connect muscle to bone. This adds greater strength.

"The skull has no diaphysis or epiphysis on its flat bones. It has a sandwich-like construction with two layers of compact bone enclosing a middle layer of spongy bone called **diploe** (dip-lo-ee). Diploe is the central layer of spongy, porous, bony tissue between the hard outer and inner bone layers of the cranium. Diploe generally reduces the weight of bone."

"I guess that's a good thing if you are talking about the weight of your skull and having strong enough neck muscles to hold your head up properly," remarked Will.

"So, the spongy layer in the skull is the diploe?" asked Grace.

"Yes, that's right," answered Dr. T.

BONE HOMEOSTASIS

Dr. T continued, "The skeleton is a metabolically active organ that undergoes continuous remodeling throughout life. Bone tissue has some very unique types of cells. There are bone-forming cells called **osteoblasts**. And there are bone-dissolving cells called **osteoclasts**. Bone remodeling results from the combined action of these bone-dissolving osteoclasts and bone-depositing osteoblasts. Bone remodeling is the process of old bone being absorbed and new bone being deposited. It repairs micro-fractures, releases vital minerals into the blood, and reshapes bones according to use or disuse."

Dr. T added, "Bone remodeling involves the removal of mineralized bone by osteoclasts followed by the formation of bone matrix through the osteoblasts that subsequently become mineralized. The remodeling cycle consists of three consecutive phases: resorption, during which osteoclasts digest old bone; reversal, when mononuclear cells appear on the bone surface; and formation, when osteoblasts lay down new bone until the resorbed bone is completely replaced. Bone remodeling serves to adjust bone architecture to meet changing mechanical needs and it helps to repair micro damages in bone matrix preventing the accumulation of old bone. It also plays an important role in maintaining plasma calcium homeostasis. This is cyclic and continues throughout our lives."

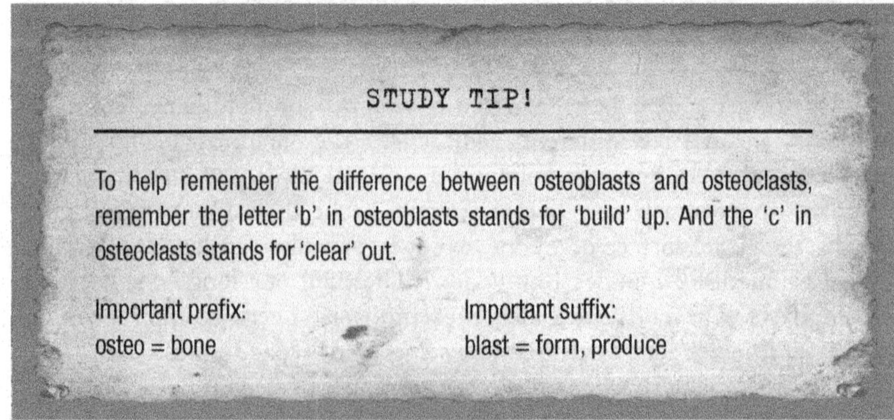

Osteoblasts (top layer) on surface of bone (pink bottom area)

osteoclasts

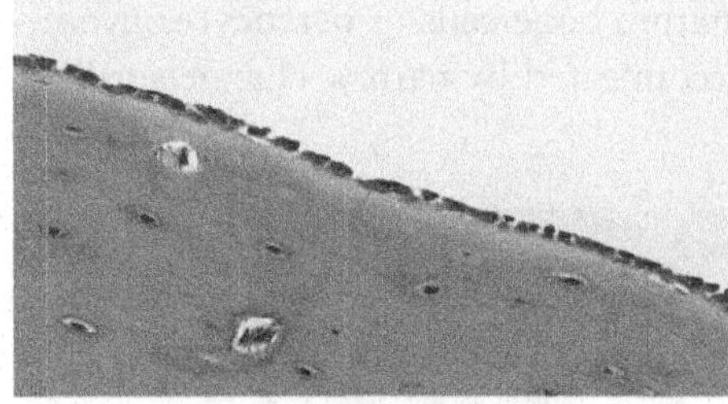

© vetpathologist, 2012. Under license from Shutterstock, Inc.

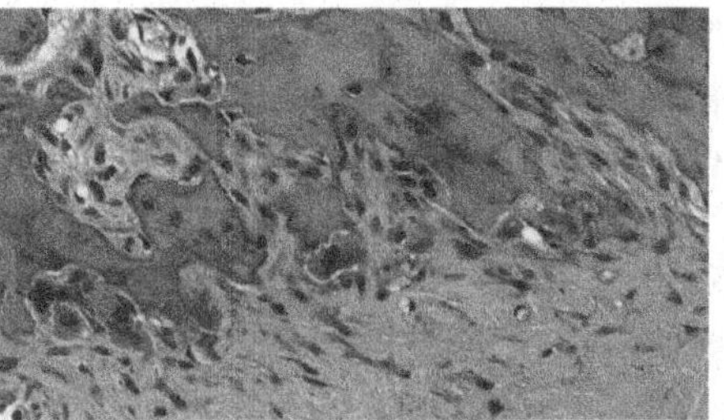

© vetpathologist, 2012. Under license from Shutterstock, Inc.

IMAGE 6.7 Osteoblasts and Osteoclasts

Dr. T explained further, "The famous German anatomist and surgeon, Dr. Julius Wolff who lived from 1836-1902, developed **Wolff's Law of Bone** which states that the architecture of bone is determined by the stresses place upon it. Bone responds in kind by growing wider and stronger. For example, a professional tennis player's racket arm bones are typically more robust than the other arm. In fact, people engaged in heavy manual labor or who are athletic usually have bones with greater density. Essentially, we should remain active in our lives doing activities that require the bones to be under positive stress. These activities should cause some jarring or weight-bearing stresses such as running, walking, and jumping rope rather than no jarring such a stationary biking, an elliptical machine, and swimming. Weight lifting is also good for the bones because it is a weight-bearing exercise."

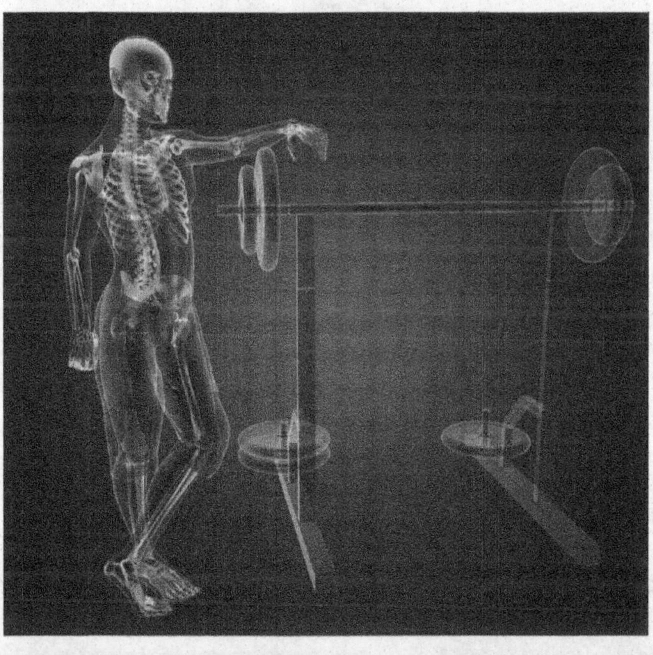

©videodoctor, 2014. Under license from Shutterstock, Inc.

IMAGE 6.8 Weight-bearing Exercise is a Great Way to Maintain Bone Throughout Your Life

BONE MICROANATOMY

"When it comes to actual bone tissue, mature bone cells or **osteocytes** live in tiny cavities called **lacunae** that are interconnected by narrow channels called **canaliculi**."

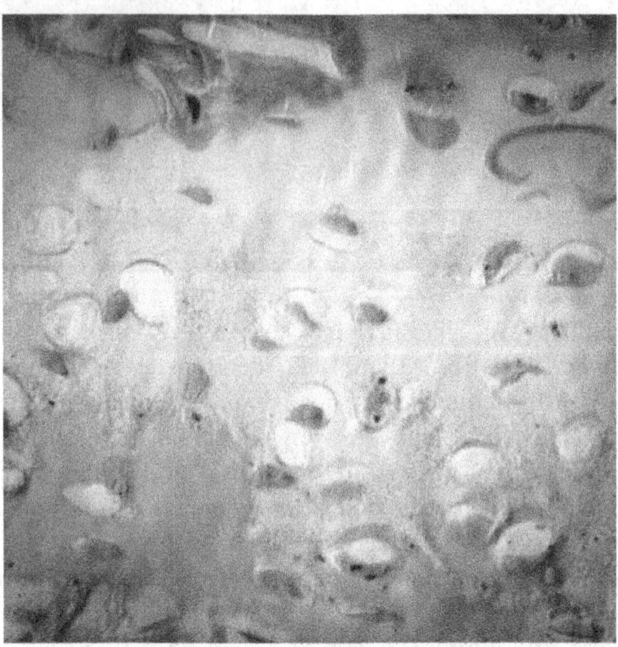

©Pan Xunbin, 2014. Under license from Shutterstock, Inc.

IMAGE 6.9 Osteocytes Inside Lacunae 40 121623757 X

STUDY TIP!

Think of it like this, if a person were an osteocyte, they would live in a home or lacunae. Houses are interconnected by streets called canaliculi.

FUNCTIONS OF BONES

Dr. T stated, "Bones have many functions, including the following:

Support: Bones provide a framework for the attachment of muscles and other tissues.

Protection: Bones such as the skull and rib cage protect internal organs from injury.

Movement: Bones enable body movements by acting as levers and points of attachment for muscles.

Mineral storage: Bones serve as a reservoir for calcium and phosphorus, essential minerals for various cellular activities throughout the body.

Blood cell production: The production of blood cells, or hematopoiesis, occurs in the red marrow found within the cavities of certain bones.

Energy storage: Lipids, such as fats, stored in adipose cells of the yellow marrow serve as an energy reservoir."

COMPACT BONE

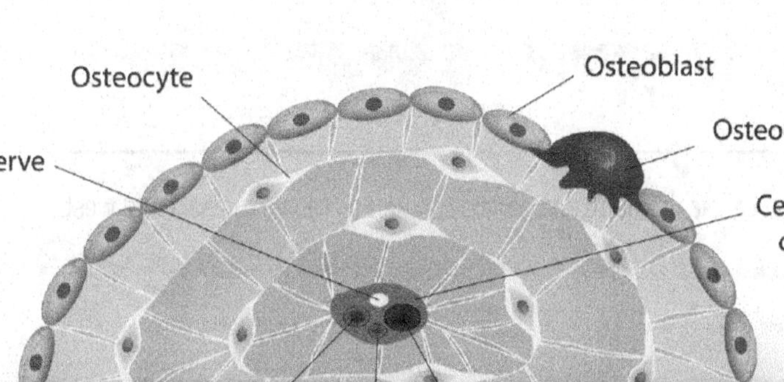

IMAGE 6.10 An Osteon

Dr. T continued, "If you take a slice of normal bone (compact bone) and study it under a microscope like this specimen… you can see certain common structures. There are concentric rings called **lamellae** arranged around a **central canal** that provides an opening for blood vessels and nerve fibers that serve the needs of the osteon's cells. The lamella are not all created equal meaning not all the lamellae in compact bone are part of osteon's. The central canal and associated lamellae rings together make up an **osteon** (or **Haversian system**)—the main structural unit of bone. Each osteon is an elongated cylinder oriented parallel to the long axis of the bone. Functionally, osteons are weight-bearing pillars. In other words, an osteon is a group of hollow tubes of bone matrix, one placed outside the next like the growth rings of a tree trunk. Horizontal canals called **Volkmann's canals** provide further conduits for blood vessels."

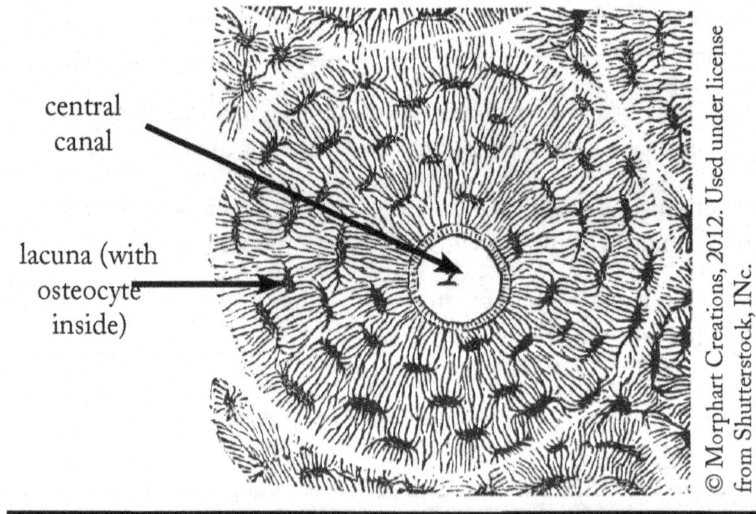

IMAGE 6.11 Osteon, Central Canal, Lamellae, and Volkmann's Canal

STUDY TIP!

Central canals run north and south. Volkmann's canals run east and west.

SPONGY BONE

Dr. T continued, "So, there are two types of bone tissue: compact and spongy. We have already talked about compact or 'hard' bone. Now lets talk about spongy bone. Spongy (cancellous) bone is lighter and less dense than compact bone. Spongy bone consists of plates (trabeculae) and bars of bone adjacent to small, irregular cavities that contain red bone marrow. The canaliculi connect to the adjacent cavities, instead of a central Haversian canal, to receive their blood supply. It may appear that the trabeculae are arranged in a haphazard manner, but they are organized to provide maximum strength similar to braces that are used to support a building. The trabeculae of spongy bone follow the lines of stress and can realign if the direction of stress changes."

Self-Check – Answer these questions now before proceeding:

7. Distinguish between a long bone's diaphysis and epiphysis.

____

____

____

BONE ANATOMY

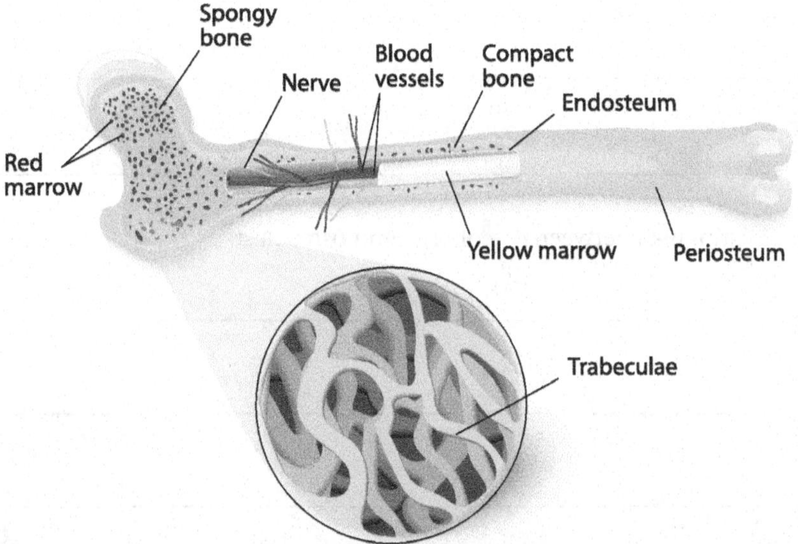

IMAGE 6.12 Spongy Bone and Trabeculae

8. Describe the membranes that make up a bone.

____

____

____

9. What is diploe?

____

____

____

10. Explain Wolff's Law of Bone. Give supporting examples.

____

____

____

11. What are osteocytes? Where are they located specifically?

____

____

____

12. Distinguish between osteoblasts and osteoclasts.

____

____

____

13. Draw a section of compact bone. Label the following structures: osteon, central canal, lamellae, and Volkmann's canal on your diagram.

FRACTURES

Dr. T belted out, "There is an apparent depression fracture in the victim's skull. This usually indicates a wound from a gunshot or head trauma like a blow from a hammer. Either way—it's foul play and that places this case in the category of homicide."

Dr. T continued, "Bone fractures are classified by several factors including the position of the bone ends after fracture, completeness of the break, orientation of the bone to the long axis, and whether or not the bone's ends penetrate the skin. Here are some of the more common fracture types…"

TABLE 6.3 Types of Fractures

Type of Fracture	Example
Depression	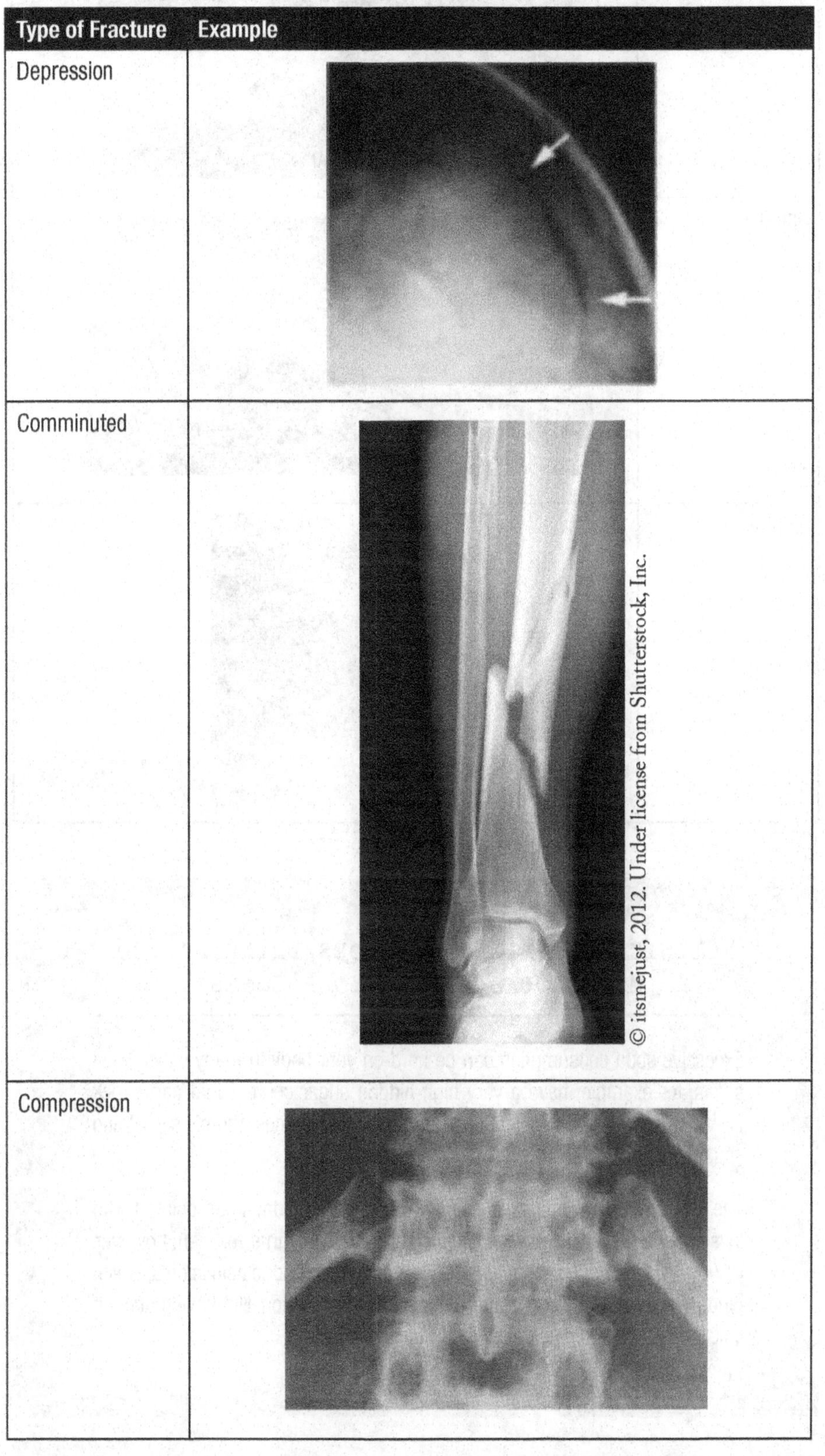
Comminuted	
Compression	

© itsmejust, 2012. Under license from Shutterstock, Inc.

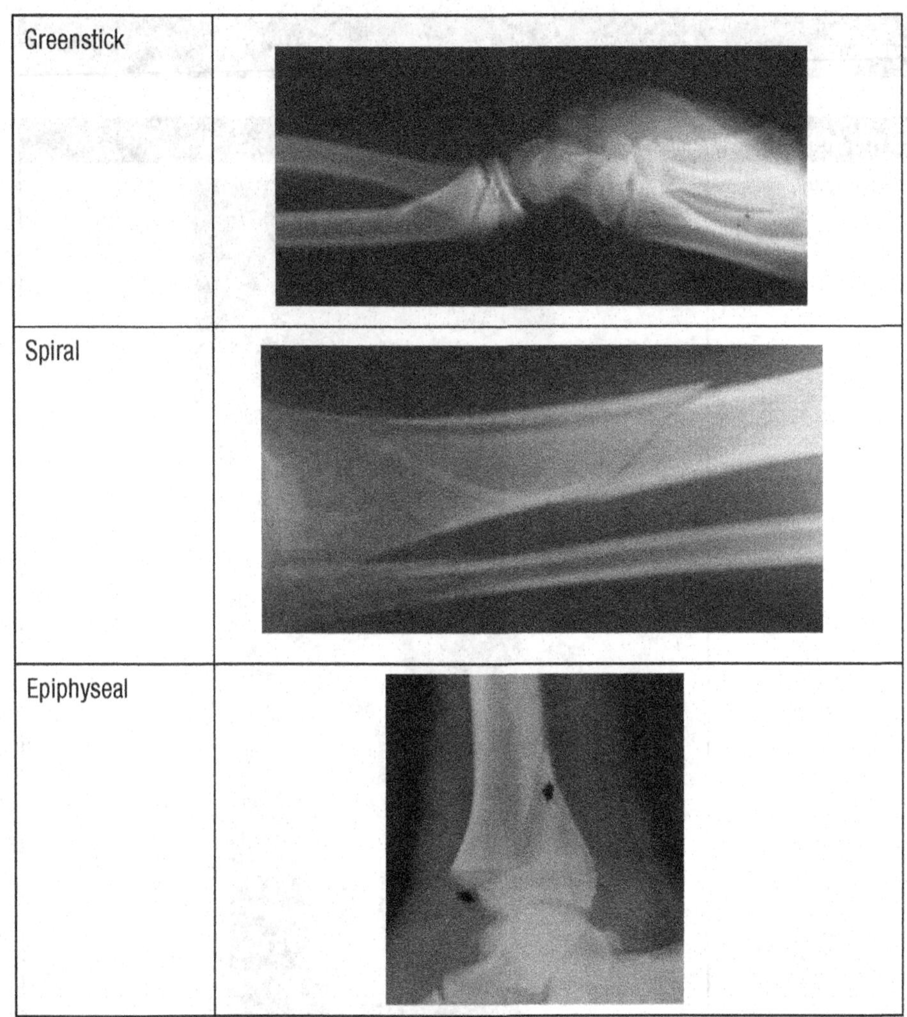

Greenstick	
Spiral	
Epiphyseal	

Excessive soda consumption can be hard on your body in many ways. Most sodas, for example, have a very high hidden sugar content that can wreck havoc on blood sugar levels. But when it comes to bones, there's something you may not know.

The consumption of more than three, 12-oz. sodas a day is associated with loss of bone density in women. This is not necessarily true for men, however. In women, the effect is thought to be due to the phosphoric acid in cola, which binds intestinal calcium and interferes with its absorption. Not having enough calcium can lead to it being leached from the bones.

IMAGE 6.13 Soda

Clinical Reflection Questions

1. What is the average size cola you consume? Is it a small 12-oz. can or a typical 20-oz. bottle? Or is it something larger like a large 32-oz. plastic cup?

   ____

   ____

   ____

2. How many of that size do you average a day? _______________________

3. If you are a woman, how does this information personally affect you? If you are a man, what women do you need to share this information with?

   ____

   ____

   ____

OTHER FACTORS AFFECTING BONES

Grace asked, "Aren't there certain diseases like osteoporosis that can fracture bones too?"

Dr. T replied, "Yes. There are some 20+ hormones that affect bone growth and health. A deficiency of growth hormone from the pituitary gland during childhood, for example, can result in pituitary dwarfism."

Dr. T continued, "There are some other significant bone diseases that you should be aware of as well, such as **rickets** (faulty mineralization in the bones of children, particularly the legs, due to insufficient amounts of vitamin D), **Paget's disease** (a disorder of that triggers abnormal bone destruction and abnormal re-growth of weak bone tissue), and **osteoporosis** (an exaggerated loss of bone tissue that causes bones to be much weaker than normal). Here's a diagram showing normal bone and bone with osteoporosis."

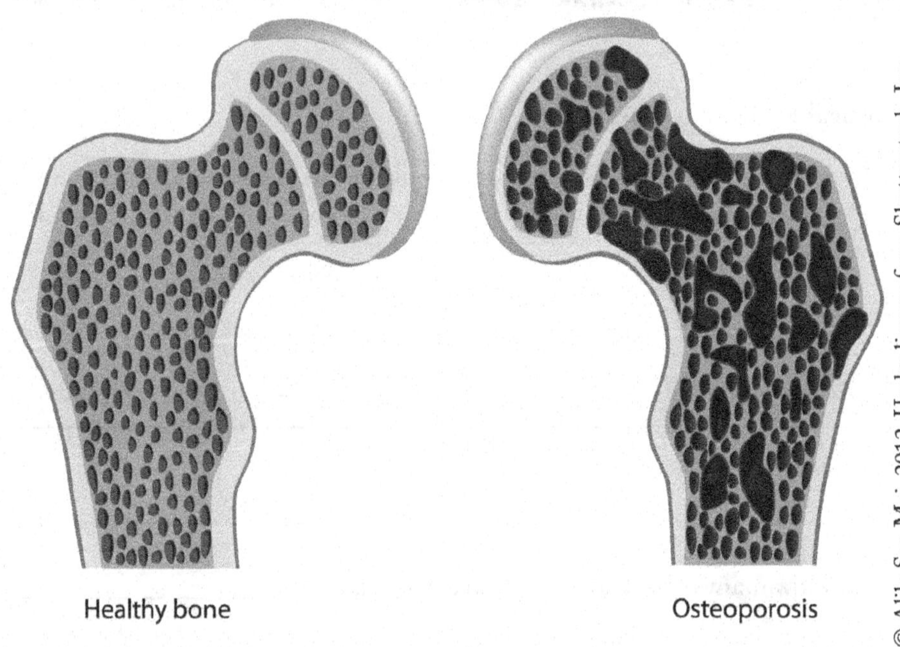

IMAGE 6.14 Healthy Bone and Bone with Osteoporosis

Grace and Will thanked Dr. T and concluded the appointment. On the way back to campus, they discussed how much they had learned about bone tissue, growth and remodeling, fractures, and bone diseases.

Self-Check—Continued

14. Explain the bone diseases rickets, Paget's disease, and osteoporosis.

____

____

____

____

____

____

15. In your own words, what have you learned that you didn't know before about bone tissues, growth and remodeling, fractures, or bone diseases?

____

____

____

____

____

____

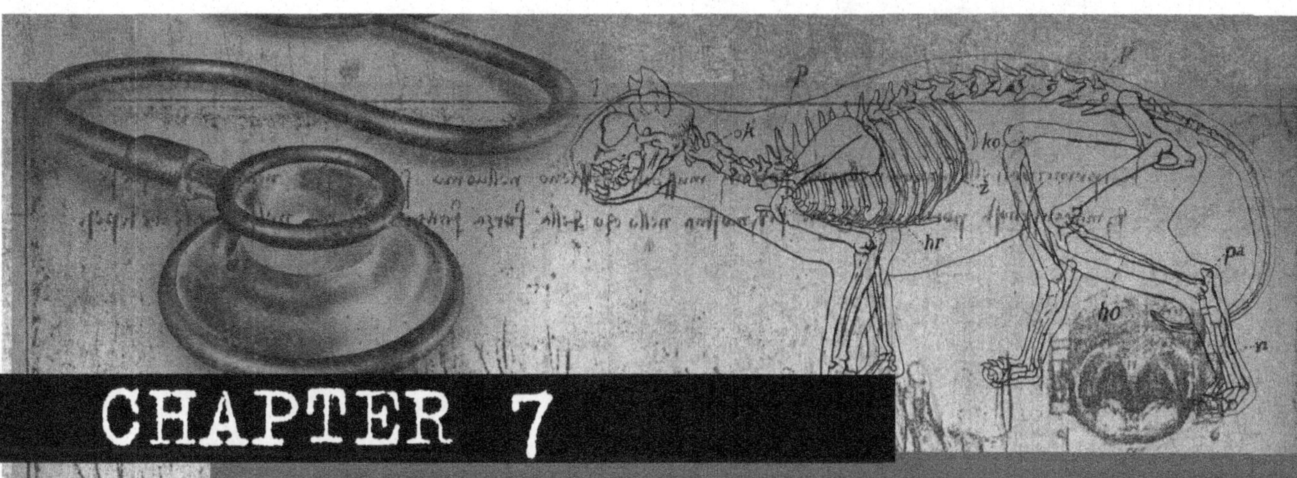

CHAPTER 7

The Skeletal System: Return to the Local Radiology Department

KEY WORDS

The study of Anatomy & Physiology involves many new vocabulary words. It is helpful to gain familiarity with these new key words, just as you would a foreign language.

acetabulum	floating ribs	obturator foramen
acromion	frontal bone	parietal bone
appendicular skeleton	glenoid fossa	patella
atlas	greater wings	pectoral girdle
axial skeleton	hamate	pedicles
axis	humerus	pelvic girdle
body of a vertebrae	iliac crest	phalanges
calcaneus	ilium	phalanx
capitate	intermediate cuneiform	pisiform
carpal	ischium	pubis
cervical vertebrae	lacrimal bone	radius
clavicle	lambdoid suture	sacrum
coccyx	lateral cuneiform	sagittal suture
coracoid process	lumbar vertebrae	scaphoid
coronal suture	lunate	scapula
crista galli	mandible bone	sella turcica
cuboid bone	mastoid process	sphenoid bone
dens	maxilla bone	spinous process
ethmoid bone	medial cuneiform	squamous suture
external auditory canal	metacarpals	styloid process
false ribs	metatarsals	superior nasal concha
femur	nasal bone	sutures
fibula	navicular	talus

tarsal	trapezium	vertebral foramen
temporal bone	trapezoid	vomer bone
thoracic cage	triquetral	zygomatic bone
thoracic vertebrae	true ribs	zygomatic process
tibia	ulna	
transverse process	vertebrae	

STUDENT STUDY GUIDE/OUTCOMES

It is helpful to have an idea of what you need to learn before proceeding. Here is a study guide to assist you:

1. Explain the two main subdivisions of the skeleton.
2. State the approximate number of bones in the body.
3. Identify the 22 bones of the skull and the significant processes.
4. Describe the 4 sutures of the skull and the bones they are associated with.
5. Describe the vertebrae of the spinal column and their characteristics.
6. Describe the bone markings found on a typical thoracic vertebrae.
7. Compare and contrast true, false, and floating ribs.
8. Compare and contrast C1 and C2 vertebrae.
9. Explain the bones of the pectoral girdle.
10. Describe the three major bone markings of the scapula.
11. Describe the bones of the upper limbs including the wrist.
12. Describe the bones of the pelvic girdle and the significant markings.
13. Compare and contrast a male and female pelvis.
14. Describe the bones of the lower limbs including the foot and ankle.
15. Label the major bone markings of the femur.

Will and Grace decided to revisit their established contact and friend, Dr. Monroe, who had helped them previously at the local hospital radiology department when they were learning regional and directional anatomy (back in Chapter 1).

Dr. Monroe greeted them whole-heartedly, "Hi you two! Welcome back to the Radiology Department. So, you are seeking information about the skeletal system?"

"Yes, we are. Can you teach us the bones?" questioned Will.

"And, can you show us some actual bones, too?" asked Grace.

Dr. Monroe continued, "Okay, sure. Let's get started! Look at this diagram first (below). The skeleton is divided into two main regions: the axial skeleton and the appendicular skeleton. The **axial skeleton** forms a core axis through the body so-to-speak. It is made up of the skull, ossicle (ear) bones, hyoid bone, vertebral column, and thoracic cage (ribs and sternum). While the **appendicular skeleton** focuses on appendages like arms, legs, as well as shoulders (pectoral girdle), and hips (pelvis)."

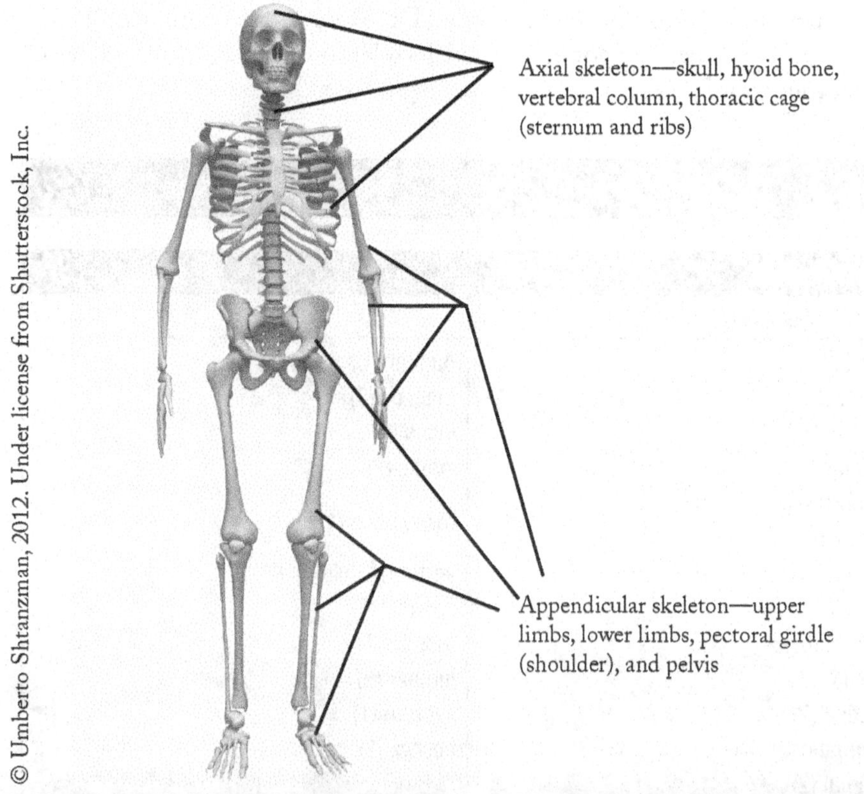

IMAGE 7.1 Axial and Appendicular Skeleton

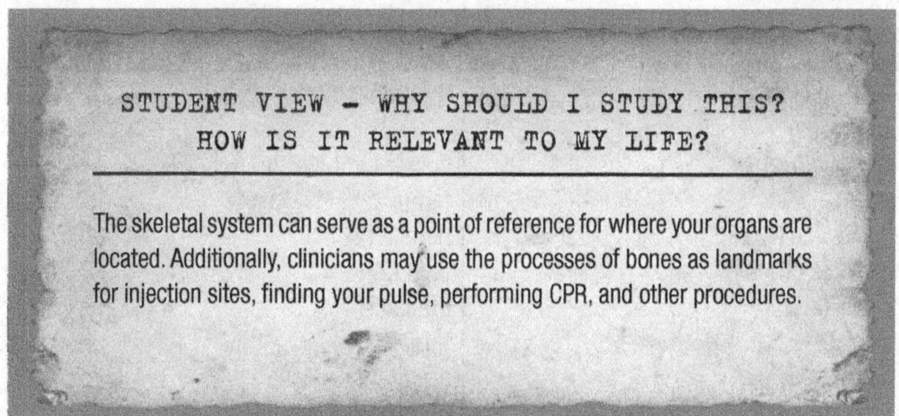

STUDENT VIEW – WHY SHOULD I STUDY THIS? HOW IS IT RELEVANT TO MY LIFE?

The skeletal system can serve as a point of reference for where your organs are located. Additionally, clinicians may use the processes of bones as landmarks for injection sites, finding your pulse, performing CPR, and other procedures.

THE BONES

"How many bones are in the body?" asked Will.

"Adults have approximately 206," answered Dr. Monroe. He continued, "Here's a list of all the bones categorized by axial skeleton or appendicular skeleton. This is your copy."

TABLE 7.1 Bones of the Adult Skeleton

Axial Skeleton	
Skull (22 bones)	
Cranial bones:	Auditory Ossicles (6):
frontal (1)	malleus (2)
parietal (2)	incus (2)
occipital (1)	stapes(2)
temporal (2)	
sphenoid (1)	hyoid bone (1)
ethmoid (1)	
	Vertebral Column (26):
	cervical (7)
Facial Bones:	thoracic (12)
maxilla (2)	lumbar (5)
palatine (2)	sacrum (1)
zygomatic (2)	coccyx (1)
lacrimal (2)	
nasal (2)	Thoracic cage (25 bones plus thoracic
vomer (1)	vertebrae):
inferior nasal concha (2)	ribs (24)
mandible (1)	sternum (1)
Appendicular Skeleton	
Pectoral girdle (4):	Hip Bones (2)
scapula (2)	
clavicle (2)	Lower Limbs (60):
	femur (2)
Upper limb (60):	patella (2)
humerus (2)	tibia (2)
radius (2)	fibula (2)
ulna (2)	tarsals (14)
carpals (16)	metatarsals (10)
metacarpals (10)	phalanges (28)
phalanges (28)	
Total = 206	

Dr. Monroe continued, "First we will study the bones of the skull. This is a very helpful diagram showing a 3D look inside the skull."

HUMAN SKULL

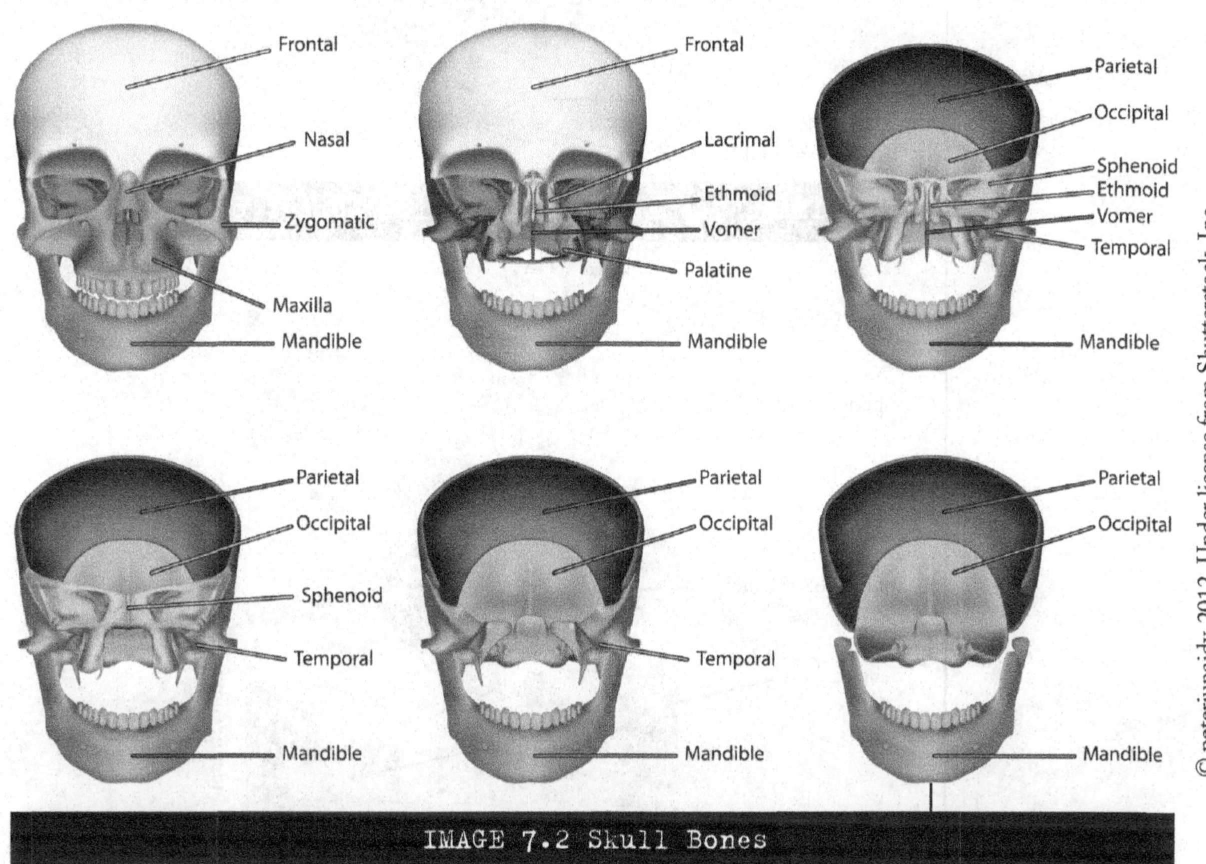

IMAGE 7.2 Skull Bones

"And, now we will look at real skulls for a closer look. It is easy to spot the **frontal** (forehead bone), **parietal** (two on top of head), **temporal** (near temples), **vomer** (bottom bone forming nasal septum), **maxilla** (upper jaw), **mandible** (lower jaw), **ethmoid** (from the front visible in the inner eye, top bone of the nasal **septum (inner wall), superior nasal concha,** and **lacrimal** (tear) bones from the anterior (front) aspect (view). Additionally, you can see the **inferior nasal concha** and a significant bone marking called the **mental foramen** (a small hole in the mandible that allows blood vessels and nerves to reach the front of the face)," continued Dr. Monroe.

NOTES

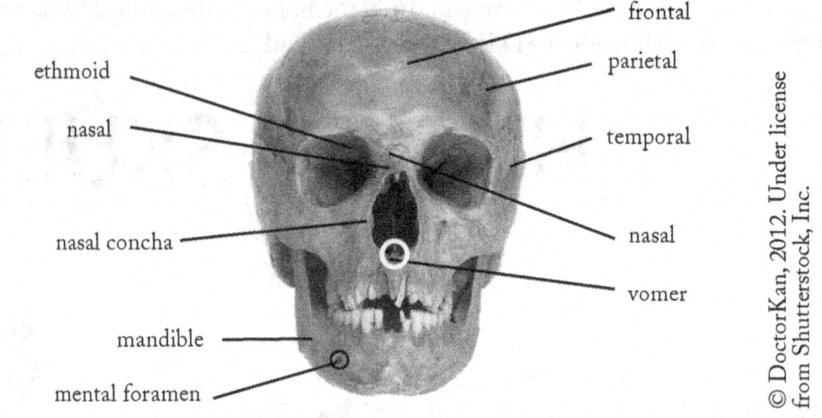

IMAGE 7.3 External Anterior Aspect of Skull

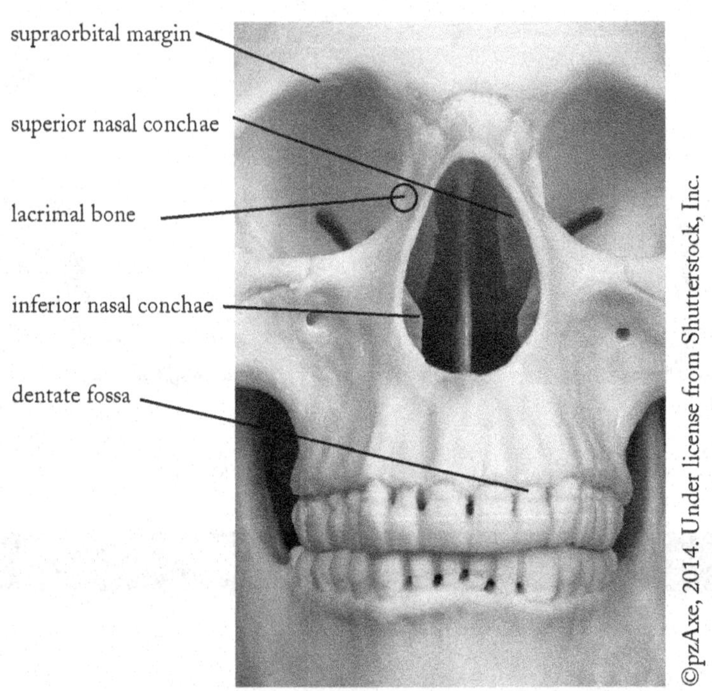

IMAGE 7.4 Close-up of Anterior Aspect of Skull

"What are the squiggly lines across the top of the skull?" asked Grace.

"Those are **sutures** or immoveable joints. When you are an infant, your head is expandable to make it through the birth canal. As adults, however, those joints are fixed and do not move. We have the **coronal suture** between the parietal bones and the frontal bone. The **squamous suture** separates the parietal bones from the temporal and sphenoid. The **lambdoid suture** separates the parietals from the occipital and the **sagittal suture** separates one parietal bone from the other down the very center of the top of the skull," commented Dr. Monroe.

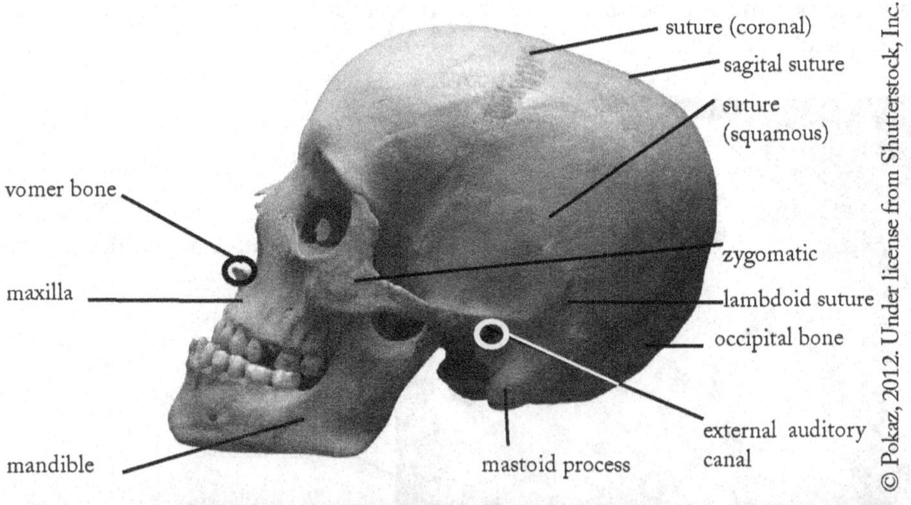

IMAGE 7.5 External Lateral Aspect

STUDY TIP!

Sometimes it helps to be goofy when creating ways to remember things, especially, sutures.

Coronal = Corona beer
Squamous = "s-shaped"
Lambdoid = lay the back of your head (occipital area) down, like a lamb, for a nap
Sagittal = median (as in Chapter 1 terminology)

"And, what's the mastoid process and external auditory canal? I didn't see that before," asked Will.

"The **mastoid process** is a large process behind the ear. Turn your head and feel behind your ear. You can feel it. And the **external auditory canal** is the opening to the ear," replied Dr. Monroe.

Dr. Monroe continued, "Now if we turn the skull upside down (photo below) and remove the mandible, we can see many more bones and significant markings lie in the **palatine** bone that helps form the roof of the mouth, the **pterygoid process** (extensions that stick out like slides off the sphenoid bone), **occipital condyles** ('knuckles' on the sides of the magnum foramen), the **magnum foramen**—the largest foramen in the skull, (the vertebral column inserts there; part of the occipital bone), and the **styloid process** (spine). Two bones actually make up the cheek—the **zygomatic** or front part of the cheek and the **zygomatic process**, which is a process of the temporal bone that creates the side of the cheek."

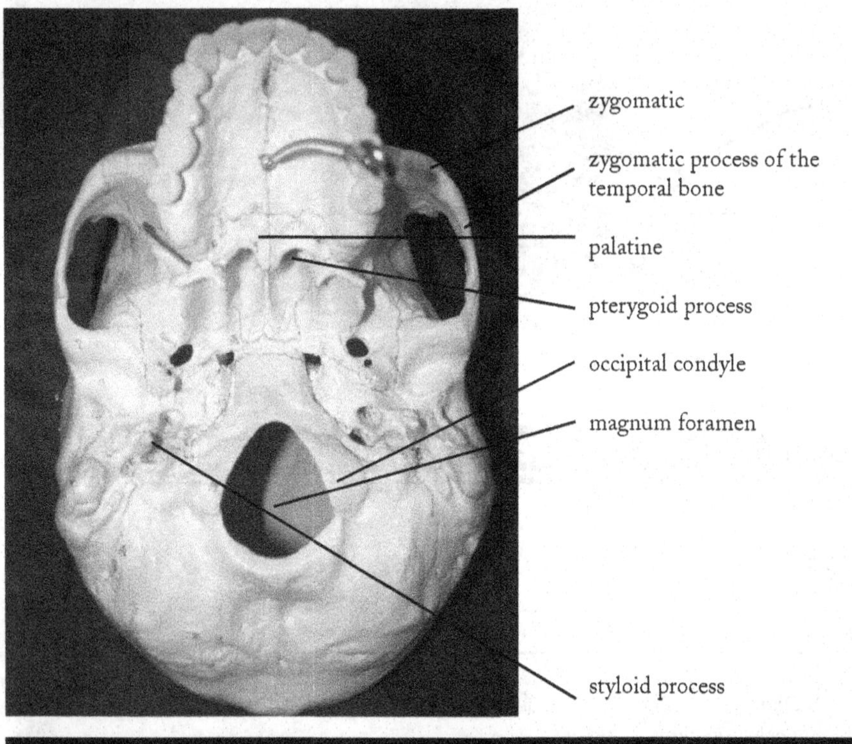

IMAGE 7.6 External Inferior Aspect

Dr. Monroe added, "And, if we remove the top of the skull and look inside… you can get a good look at the **ethmoid** from the interior aspect with its **crista galli** (sharp tip) and **cribiform plate** (perforated looking bone on the sides). You can also see the beautiful butterfly bone or **sphenoid** as well as its **lesser wings** and **greater wings** and **sella turcica** (or cradle that holds the pituitary gland). The **hypoglossal canal** is also apparent when you look closer inside the magnum foramen. It acts as a passageway for nerves and blood vessels."

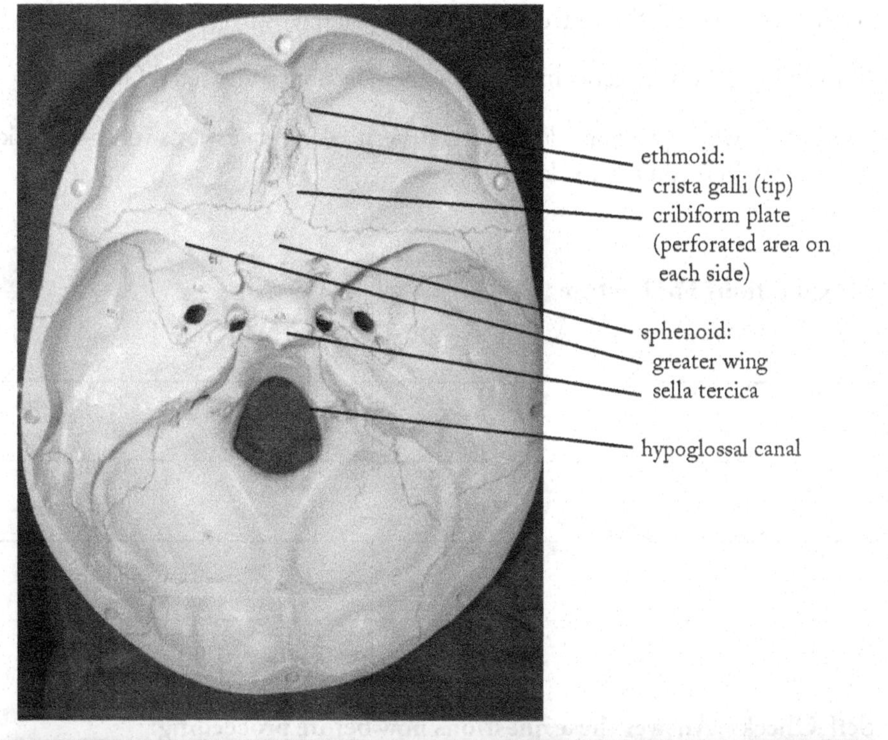

IMAGE 7.7 Interior Aspect from Superior View

IMAGE 7.8 Top of Skull

The Skeletal System: Return to the Local Radiology Department

Dr. Monroe asked, "So Grace and Will, do you have any questions?"

"I don't, but others students may have some questions," answered Grace.

Students… what questions do you have? Write down questions you would like to discuss when you get to class.

My Questions For Lecture

____

____

____

____

Self-Check – Answer these questions now before proceeding:

1. Apply what you have just seen to a real skull. Label the bones on these diagrams.

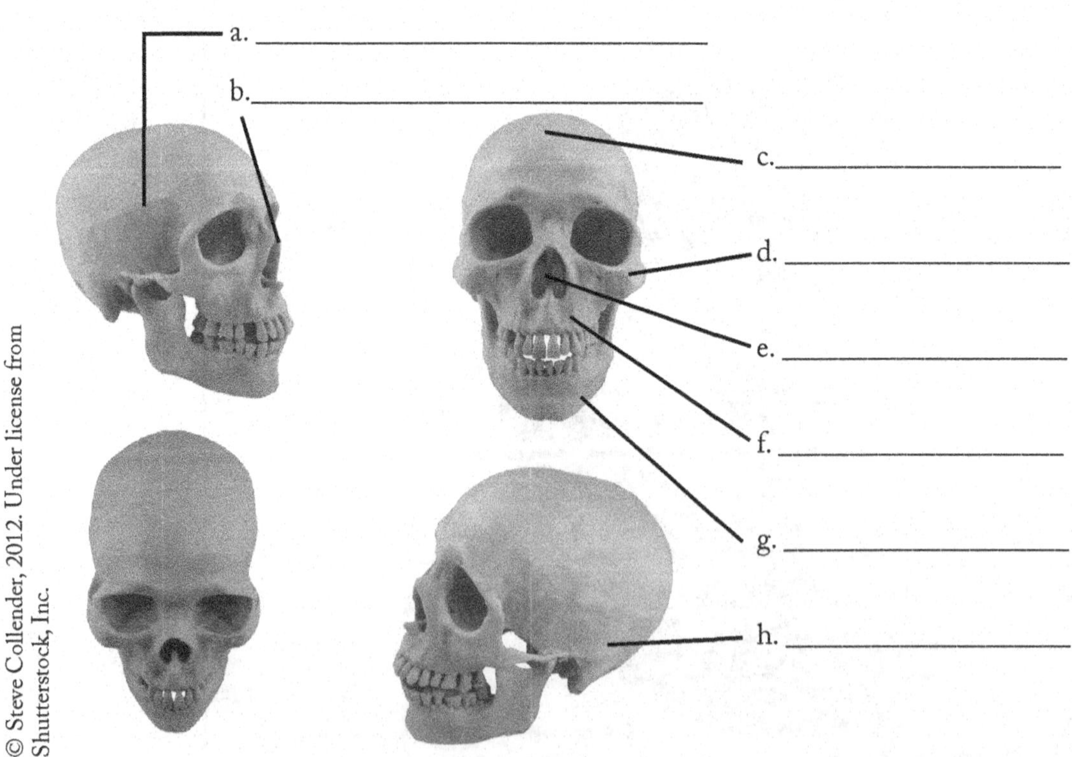

IMAGE 7.9 Blank Skulls To Label

2. (a) What bone is the sella turcica associated with?

____

(b) What does it "protect" in its cradle?

____

3. What bone is the crista galli associated with?

____

4. If you are looking at the anterior external aspect of the skull, describe some of the areas where you can see the ethmoid bone.

____

____

____

5. What is the significance of the magnum foramen?

____

____

____

6. List the name and locations (bones) of the sutures in the skull.

____

____

____

____

____

THE VERTEBRAL COLUMN

"Hey Doc, doesn't the vertebral column (or spine) support the skull and trunk of the body?" asked Will.

Dr. Monroe replied, "Yes and it also allows you to move, protects your spinal cord, and absorbs the physical stresses of activity like jumping and running. The vertebral column isn't one giant solid piece—it is made up of separate bones called **vertebrae**. The vertebrae have a fibrocartilage disc between most of them for cushioning."

"How many vertebrae?" asked Grace.

"Here—why don't you count for yourself…" Dr. Monroe propositioned. Dr. Monroe further challenged, ""Try it now too."

Self-Check—Continued

7. Count the vertebrae and record your answers below.

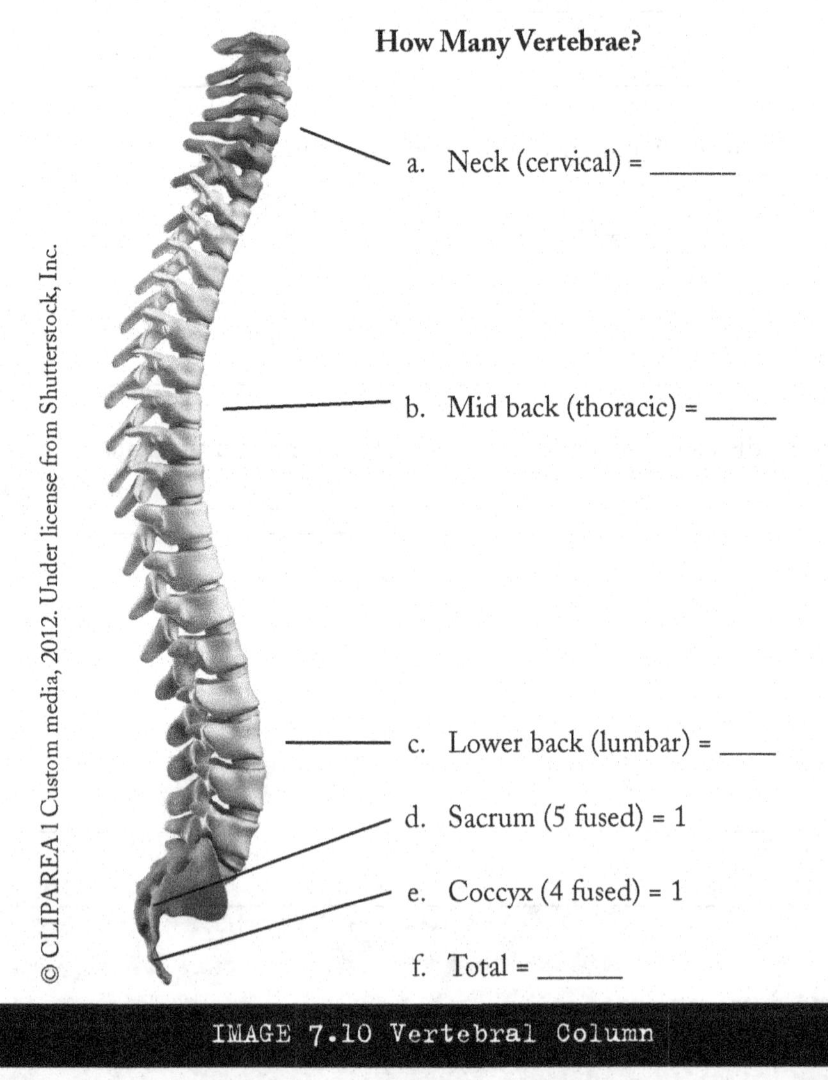

IMAGE 7.10 Vertebral Column

"I counted 7 cervical, 12 thoracic, 5 lumbar, 1 sacrum, and 1 coccyx, totaling 26. Is that right?" asked Grace.

"Yes, that is correct," confirmed Dr. Monroe.

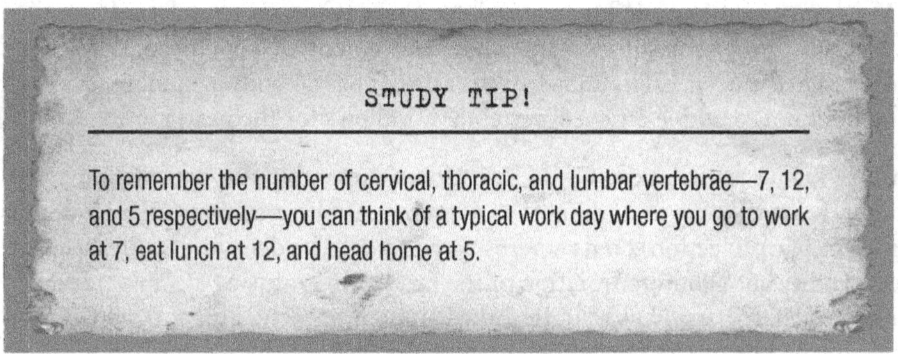

"Vertebrae have some special markings that you should be familiar with. They have a spine or **spinous process** projecting toward the posterior side of the body. The spinous procesess are not very pronounced in the cervical vertebrae but are very prominent in the thoracic vertebrae. In the lower lumbar vertebrae, they are broad and hatchet-shaped. There are **transverse processes** protruding from the lateral portion of the vertebrae. There are stalk-like **pedicles** that anchor the **body** of the vertebrae to the rest of it. The spinal cord is sheltered inside the **vertebral foramen**," continued Dr. Monroe.

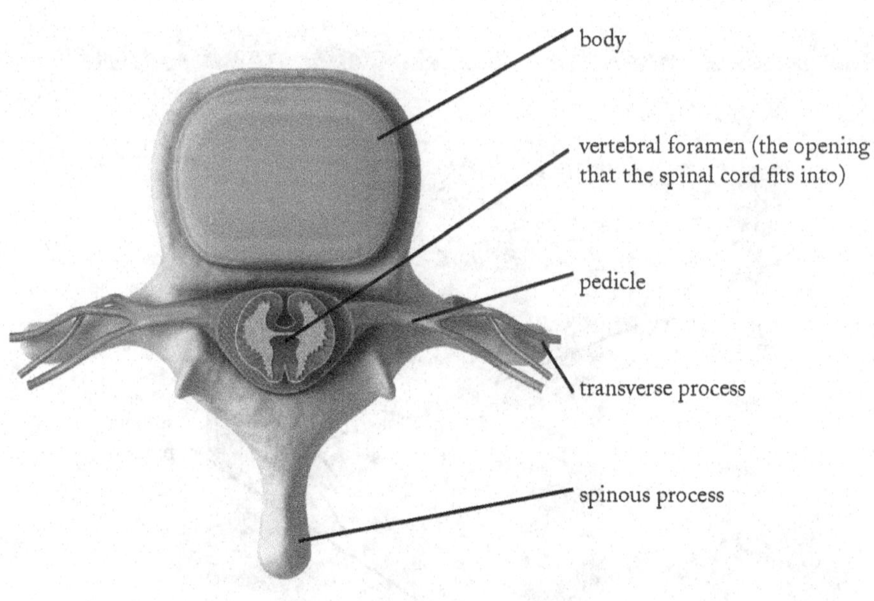

IMAGE 7.11 Vertebrae with Markings

GENERAL CHARACTERISTICS OF CERVICAL VERTEBRAE

Dr. Monroe continued, "The cervical vertebrae (C1-C7) are relatively small. Their function is to support and allow movement of your head. The first vertebra in the neck, C1, is known as the **atlas**. It has no body or spinous process and does not resemble a typical vertebra. C1 allows for the head motion 'yes.'

"C2 is known as the **axis**. Like C1, it doesn't resemble the typical vertebrae either. It actually has a marking that no other vertebrae exhibits called a **dens** (a knob-like projection) on its anterosuperior side. The dens fuses with the axis for stability but allows for rotation of the head as in gesturing 'no' in a side-to-side motion. C2-C6 have a short, forked spinous process. This fork allows for the attachment of the nuchal ligament at the back of the neck. C7 is a little different because its spinous process is extremely long and forms a prominent bump on the back of the neck that can be felt."

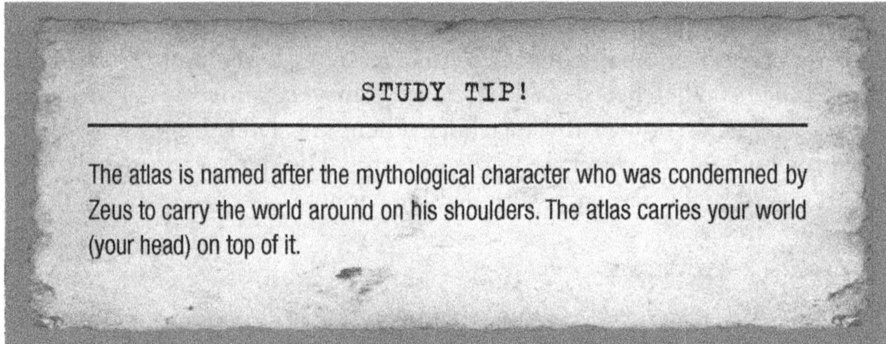

STUDY TIP!

The atlas is named after the mythological character who was condemned by Zeus to carry the world around on his shoulders. The atlas carries your world (your head) on top of it.

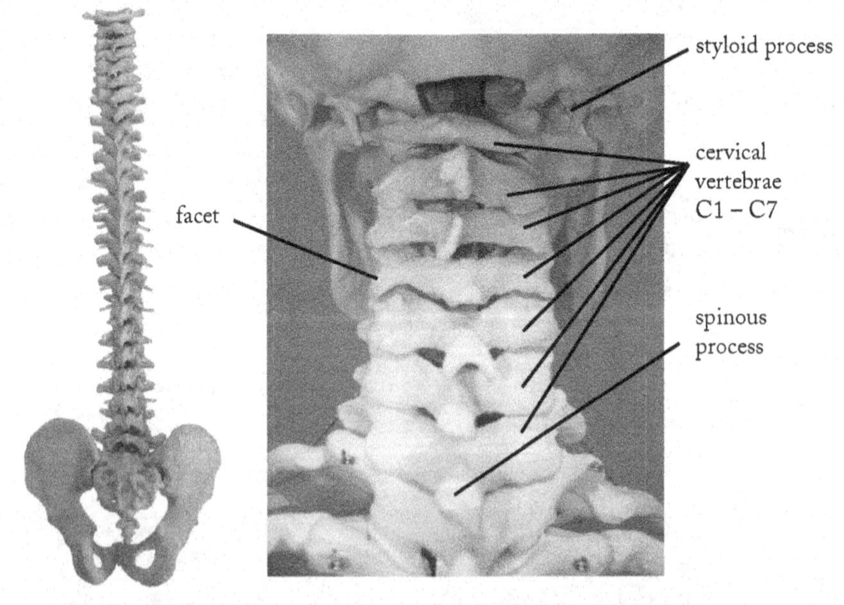

IMAGE 7.12 Cervical Vertebrae C1-C7

"What about the **thoracic vertebrae**… are they different from the cervicals?" questioned Grace.

Dr. Monroe replied, "T1-T12 are unique in several ways. One, they all attach to a rib. And no other vertebrae have ribs attached to them. Secondly, the spinous process are very exaggerated and angle sharply downward. Third, the bodies are somewhat heart-shapped and bigger than the cervical vertebrae. Overall, they vary among themselves in the way they articulate with the ribs— their facets (joints)."

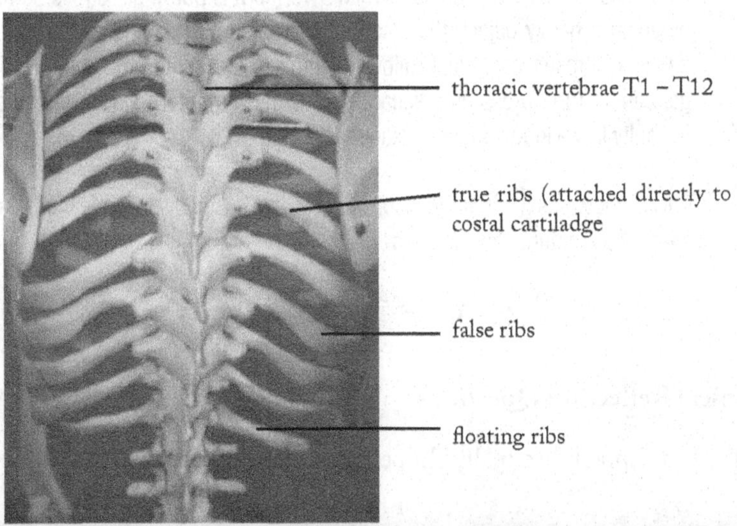

IMAGE 7.13 Thoracic Vertebrae (T1-T12)

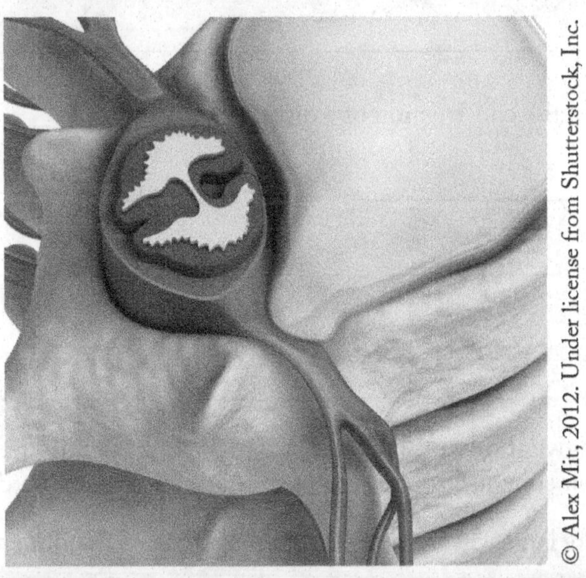

IMAGE 7.14 Herniated Disc

CLINICAL APPLICATION: IS IT REALLY A
SLIPPED DISC?

Have you heard the term "slipped disc" when referring to a serious back injury? What does that really mean? Is it an accurate description for disc problems?

Intervertebral discs are cartilaginous pads between vertebrae. They consist of an inner gel-like center called the nucleus pulposus surrounded by one of the tougher types of cartilage—fibrocartilage. The outer fibrocartilage ring is known as the annulus fibrosus. When a vertebra is put under stress perhaps by lifting a heavy object, the disc expands laterally. Under great stress, the annulus fibrosus can crack allowing the gel-like center to ooze out. This gel can, in turn, press on adjacent nerves causing intense pain, numbness, difficulty in walking, or loss of bladder control.

This is known as a herniated disc, which is also called by several other names such as a "ruptured" or "slipped" disc.

Clinical Reflection Questions

1. Is a slipped disc really slipped? Explain your answer.

2. Explain the construction of a disc. _______________________

LUMBAR VERTEBRAE

"So there are five **lumbar vertebrae**," stated Will. He continued, "From what I can tell, their spinous process are not so sharp but kind of hatchet-shaped."

"They sure are large vertebrae compared to the others," commented Grace.

"Can you think of why they need to be so big?" asked Dr. Monroe.

"Well, I guess they have to support all our weight," answered Will.

"Right you are!" confirmed Dr. Monroe.

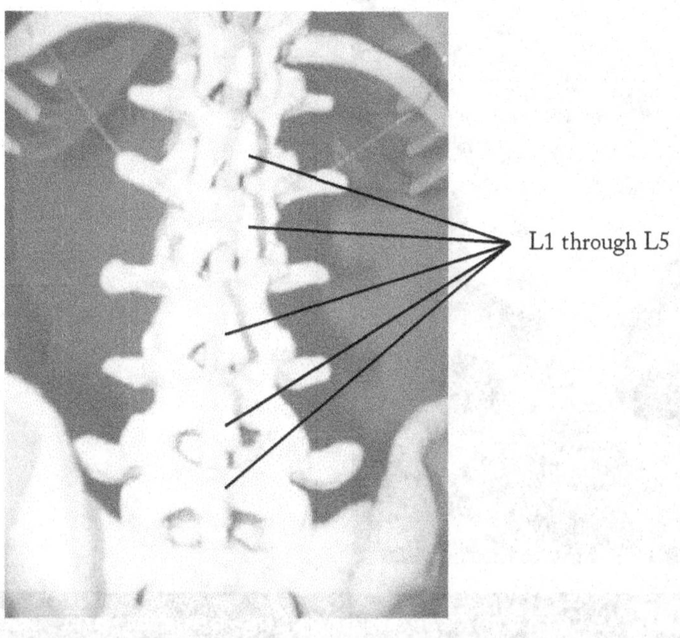

IMAGE 7.15 Lumbar Vertebrae L1–L5

THE SACRUM

"And this used to be five individual vertebrae at one time?" questioned Grace looking at the solid sacrum.

"Yes, in children there are 5 individual sacral vertebrae which begin to fuse around age 16 and are fully fused by age 26," instructed Dr. Monroe. He continued, "When fused, they form a bony plate—the sacrum is the strongest bone of the vertebral column."

THE COCCYX

"Is it similar for our coccyx?" asked Will.

"Yes, in children we see 4 or 5 small individual vertebrae that fuse sometime between age 20 and 30," confirmed Dr. Monroe.

"It seems useless. Why do we have a tailbone anyway?" asked Grace.

Dr. Monroe laughed and replied, "Well, it's not completely useless… it serves as a point of attachment for muscles that form the pelvic floor."

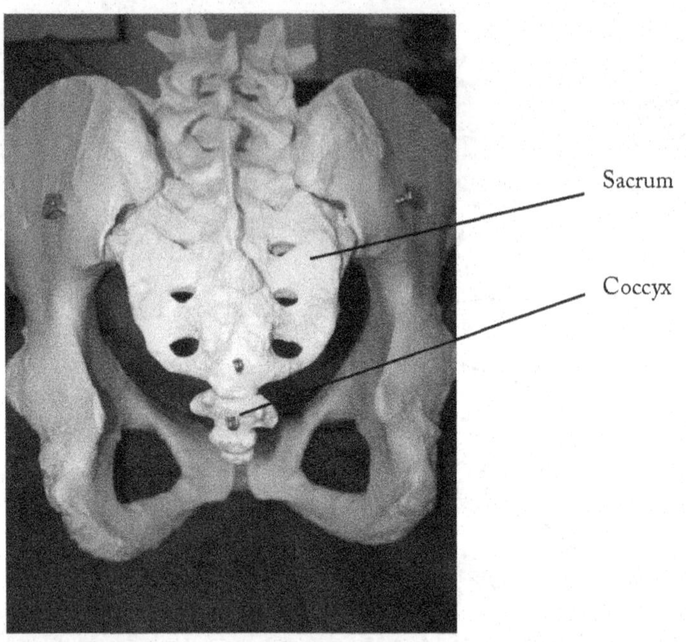

IMAGE 7.16 Anterior Aspect Showing Sacrum and Coccyx

THE THORACIC CAGE

Dr. Monroe continued, "Your **thoracic cage** is an enclosure that is a great protector of your heart, lungs, and other internal organs. It consists of your thoracic vertebrae, ribs, and sternum."

"I just have to know… is it true that us guys have one less rib than girls?" Will asked intensely.

"No, women and men are the same… 12 ribs," answered Dr. Monroe. He continued, "The first 7 ribs are known as **true ribs** because they are connected directly to the sternum by hyaline cartilage. The hyaline cartilage in the rib area is also known as costal cartilage. The remaining 5 ribs are called **false ribs** because they actually suspend via extra cartilage to rib 7 and thus are not directly connected to the sternum. And the anterior ends of the last two false ribs do not connect to anything—they are called **floating ribs**."

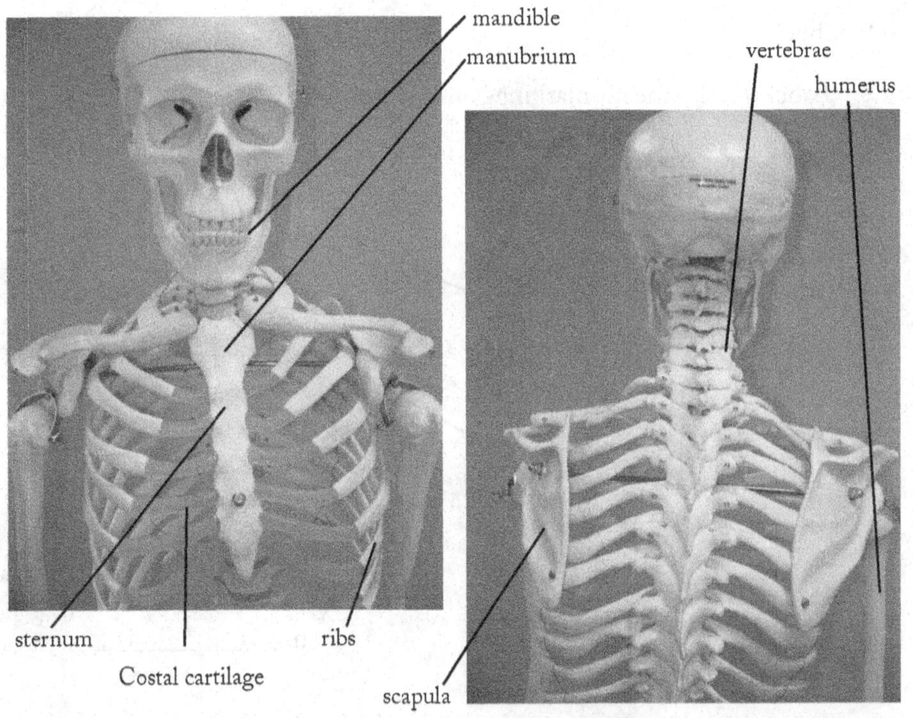

IMAGE 7.17 Thoracic Cage

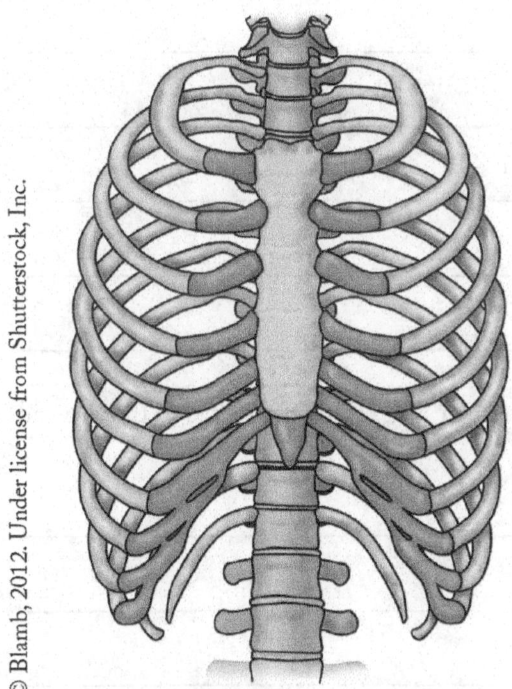

true ribs 1-7 (all connect via cartilage directly to the sternum)

false ribs 8-12 connect to the cartilage of true rib 7, they are not directly connected to the sternum

ribs 11-12 do not connect at all, they are floating ribs

IMAGE 7.18 True and False Ribs

Self-Check

8. Label the significant markings on this vertebra.

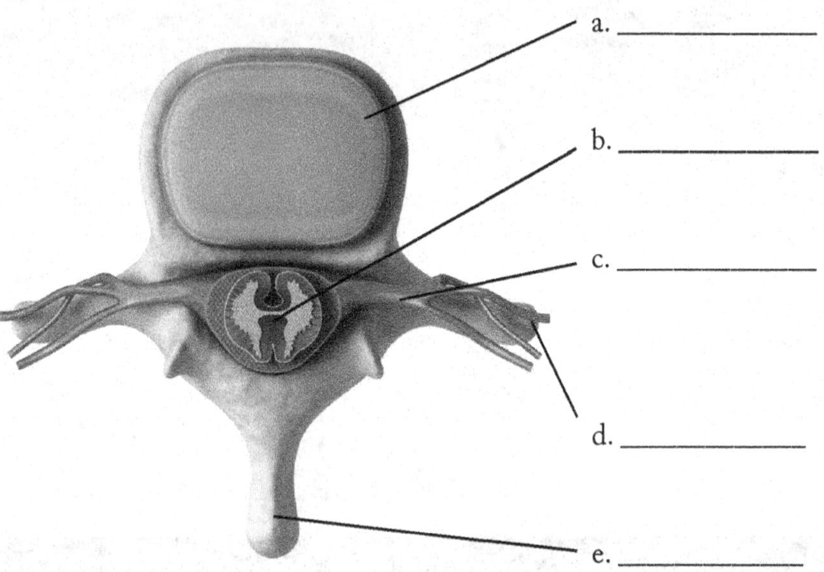

9. Describe the atlas and the axis. ________________________________

____

____

____

____

10. Compare and contrast the thoracic and lumbar vertebrae to the cervical vertebrae.

____

____

____

____

11. Explain the statement, "The sacrum is 1 vertebra." _________________

____

____

____

____

____

12. Differentiate between true and false ribs. Also, explain how the floating ribs fit into the picture.

____

____

____

____

____

THE PECTORAL GIRDLE

Dr. Monroe continued, "The **pectoral girdle** or shoulder connects the arm to the axial skeleton. It is made up of only two bones—the scapula and the clavicle, shown here on this X-ray."

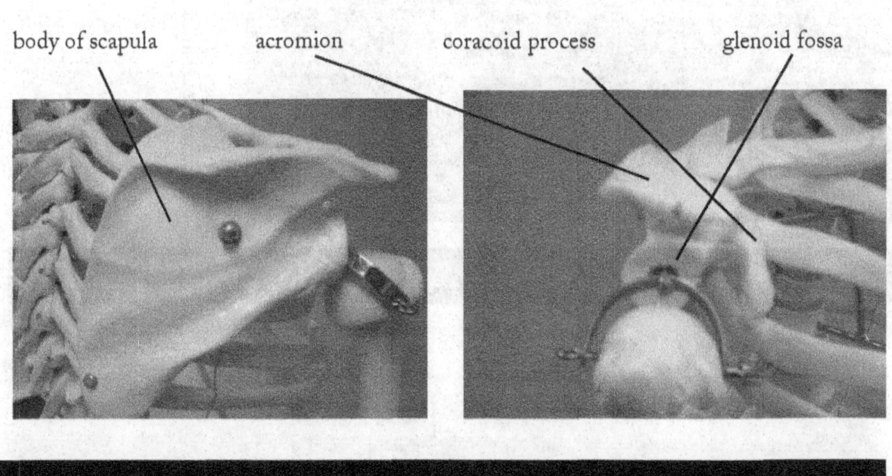

IMAGE 7.19 Pectoral Girdle

"Isn't the **clavicle** also known as your collarbone?" asked Grace.

"It is the same thing," answered Dr. Monroe. Then he continued, "It braces the shoulder and helps keep the upper limbs from folding in toward the midline of the body. It's somewhat s-shaped and easily seen and felt through the skin near your lower neck on the front part of your body."

"And, isn't your **scapula** the same thing as your shoulder blade?" questioned Grace again.

"Correct" answered Dr. Monroe again patiently. He continued, "Because the scapula has a triangular appearance, it has been said to resemble a spade or shovel. It glides over the ribs as the arm and shoulder move, attached only by muscles to the thorax (chest). It has 3 significant markings that I would like to point out. (1) The **acromion** which forms the apex (tip) of the scapula. (2) The **coracoid process** which is shaped vaguely like a pretzel twist, protrudes toward the anterior of the body and forms the attachment for tendons of the biceps brachii and other muscles of the arm. And (3) The **glenoid fossa** is a shallow socket-type cavilty which articulates with the head of the humerus."

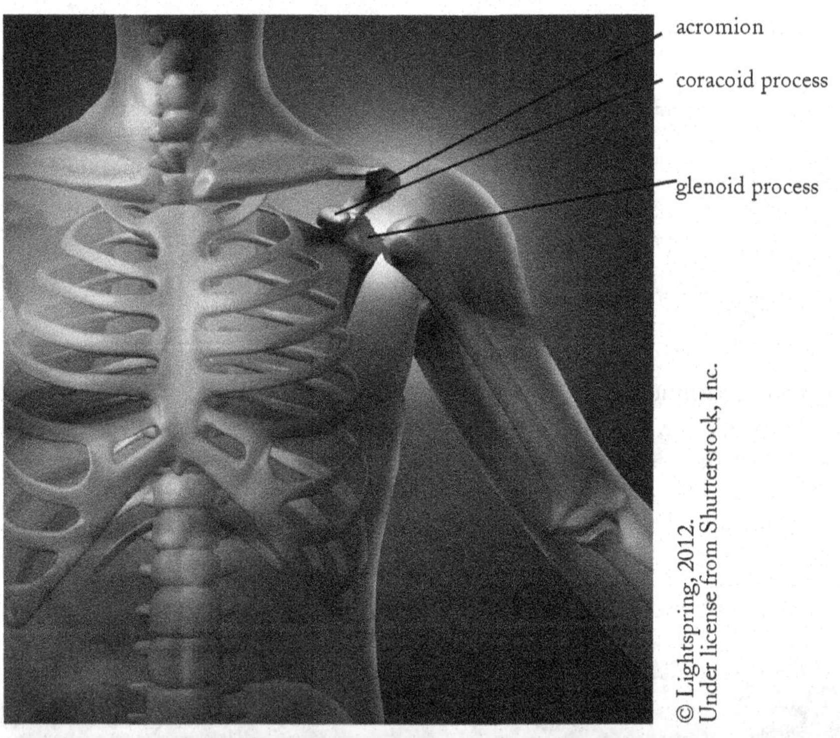

IMAGE 7.20 Significant Markings of the Scapula

THE UPPER LIMBS

"Now let's look at the arms," directed Dr. Monroe.

"I remember you said earlier that there are 30 bones per upper limb. So, that's 60 for both arms. That is a lot when you remember that the entire adult body is 206 bones," commented Grace.

"So, the upper arm is the **humerus,**" added Will.

"And the two forearm bones are the radius and ulna. But how do you tell them apart exactly?" asked Grace.

Dr. Monroe explained, "The **radius** is in line with the thumb. It also rotates over the other bone when you turn your thumb downward. Additionally, it has a distinct radial tuberosity on its medial surface, which is where the distal tendon of the biceps brachii terminates. Notice the radial tuberosity and other bone markings (see Image 7.22). And the **ulna,** of course, is the other bone, which is slightly longer (see Image 7.23 for more detail)."

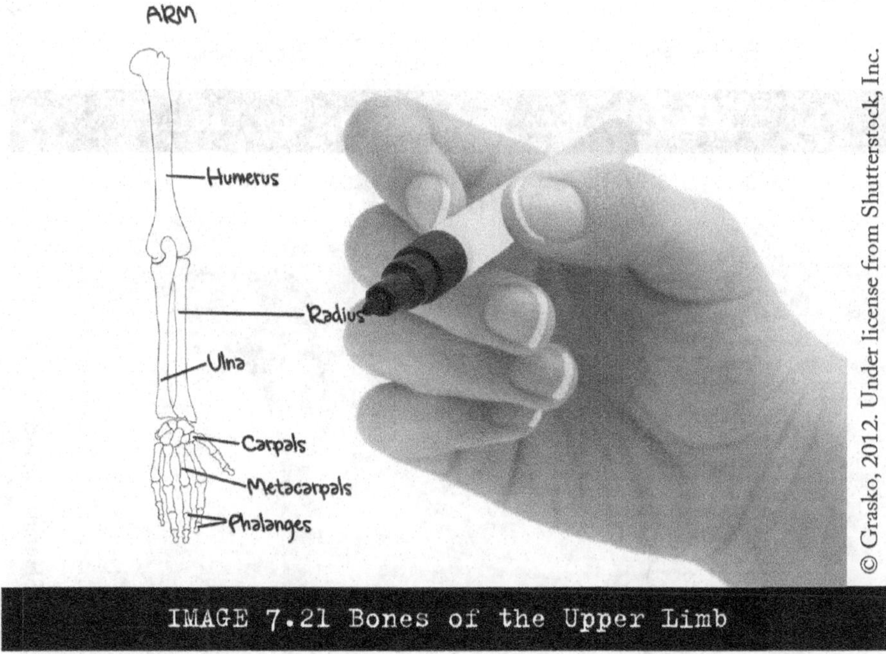

IMAGE 7.21 Bones of the Upper Limb

RIGHT RADIUS

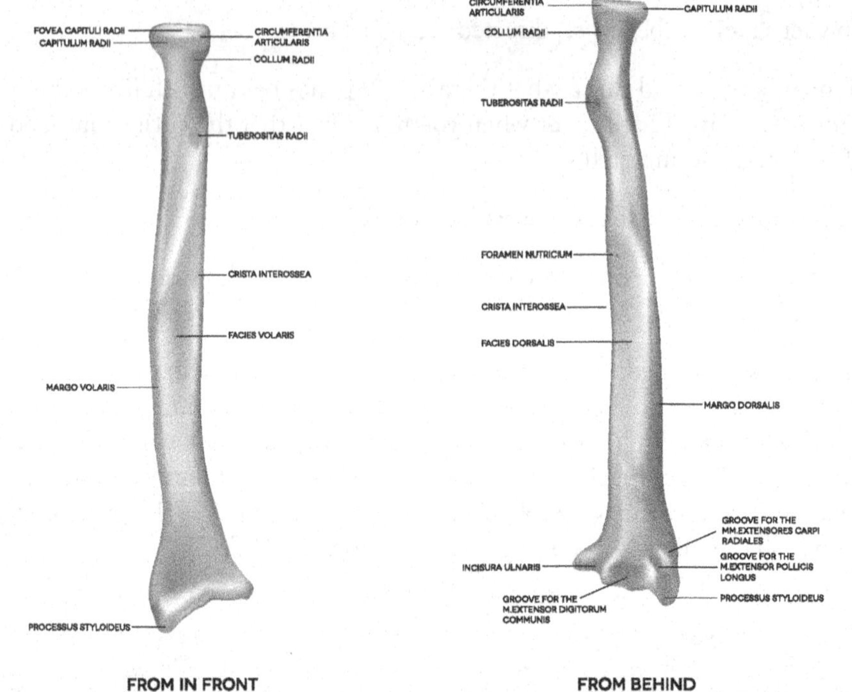

IMAGE 7.22 Bone Markings of the Radius

©GRei, 2014. Under license from Shutterstock, Inc.

RIGHT ULNA

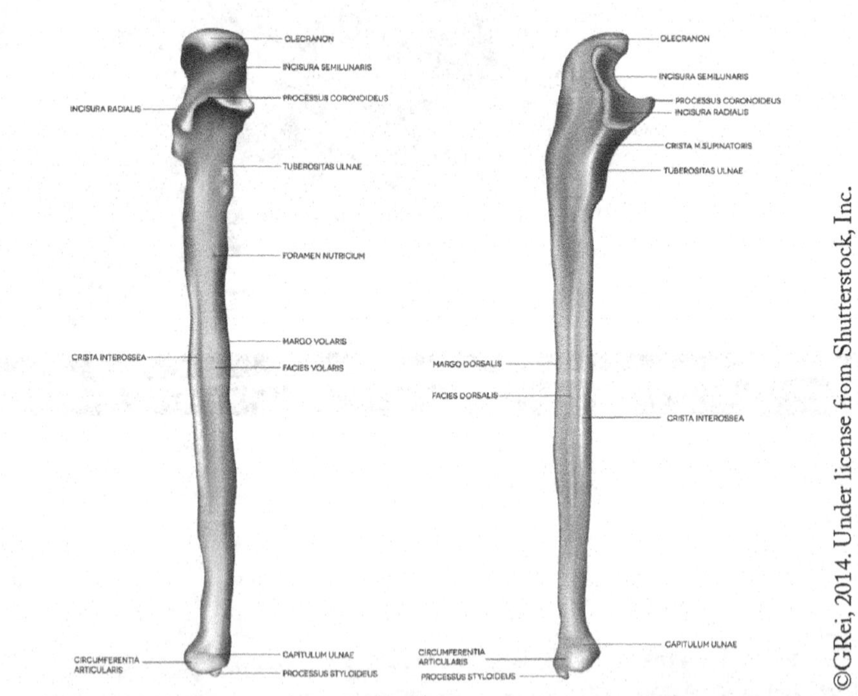

IMAGE 7.23 Bone Markings of the Ulna

©GRei, 2014. Under license from Shutterstock, Inc.

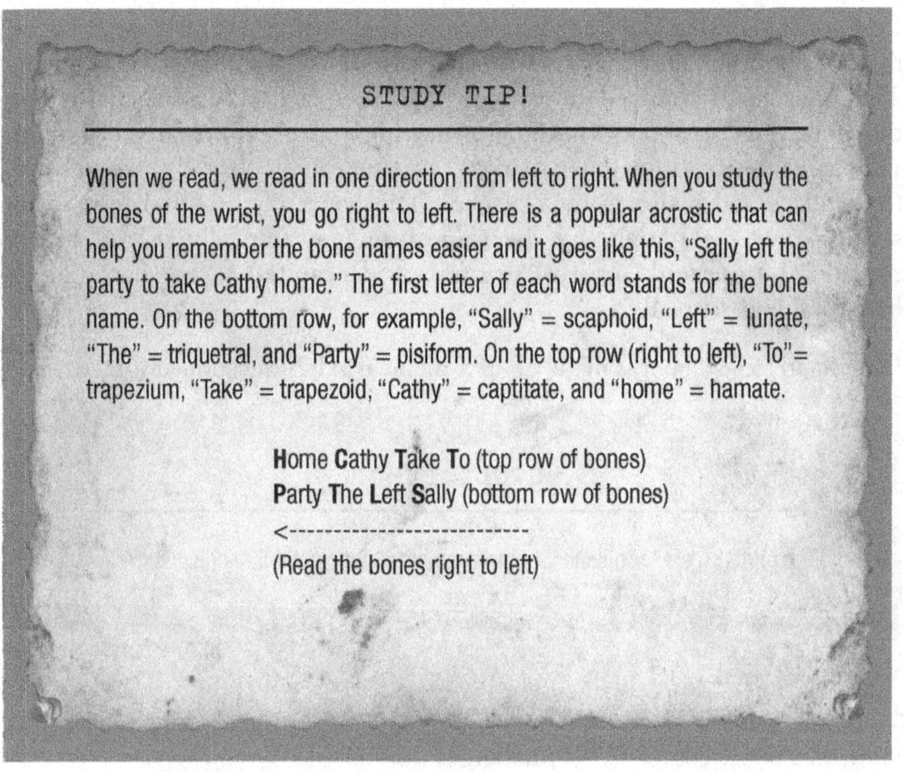

When we read, we read in one direction from left to right. When you study the bones of the wrist, you go right to left. There is a popular acrostic that can help you remember the bone names easier and it goes like this, "Sally left the party to take Cathy home." The first letter of each word stands for the bone name. On the bottom row, for example, "Sally" = scaphoid, "Left" = lunate, "The" = triquetral, and "Party" = pisiform. On the top row (right to left), "To"= trapezium, "Take" = trapezoid, "Cathy" = captitate, and "home" = hamate.

Home **C**athy **T**ake **T**o (top row of bones)
Party **T**he **L**eft **S**ally (bottom row of bones)

<----------------------------

(Read the bones right to left)

Bones of human hand and wrist

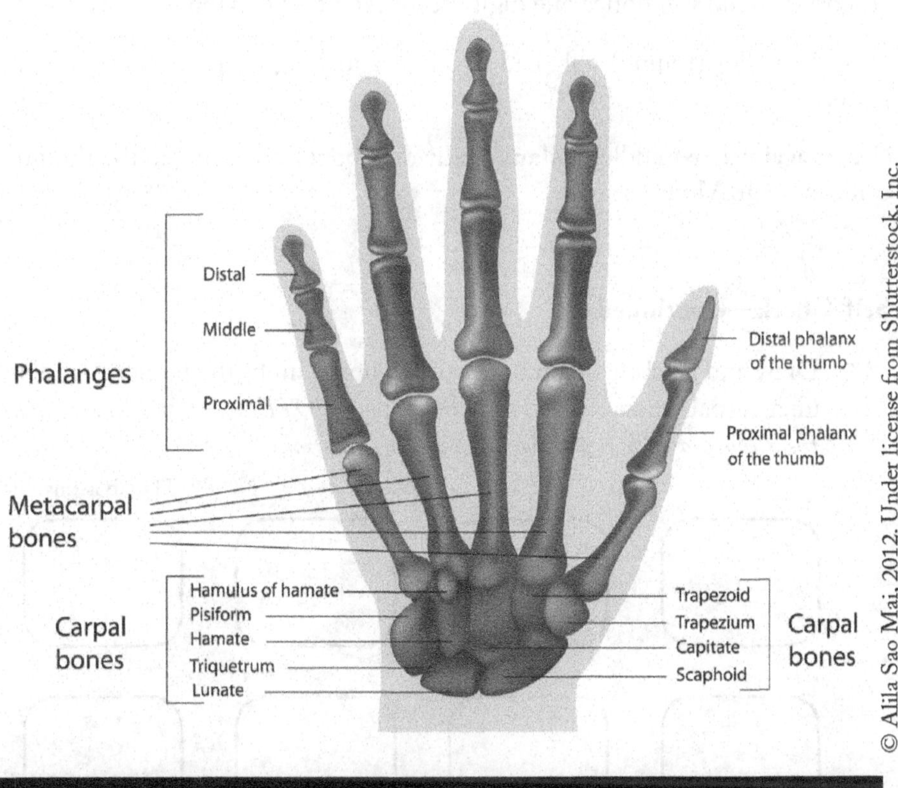

IMAGE 7.24 Bones of the Hand and Wrist

Dr. Monroe continued, "The wrist is the **carpal** area. There are 8 wrist bones that are short and allow for side-to-side, as well as up and down movement. They are actually arranged in two rows of 4 bones each, stacked one on top of the other. The carpal bones of the proximal row starting at the lateral (thumb) side are the **scaphoid** (boat-shaped), **lunate** (moon-shaped), **triquetrum** (triangular), and **pisiform** (pea-shaped). The bones of the distal row, again starting near the thumb, are the **trapezium** (geometrically shaped like a trapezium), **trapezoid** (shaped like a trapezoid), **capitate** (head-shaped), and **hamate** (hook-shaped)."

"The hand looks like it is made up of 5 **metacarpals**?" questioned Grace.

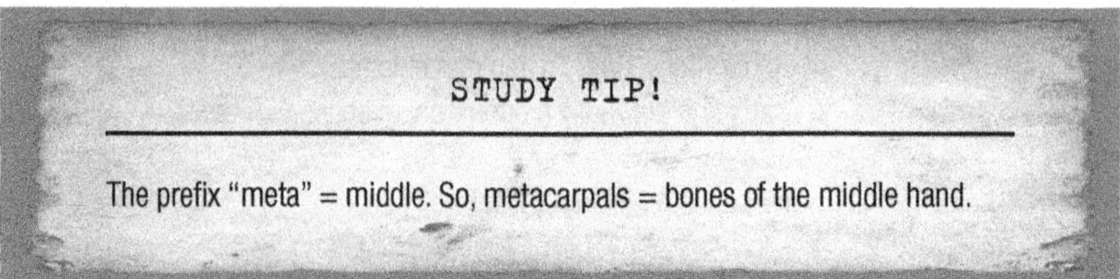

"It is," replied Dr. Monroe. He challenged, "And can you count how many individual bones comprise the **phalanges** (digits)?"

"Um… 14," answered Grace.

"Right! And did you notice one digit was different?" Dr. Monroe asked.

"Yes, the pollex (thumb) only had two bones making it up," answered Grace again.

"Right again! The middle **phalanx** (individual bone) is missing in the thumb," confirmed Dr. Monroe.

Self-Check—Continued

13. Draw and label the main bones of the upper limb (the humerus, radius, ulna, carpals, metacarpals, and phalanges). *NOTE: you do not have to label the 8 individual bones of the wrist on this question.*

NOTE: Thumb is this side

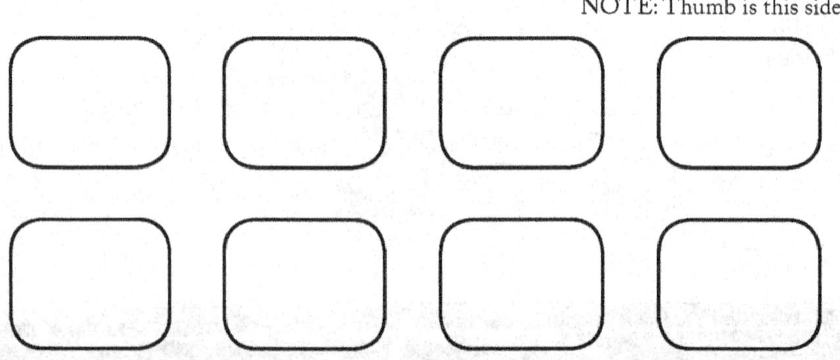

14. Assuming that these shapes represent the 8 bones of the wrist, label them with the correct bone names.

15. List and describe the two bones that comprise the pectoral girdle.

____

____

16. What are the three significant markings on the scapula and where are their locations (describe and explain).

____

____

____

____

____

____

THE PELVIC GIRDLE AND LOWER LIMBS

Dr. Monroe continued, "Now that we've talked about the top half of the body, let's head south to the pelvic girdle and lower limbs. We still have a lot of bones to discuss, so let's start with the pelvis (also referred to as the pelvic girdle). Collectively, the **ilium, ischium, pubis, sacrum,** and **coccyx** make up the pelvic girdle. Additionally, there are significant bone markings such as the **acetabulum** (hip socket), **iliac crest** (crest at the top of the ilium), and **obturator foramen** (large holes in the ischium). There is also a large piece of cartilage connecting the bottom of the pelvic girdle together in the pubis (pubic bone) area, called the **pubic symphysis**."

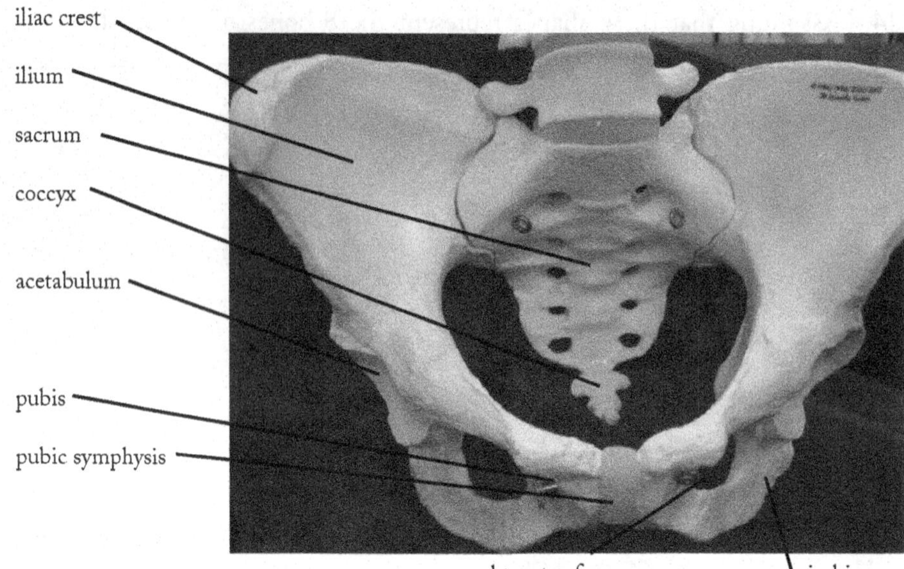

IMAGE 7.25 Pelvic Girdle, Anterior Aspect

STUDY TIP!

When trying to remember the obturator foramen, it helps if you think about placing an emergency call to 9-1-1… when the dispatcher answers the phone "Hello," you would then say, "Hello obturator…" instead of operator. The similarity in the two words should create a memorable phrase for life.

Dr. Monroe expounded, "The pelvis is the most sexually dimorphic portion of the skeleton, meaning that its anatomy differs the most between sexes. Here's some information from our forensics department that I keep handy that summarizes the differences between male and female pelvic girdles."

TABLE 7.2 Male and Female Pelvic Differences

Feature	Male Pelvis	Female Pelvis
Overall	Heavier; thicker	Lighter than male pelvis
Size	Taller and narrower overall	Shorter and wider
Coccyx	Coccyx 'hooks' forward and fixed	Coccyx straight and slightly moveable
Acetabula	Distance between the acetabula lesser than female due to narrow hips; actual sockets are larger	Distance between the acetabula greater than male due to wider hips; sockets are smaller than in males
Pubic arch	Pubic arch is at a sharper angle	Less angle on the pubic arch

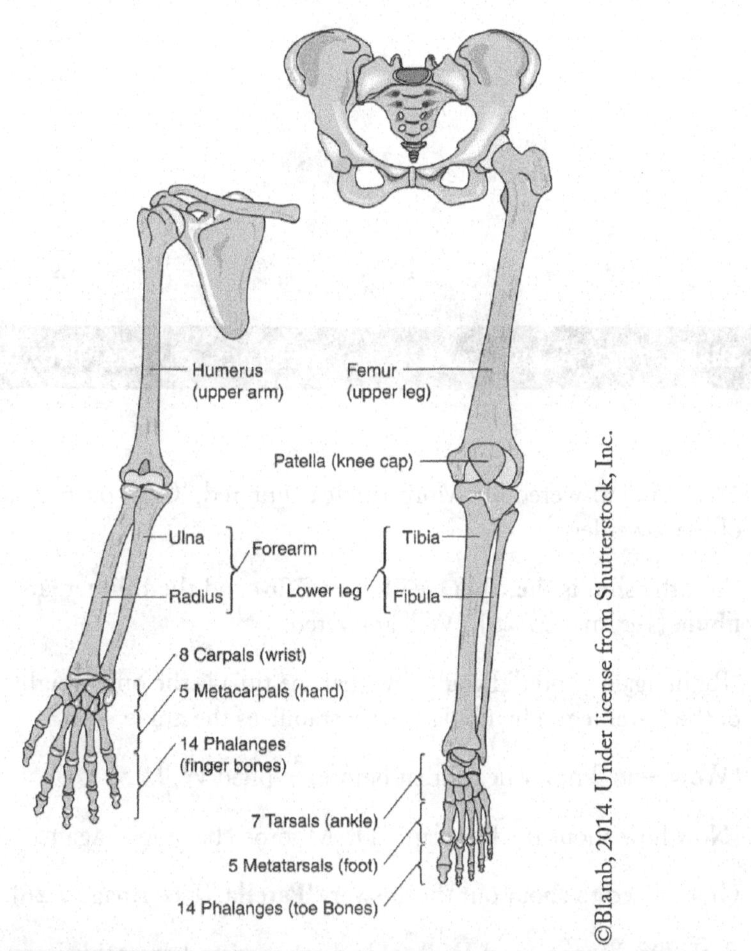

IMAGE 7.26 Leg Bones (shown right side of this image)

The Skeletal System: Return to the Local Radiology Department

"I know the name of the thigh bone," exclaimed Will. "It's the **femur!**" (see Image 7.26). The femur has many significant bone markings… note the head, neck, greater trochanter, and lesser trochanter on the proximal femur (see Image 7.27). At the distal end (anterior aspect), you will find the epicondyles (small projections) and on the posterior side you will see the medial and lateral condyles (they are markings that look like smooth knuckles)."

The Femur

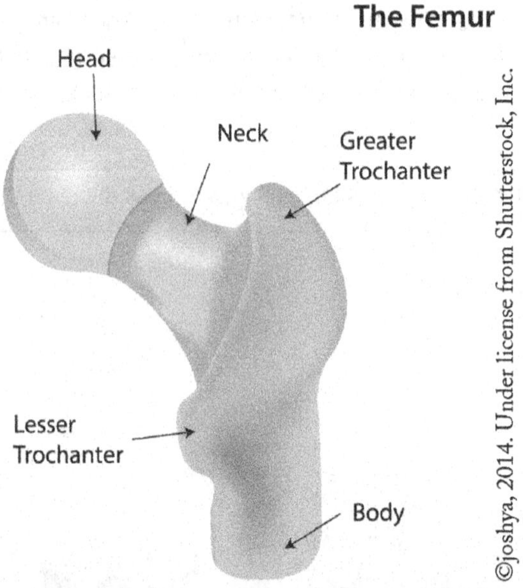

IMAGE 7.27 Significant Bone Markings of the Proximal Femur

"Yes, it is," answered Dr. Monroe. He countered, "Can you also name the bones of the lower leg?"

"Yes, the shin is the **tibia** (see Image 7.26) and the thinner lateral bone is the **fibula** (see Image 7.26)," Will answered.

"Right again. And did you know that the tibia is the only weight-bearing bone of the lower leg? The fibula merely stabilizes the ankle."

"Wow—no. I never heard that before," replied Will.

"Now how about the kneecap?" Dr. Monroe challenged again.

Grace raced to shout out the answer, **"Patella!"** (see Image 7.26).

Both Dr. Monroe and Will looked stunned at her enthusiasm and laughed. Then Dr. Monroe continued, "Yes, right you are. And the triangular patella is one of those bones that is actually held in place by the tendon of the knee. This is a bit unusual. Can you think of why?"

"I think it's because usually ligaments hold bone in place. And tendons hold muscle to bone. But with the patella, suddenly we have a tendon holding the bone in place," explained Will.

"Ah—indeed. That is nice reasoning Will," confirmed Dr. Monroe.

ANKLE AND FOOT

Dr. Monroe continued, "And now we're at the **tarsal** area (ankle). Does anyone remember how many ankle bones there are?"

"I believe it was 7," replied Grace,

"Yes, correct," replied Dr. Monroe.

STUDY TIP!

To remember that carpals are the wrist bones and tarsals are the ankle bones, simply remember that "c" comes before "t" in the alphabet. Thus, the top of your body comes first just like "c" in the alphabet.

Dr. Monroe continued, "The ankle bones are arranged much like the wrist in two rows. The proximal row has 3 bones—the **calcaneus** also known as the heel (the largest ankle bone), the **talus** (or second largest bone), and the **navicular.** The distal (bottom) row has 4 bones from medial to lateral—**medial, intermediate,** and **lateral cuneiforms** as well as the **cuboid** (the largest in this group/row of bones)."

"Are there five metatarsals?" questioned Grace.

"Yes," answered Dr. Monroe.

"And are the toes similar in bone count to the fingers?" questioned Will.

"Yes. Like the pollex (thumb), the hallux (greater toe) has only two phalanx, all the other toes have three," confirmed Dr. Monroe.

NOTES

IMAGE 7.28 Bones of the Foot and Ankle

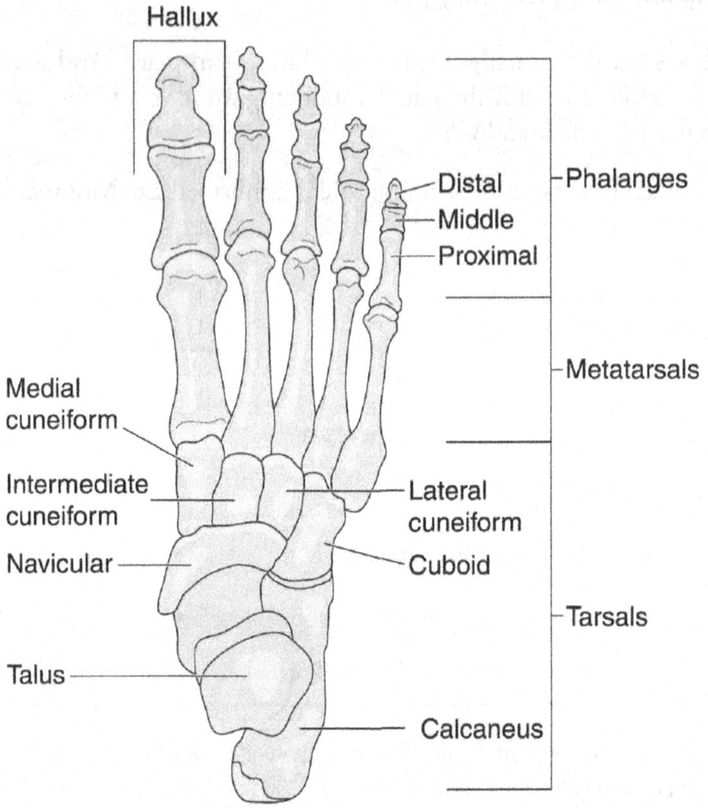

Self-Check – Continued

17. Label the bones and significant markings of the pelvis on the photo below.

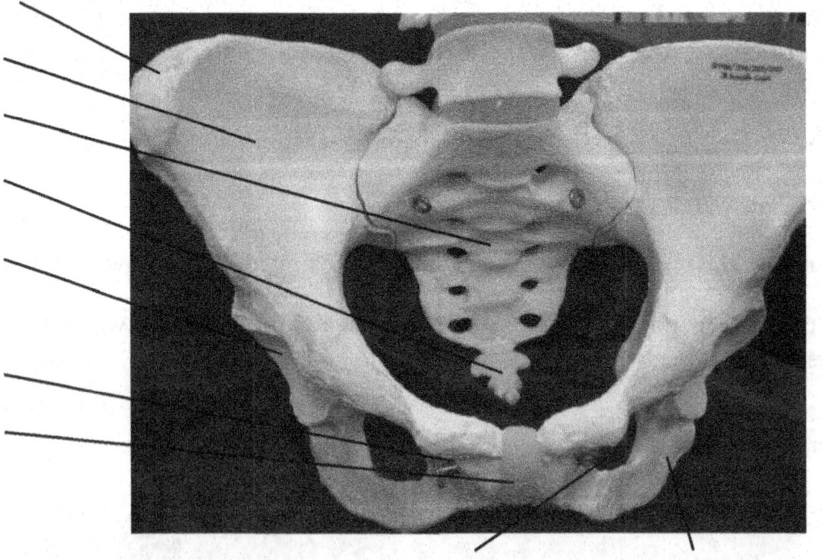

18. Describe the bones of the lower leg.

____

____

19. Compare and contrast the male and female pelvic girdles.

____

____

____

____

____

____

WRAP-UP

"I guess we have finished the bones now so we should let you get back to work," said Will as he addressed Dr. Monroe.

Grace looked at Dr. Monroe and said in an appreciative tone, "Again, we have learned so much from you. Thank you for your expertise and your time."

"My pleasure, you two and keep practicing the bones until you learn them thoroughly!" finalized Dr. Monroe.

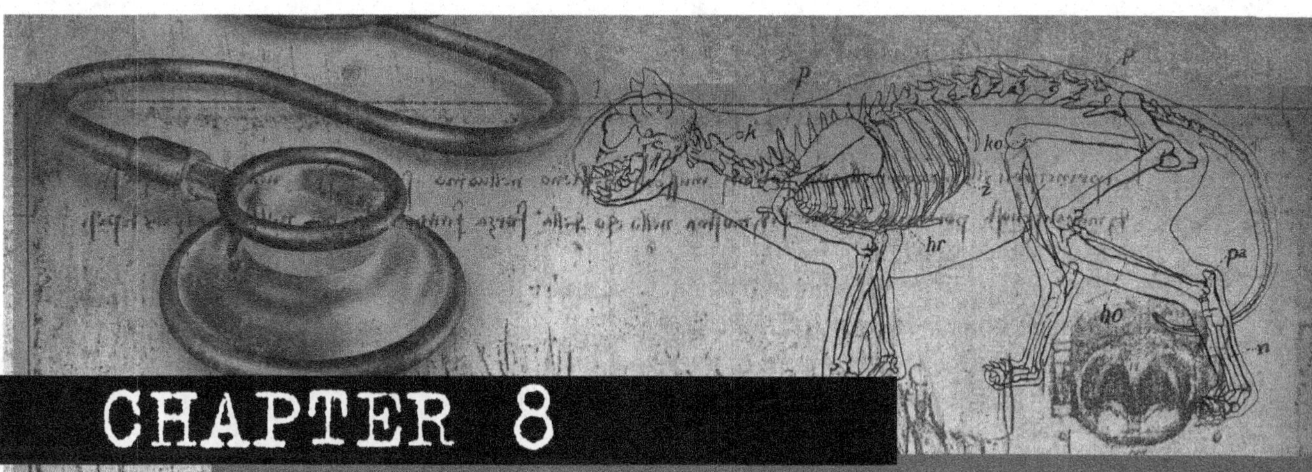

CHAPTER 8

Joints: Working with a Physical Therapist to Better Understand Joint Movement, Function, and Disease

KEY WORDS

The study of Anatomy & Physiology involves many new vocabulary words. It is helpful to gain familiarity with these new key words, just as you would a foreign language.

abduction

adduction

amphiarthroses

articulation

bursae

cartilaginous joint

circumduction

depression

diarthroses

dorsiflexion plantar

elevation

eversion

extension

fibrocartilage

fibrous joint

flexion

gomphoses

hyperextension

inversion

opposition

pronation

protraction

range of motion (ROM)

retraction

supination

sutures

symphyses

synarthroses

synchondroses

syndesmoses

synovial fluid

synovial joint

tendon sheath

It is helpful to have an idea of what you need to learn before proceeding. Here is a study guide to assist you:

1. **Explain an articulation (joint).**
2. **Classify joints based on structure and function. Describe their characteristics.**
3. **Demonstrate common body movements.**
4. **Identify and describe the main joints of the elbow, knee, hip, and shoulder.**
5. **Explain the basis of common joint diseases such as arthritis.**

Grace and Will are both extremely interested in learning more about joints. Grace has an aunt with rheumatoid arthritis. Will has knee issues. They have made an appointment to talk with Elijah "Eli" Crane, a physical therapist that focuses on limb mobility.

INTRO TO JOINTS

"Hey guys! I'm Eli. What kind of things would you like to know more about?"

"We'd like to know more about the joints of the human body and how they work as well as diseases that threaten joints… and maybe how you treat each condition," explained Will.

"Okay… well, joints are articulations (where two bones meet) that link the bones of the skeletal system together so you can move. The name of the joint is typically derived from the names of the bones involved. For example, the radioulnar joint is where the radius meets the ulna," Eli explained. He continued, "They are classified by the way adjacent bones are bound to one another. For example, is the joint freely moveable or not moveable at all. There are actually three major categories of joints: **fibrous**, **cartilaginous**, and **synovial**."

STUDENT VIEW – WHY SHOULD I STUDY THIS? HOW IS IT RELEVANT TO MY LIFE?

Perhaps you know someone that suffers from a debilitating rheumatoid arthritis? Or you yourself have early symptoms of osteoarthritis from physical wear and tear on your body? Understanding your joints can help you understand not only how they function but how to better care for them or help a loved one suffering from arthritis.

Eli continued, "Fibrous joints are also known collectively as synarthoses (NOTE: synarthroses is plural; synarthrosis is singular/one)."

"Yeah, we have," answered Will.

"Okay, good. That's an example of a **fibrous joint**. Typically, the gap between two skull bones like the frontal and parietal ossifies (hardens) and the two bones become one bone. Immoveable joints can also be cartilaginous."

"We learned all the names of the sutures already when we visited Dr. Monroe at the Radiology Department at the local hospital here," revealed Grace.

Self-Check – Answer this question now before you continue:

1. Let's review. Write the names of the four skull sutures and the bones (learned in previous chapter) involved with each suture here.

____

____

____

Eli continued, "Fibrous joints are also known as **synarthoses**."

"What makes it a synarthrosis exactly?" questioned Will.

"Glad you asked," replied Eli. "Adjacent bones bound by collagen fibers. These collagen fibers usually emerge from one bone, run across the space between the two bones and penetrate into the other bone like webbing. The attachment of a tooth to its socket is an example of another type of fibrous joint called a **gomphoses**. A tooth is held in its socket by a fibrous periodontal ligament. This ligament is made up of collagen fibers that extend from the maxilla and mandible into the dental tissue. This type of joint does allow for some movement under the stress of mastication. This also allows us to sense how hard we are biting."

Tooth anatomy

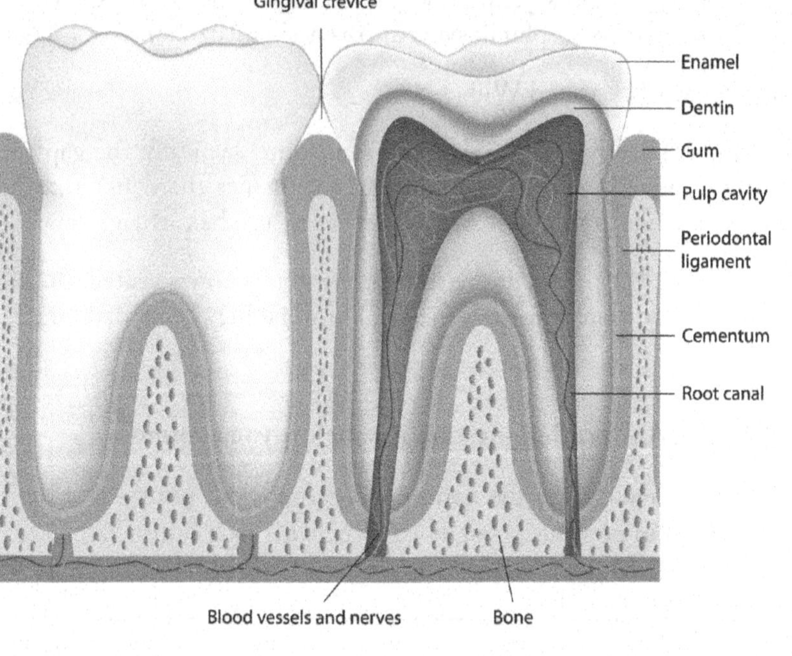

IMAGE 8.1 Gomphoses with Periodontal Ligament

Eli continued, "The collagen fibers (ligaments) are long and especially moveable between the diaphysis (shafts) of the radius and ulna. Fibrous joints of this nature are known as **syndesmoses**. Another example of a syndesmoses is the distal ligament connecting the ends of the tibia and fibula."

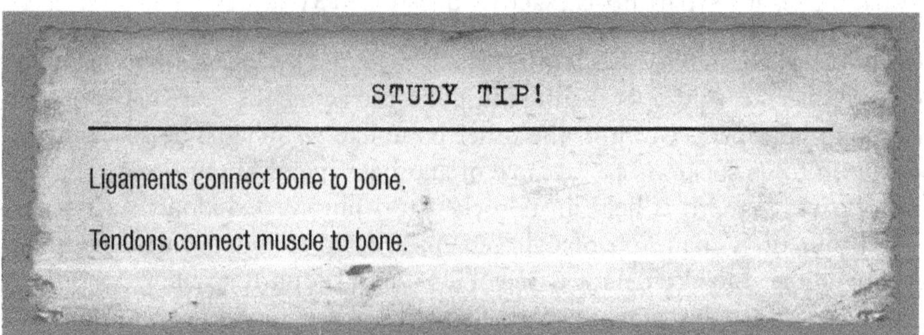

CARTILAGINOUS JOINTS

Eli continued, "A **cartilaginous joint** is also known as an **amphiarthrosis**. These are slightly moveable joints."

"If cartilaginous joints are made out of cartilage, then I think I can come up with an example," exclaimed Will.

"Okay, what?" challenged Eli.

"I think how the ribs connect to the sternum via hyaline cartilage," answered Will.

"Yes, that's certainly one example of a cartilaginous joint known as a **synchondroses**."

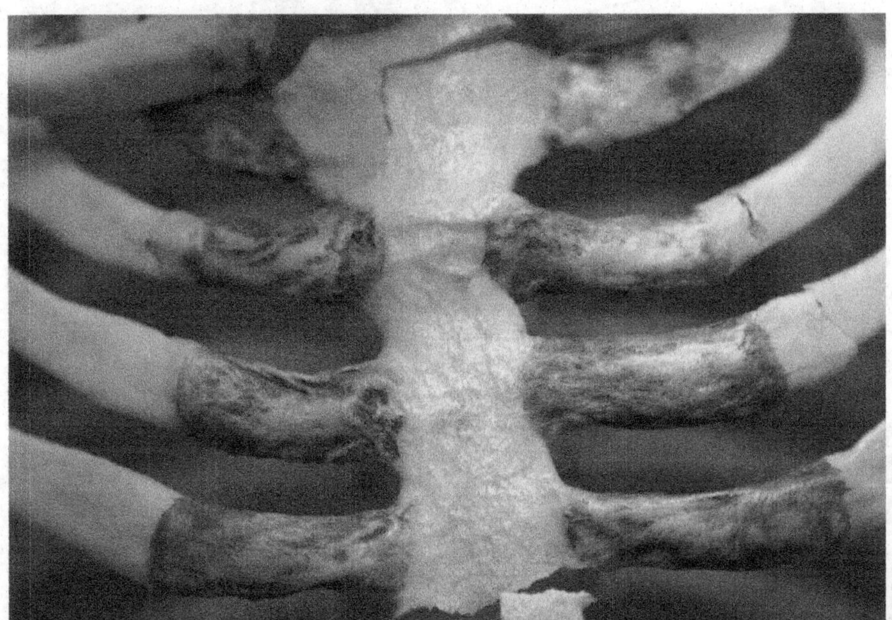

© Ryan Kelm, 2012. Under license from Shutterstock, Inc.

IMAGE 8.2 Synchondroses Consisting of Ribs, Hyaline Cartilage, and Sternum

Eli interjected, "But… there is another type of cartilaginous joint. And it's called a **symphysis**."

"As in the pubic symphysis?" questioned Will.

Eli shot back "Actually, yes. That joint is **fibrocartilage** (the toughest type of cartilage) which is what constitutes a symphyses joint. Another example is the fibrocartilage discs between the vertebrae in the spinal column known as intervertebral discs."

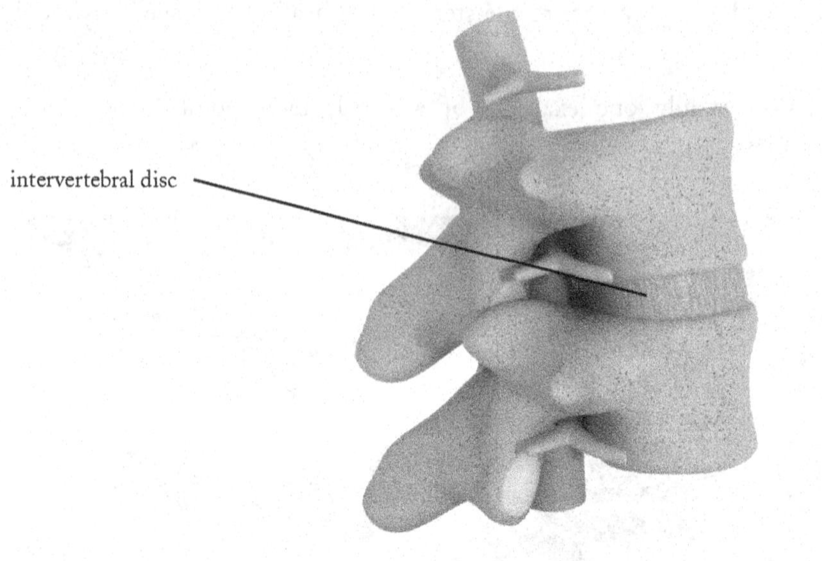

IMAGE 8.3 Symphysis

SYNOVIAL JOINTS

Eli continued, "Probably the most familiar types of joinst are synovial joints also known collectively as **diarthroses**. Generally, these are joints are freely moveable. In my line of work, these are the joints I am usually helping people with the most."

"Is the knee a synovial joint?" asked Grace.

"Yes it is. So are your elbow and knuckles. There are many examples, actually," replied Eli.

Eli continued, "Synovial joints are the most complex type of joint. With these types of joints, the ends of two bones are covered in articular cartilage (cartilage that covers the ends of long bones; usually hyaline) and lubricated with a slick lubricant called **synovial fluid**. In the knee specifically, a synovial membrane helps to lubricate the joint with synovial fluid. There is also a meniscus, a rubbery C-shaped disc, that helps to cushion your knee. A joint capsule wraps the joint to retain this synovial fluid. Another capsule known as the fibrous capsule wraps the joint capsule. There are some friction reducing structures that are also known to occur in these types of joints called **bursae** (fibrous sacs filled with synovial fluid as in the shoulder, where it acts like a cushioned pad) and **tendon sheaths** (elonagated bursae wrapped around tendons, especially in the hands and feet)."

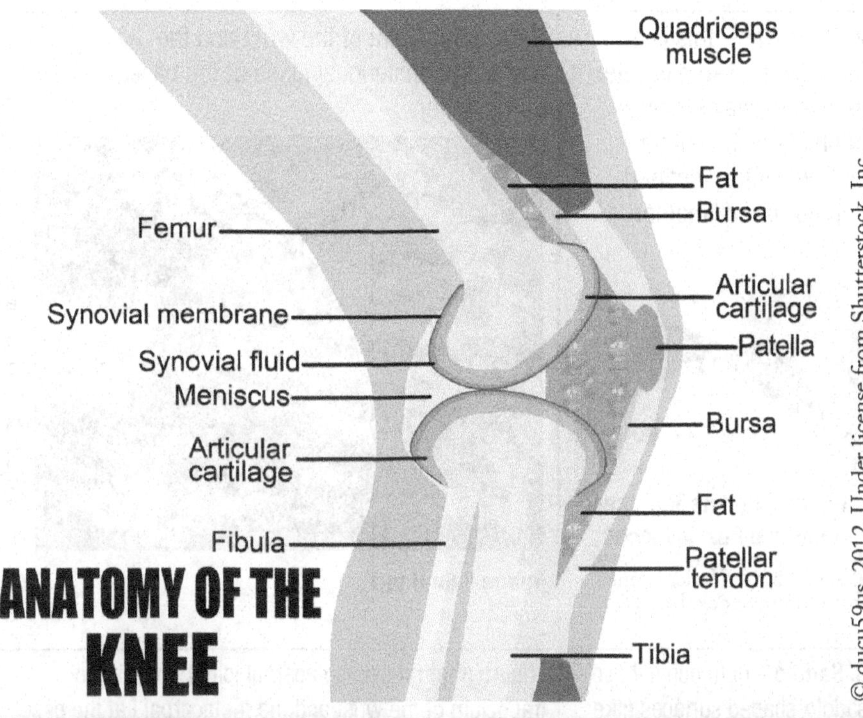

IMAGE 8.4 Knee with Synovial Fluid and Bursae

CLASSES OF SYNOVIAL JOINTS

Eli continued, "There are actually six classes of synovial joints. Here's a brochure that I give to my patients to help them understand this concept better."

TABLE 8.1 Eli's Brochure Showing the Six Classes of Synovial Joints

Class of Synovial Joint	Example
1. Ball-and-Socket – the only multi-axial joints in the body; one bone (the humerus or femur) has a smooth head that fits into a cup-like socket (glenoid fossa in the arm and acetabulum in the hip) 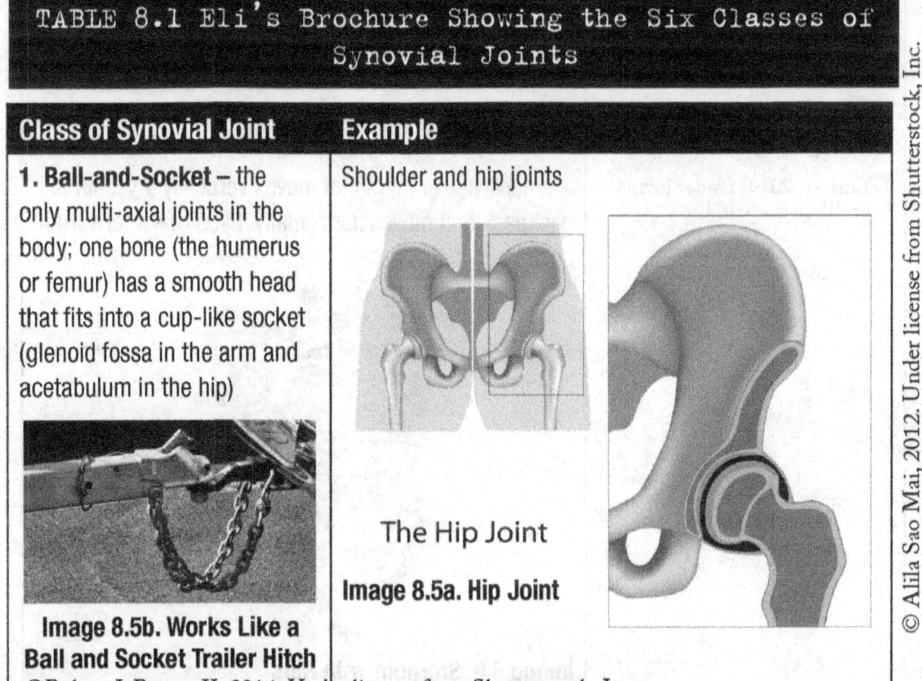 **Image 8.5b. Works Like a Ball and Socket Trailer Hitch**	Shoulder and hip joints The Hip Joint **Image 8.5a. Hip Joint**

2. Condylar – this joint fits together like a puzzle; an oval convex surface of one bone fits into a complimentary-shaped depression on another

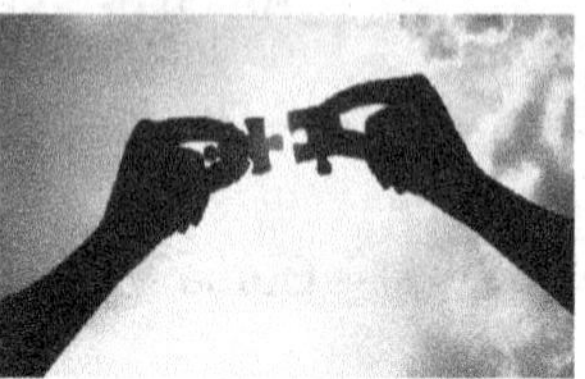

Image 8.6b Fits Together Like Two Puzzle Pieces

©2jenn, 2014. Under license from Shutterstock, Inc.

Radiocarpal joint of the wrist and the metacarpophalengeal joints at the bases of the fingers

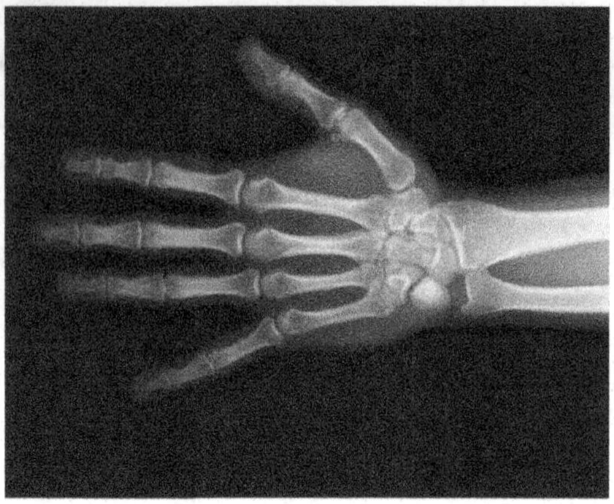

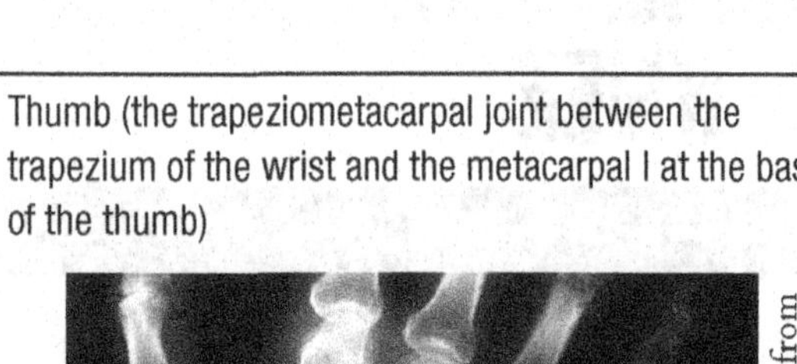

© itsmejust, 2012. Under license from Shutterstock, Inc.

Image 8.6a Hand

3. Saddle – both bones have saddle-shaped surfaces (like the front to the rear curvature of a horse's saddle) and are concave in one direction; biaxial joint

Image 8.7b Like a Rider's Legs Over a Saddle

©YanLev, 2014. Under license from Shutterstock, Inc.

Thumb (the trapeziometacarpal joint between the trapezium of the wrist and the metacarpal I at the base of the thumb)

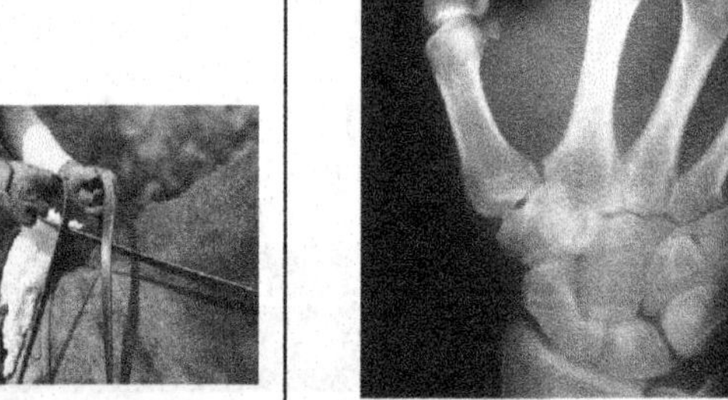

© Mats, 2012. Under license from Shutterstock, Inc.

Image 8.7a. Hand 2

Sternoclavicular joint (that moves vertically if you lift a backpack and moves horizontally if you reach forward to push open a door)

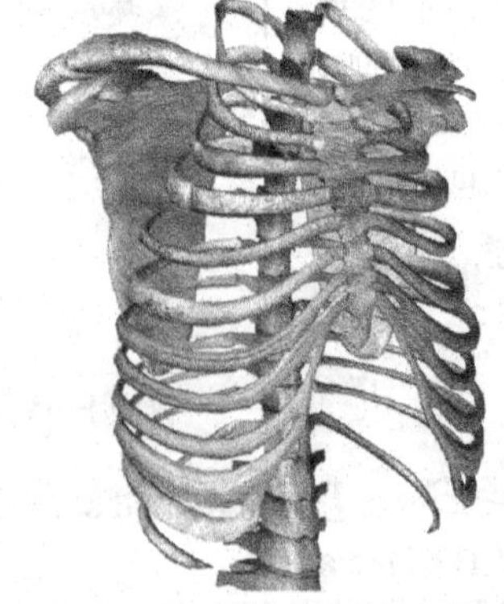

© YorkBerlin, 2012. Under license from Shutterstock, Inc.

Image 8.8. Sternum with Ribs

4. Plane or gliding joint – the bones slide over one another with relatively limited motion  **Image 8.9b Works Like a Board Sliding Across Another** ©indigolotos, 2014. Under license from Shutterstock, Inc.	Carpals of the wrist Tarsals of the ankle 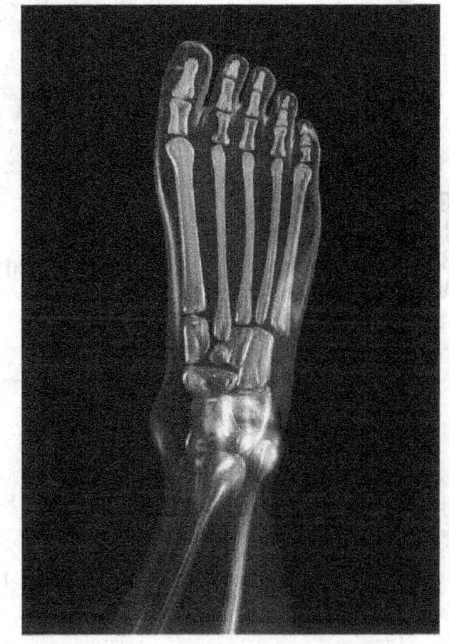 © BioMedical, 2012. Under license from Shutterstock, Inc. **Image 8.9a. Ankle**
5. Hinge – like a door hinge  **Image 8.10b Works Like a Door Hinge** ©Pablo Hidalgo, 2014. Under license from Shutterstock, Inc.	Elbow Knee 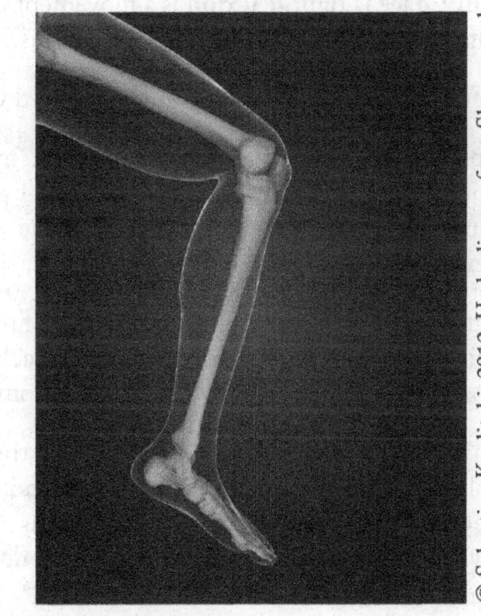 © Sebastian Kaulitzki, 2012. Under license from Shutterstock, Inc. **Image 8.10a. Knee Joint** Finger and toe joints

6. Pivot – like a wheel turning on its axle  **Image 8.11b Hoop Moving Around a Person is Like A Ring of Bone Moving Around Another Bone** ©Pablo Hidalgo, 2014. Under license from Shutterstock, Inc.	Radioulnar joint (radius crosses over ulna during rotation of the lower arm) © Robnroll, 2012. Under license from Shutterstock, Inc. 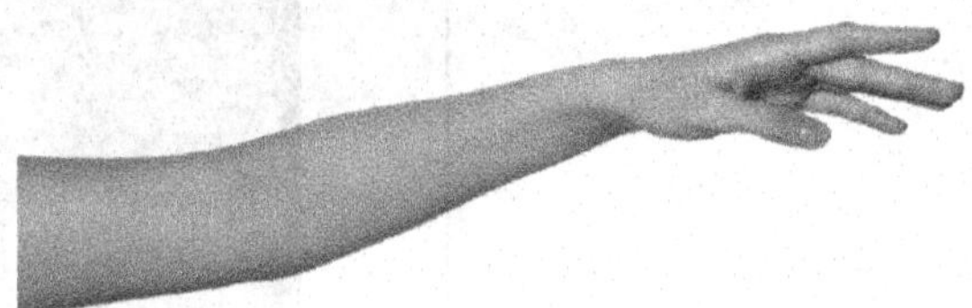 **Image 8.11a. Radius Crossing-over Ulna**

MOVEMENTS OF SYNOVIAL JOINTS

"Do you know what flexion is? Could you show me what flexion of the back looks like?" questioned Eli.

At the same time, Will bent backward and Grace bent forward. They looked at each other and giggled. They knew that someone was right and someone was wrong! "I'm right, aren't I?" quipped Will.

"Actually Grace is right. **Flexion** is a movement that decreases a joint angle. So bending forward at the waist accomplishes this," stated Eli.

"What's the opposite of this movement?" asked Grace.

Eli replied, "That would be extension and hyperextension. In **extension**, you would bring the body back to anatomical position. And if you continue with further extension of the joint beyond the anatomical position, you have **hyperextension**."

Eli continued, "Let me introduce you to one of my patients, John. He has been doing light rehab exercises to strengthen his back. Today I am going to test his range of motion in his back to see how he is coming along."

They all walked toward an elderly man. As they all approached, Eli spoke, "Hi, John. I need to test your range of motion in your back today. I'd like you to demonstrate flexion by bending forward as much as you can. It's okay to go slowly. I also need you to stand back up to demonstrate extension. Okay… good job, thanks."

"Now we're going to check on Laurie who has also had a back injury." Eli prompted them to follow him. He continued, "Hi, Laurie. Today I need to have you bend backward as much as you are able. Okay, still not too much range on that particular move. Thank you, we'll get you some ice and heat therapy now to help that along."

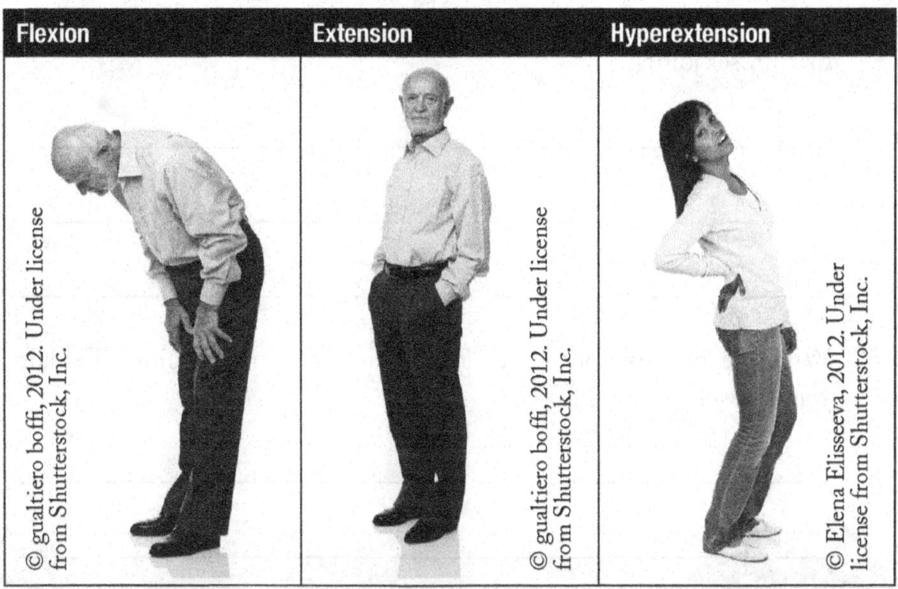

IMAGE 8.12 Flexion, Extension, and Hyperextension

Self-Check – Continued

2. Explain the three types of fibrous joints and where they are located.

   ____

   ____

   ____

3. What is another name for a fibrous joint in general?

   ____

4. What are the two main types of cartilaginous joints? Give examples of both.

   ____

   ____

   ____

5. What is another name for cartilaginous joints in general?

   ____

6. Explain the difference between synarthroses, amphiarthroses, and diarthroses joints.

____

____

____

7. What are two friction-reducing structures in synovial joints? Explain your answer.

____

____

____

8. List the six classes of synovial joints with an explanation and one example of each.

____

____

____

____

____

____

____

____

____

RANGE OF MOTION

"What I specialize in as a physical therapist is helping people restore range of motion to injured joints. One aspect of joint performance and physical assessment of a patient is joint flexibility (also known as **range of motion or ROM**) or the degrees through which a joint can move," remarked Eli.

Eli continued, "The knee, for example, can flex through an arc of 130 to 140 degrees. The elbow, however, cannot straighten out beyond 180 degrees because joint movement is limited by the shape of the bone's surfaces. Range of motion figures into a person's functional independence and quality of life. It is also an important consideration not only in monitoring the progress of a patient's rehabilitation but also in their overall clinical diagnosis and in judging readiness to return a person to athletics or dance."

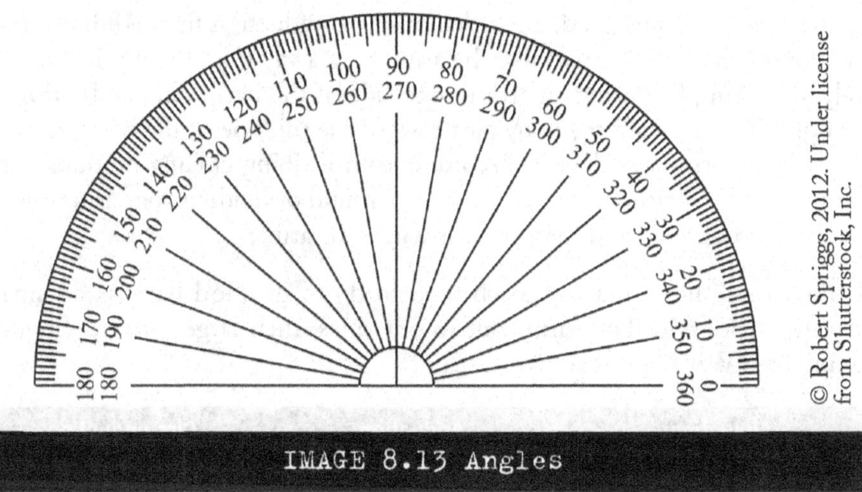

IMAGE 8.13 Angles

ROTATION

Eli added, "Lets talk about rotation now. Rotation is the turning of a bone around its own long axis. It is the only movement allowed between the first two cervical vertebra and is common at the hip and shoulder joints. Rotation may be directed toward the midline or away from it. For example, in medial rotation of the thigh, the femur's anterior surface moves toward the median plane of the body; lateral rotation is the opposite movement."

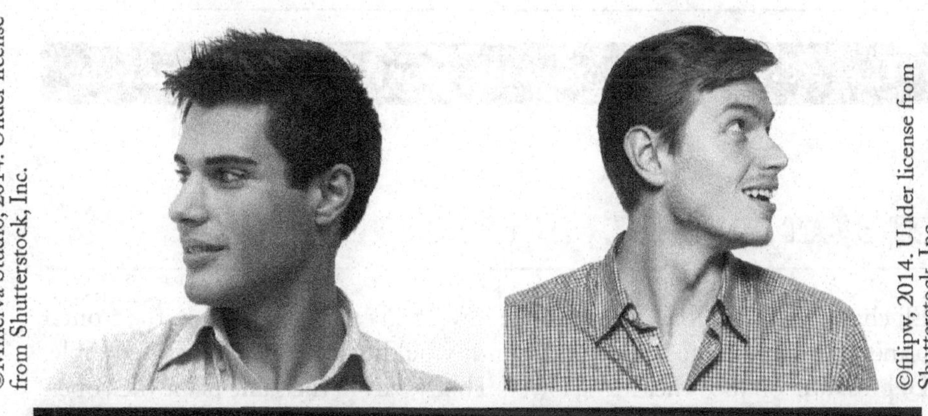

IMAGE 8.14 Head Rotated to the Left and to the Right

ABDUCTION, ADDUCTION, AND CIRCUMDUCTION

"Would you guys like to learn some more movements—maybe even practice trying some of these movements?" asked Eli.

"Sure—we're game!" answered Will. Grace nodded her head in agreement.

Eli continued, "Okay, good. Let's demonstrate abduction first. **Abduction** is the movement of a body part like the arm or leg away from the midline of the body. For example, raising an arm to one side of the body. And **adduction** is the opposite… moving the body part toward the midline of the body, such as lowering the arm back down. **Circumduction** is doing circling motions with a limb, in effect twirling a finger round and round or holding your arms out to the side of your body and doing circles with your arms."

"I think that's like what my coach used to do to us. He'd have us do small forward circles with both arms out to our sides, then large forward circles," commented Will.

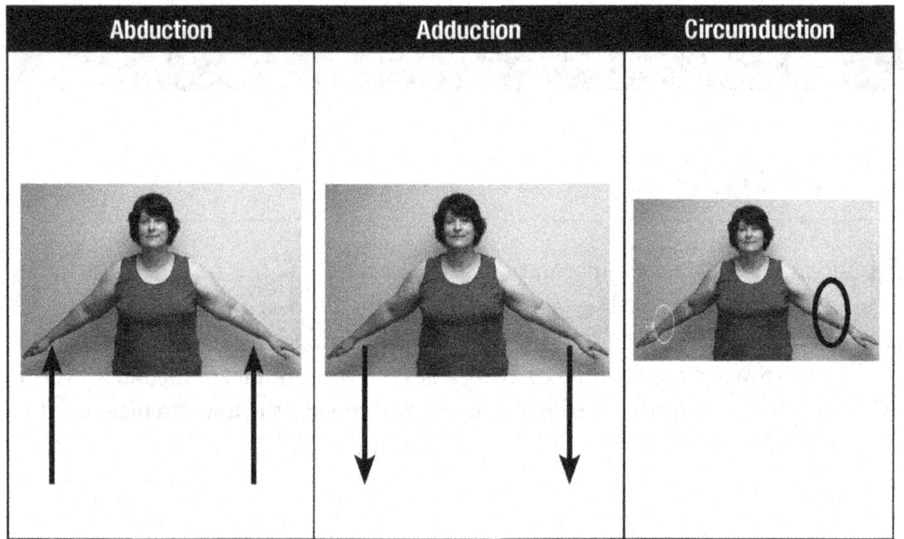

IMAGE 8.15 Abduction, Adduction, and Circumduction

ELEVATION AND DEPRESSION

Eli challenged, "Try this. **Elevation** raises a body part vertically in the frontal plane. For example, to close your mouth would be to elevate your mandible. **Depression,** on the other hand, lowers a body part in the same plane (i.e., open your mandible)."

"So, elevation is really just a nice way to tell someone to be quiet or shut their mouth," commented Grace as she laughed.

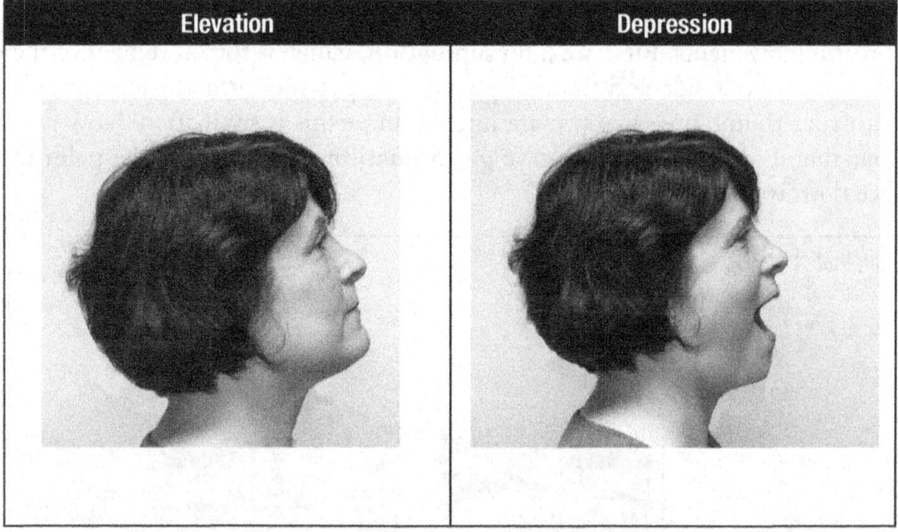

IMAGE 8.16 Elevation and Depression

PROTRACTION AND RETRACTION

Eli continued, "Since we're on a roll—here's something else easy to demonstrate using your mouth. Stick out your chin—go ahead, just jut it right out in front of the rest of your face... that's protraction of the mandible. **Protraction** is the anterior movement of a body part in the transverse or horizontal plane. **Retraction** is posterior movement. So tuck your mandible in—go ahead... make a double chin."

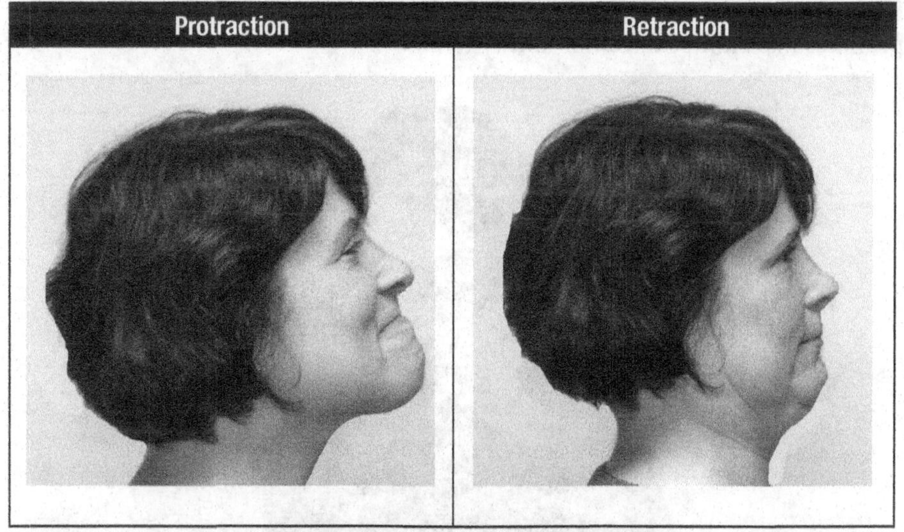

IMAGE 8.17 Protraction and Retraction

SUPINATION AND PRONATION

"So, now we are going to try supination and pronation, which are primarily forearm movements. First, we'll do **supination,** which is the movement of the arm out by your side and thumb upwards. So, put your arm out to your side with your thumb up—like you are hitchhiking—this is supination. Now turn your thumb downward and you've got **pronation,** which causes the palm to face downward."

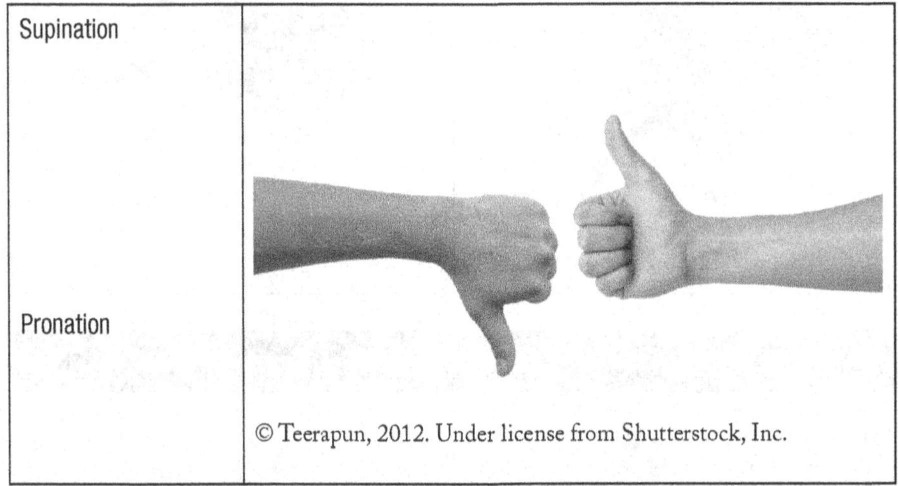

© Teerapun, 2012. Under license from Shutterstock, Inc.

IMAGE 8.18 Supination and Pronation

OPPOSITION

"We have opposable thumbs—this separates us from animals with paws. **Opposition** is the ability to move the thumb to touch any fingertip. Other primates like chimps have this ability too."

© vgstudio, 2012. Under license from Shutterstock, Inc.

IMAGE 8.19 Opposition of Thumb to Index Finger

DORSIFLEXION AND PLANTAR FLEXION

"Next, we're going to do some special movements of the foot. So, put your foot out," barked Eli.

"Yikes, my balance is horrible. It's hard to stand on one leg while I put my other foot out," Will said as he wobbled.

"Me too, I'm kinda wobbling," Grace confessed.

"Now point your toes up. This is **dorsiflexion**. Next, point your toes down. This is **plantar flexion**."

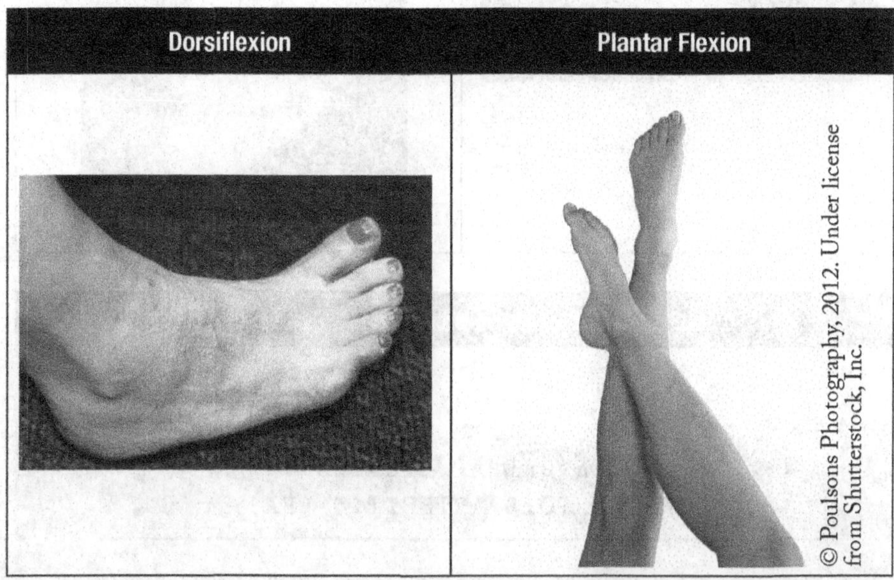

IMAGE 8.20 Dorsiflexion and Plantar Flexion

INVERSION AND EVERSION

"And now, lets try inversion. Put your foot out again. Turn the sole of your foot inward medially. This is **inversion**."

"Let me guess, turn the sole of your foot outward next and you've got eversion?" asked Grace.

"Yes, correct. **Eversion** is turning the sole of your foot outward laterally."

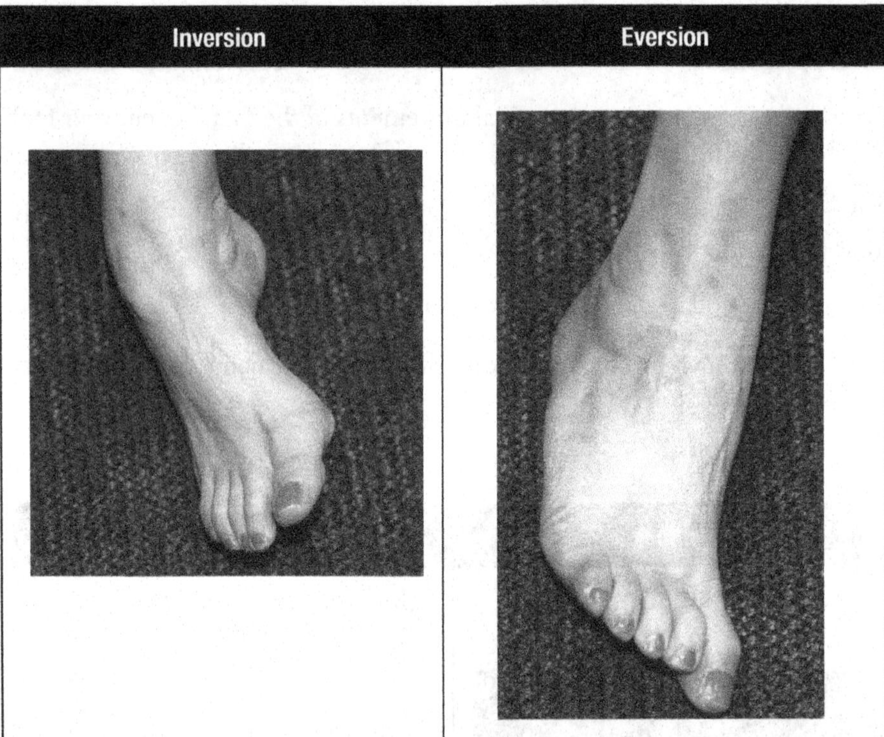

IMAGE 8.21 Inversion and Eversion

CLINICAL APPLICATION:
RHEUMATOID ARTHRITIS (RA)

There are generally two main types of arthritis—osteoarthritis and rheumatoid arthritis. This clinical spotlight will focus on the more serious of the two—rheumatoid arthritis also called RA, which results from an autoimmune attack against joint tissues. Antibodies released by our bodies typically fight infections but in the case of RA, antibodies fail to recognize the body's own tissues and also attack synovial membranes in the joints causing joint destruction, disfigurement, crippling, and severe pain and inflammation.

The name of this disease is based on the fact that its symptoms tend to come and go, flaring up periodically. It affects women more than men and can begin early in life (in some cases in the teenage years but typically onset begins between 30-40).

There is no cure but treatments like hydrocortisone and other steroids can slow it down. However, long-term use of steroids can weaken bone so aspirin is usually the first choice of treatment options. Physical therapy is often helpful as well to preserve as much range of motion as possible.

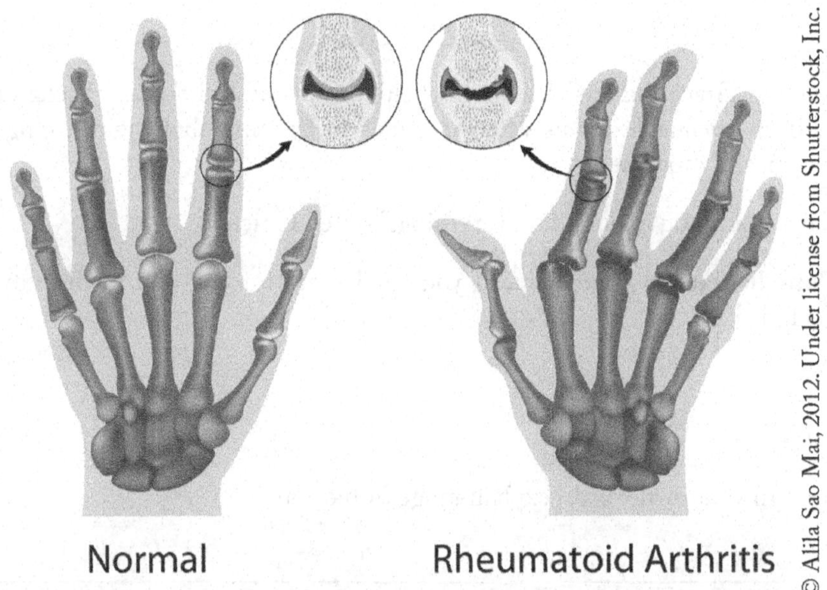

IMAGE 8.22 Rheumatoid Arthritis

Clinical Reflection Questions

1. In your own words, explain what causes RA.

____

____

____

2. What are the symptoms of the disease? What age does it typically strike?

____

____

____

3. What are treatment options?

____

____

____

WRAP-UP

Will and Grace thanked Eli for his help, "Your physical therapy background has been invaluable for us when it comes to learning about joints, range of motion, and movements."

"Can't thank you enough for everything," stated Grace.

"Don't hesitate to contact me if you need more information in the future," concluded Eli.

Self-Check – Continued

9. In your own words, explain range of motion.

    ____

    ____

    ____

10. Describe what the movement looks like for the following joint motions:

    a. gliding

    b. supination

    c. pronation

    d. opposition

    e. abduction

    f. adduction

    g. circumduction

    h. elevation

    i. depression

    j. retraction

    k. protraction

    l. dorsiflexion

    m. plantar flexion

    n. inversion

    o. eversion

CHAPTER 9

A Visit With an Athletic Trainer to Learn About Muscles

KEY WORDS

The study of Anatomy & Physiology involves many new vocabulary words. It is helpful to gain familiarity with these new key words, just as you would a foreign language.

Achilles tendon	gastrocnemius	rectus femoris
actin	gluteus maximus	sarcomere
action potential	gluteus medius	sartorius
biceps brachii	glycemic index	semimembranosus
biceps femoris	gracilis	semitendinosus
bipennate	iliopsoas	skeletal muscle
brachialis	iliotibial tract	smooth muscle
brachioradialis	isometric	soleus
cardiac muscle	isotonic	sternocleidomastoid
circular	latissimus dorsi	sternohyoid
convergent	levator labii superioris	temporalis
coupling	masseter	teres major
deltoid	motor unit	teres minor
endomysium	multipennate	tetanus
epicranius	myofibrils	tibialis anterior
epimysium	myosin	trapezius
excitation-contraction	myosin-light chain kinase	triceps brachii
extensor carpi radialis longus	occipitalis	tropomyosin
external oblique	orbicularis oculi	twitch
fascicle	orbicularis oris	unipennate
fibularis longus	parallel	vastus lateralis
flexor carpi radialis	pectineus	vastus medialis
flexor carpi ulnaris	pectoralis major	zygomaticus major
frontalis	perimysium	zygomaticus minor
fusiform	platysma	
galea aponeurotica	rectus abdominis	

It is helpful to have an idea of what you need to learn before proceeding. Here is a study guide to assist you:

1. State the three overall types of muscles, their locations, and whether or not they are voluntary or involuntary.
2. Explain the glycemic index and its relevance to blood sugar.
3. Explain muscle structure including the three levels of muscle organization—epimysium, endomysium, and perimysium.
4. Explain a fascicle.
5. State the seven common fascicle arrangements of muscle, their shape, and give an example of each.
6. Identify the major muscles of the human body and their functions.
7. Explain the macroanatomy, microanatomy and physiology of skeletal muscle.
8. Describe the sliding filament theory.
9. Explain in your own words twitch versus tetanus.
10. Explain how motor neurons stimulate muscle to contract.

Grace and Will have been asked by their college Anatomy instructor to gather as much knowledge as they can about muscles. They have also been tasked with learning the major muscles of the human body. They have scheduled an appointment with an athletic trainer, named Glenn, who works at the recreational center at their college.

Grace and Will arrive at the rec center to meet with Glenn. Will gives a shout out to Glenn, "Hey, Glenn. Thanks for agreeing to talk with us today about muscles. We understand you're an expert on the subject?"

"And… Hi, I'm Grace. I don't believe we've officially met."

"Yeah. Great to have you here. What would you like to start with?" asked Glenn.

"Well, we thought maybe an introduction to muscles first. Then later maybe teach us some of the major muscles. We know there are over 600 in the adult body so if you could help us with the major muscles, that would be helpful," answered Will.

Glenn added, "I think it's helpful to understand that our skeletal muscles are nearly half of our body's weight. Besides skeletal muscle, do you know the other two types of muscle tissue?"

Self-Check – Answer this question now before you go on:

1. What are the three overall types of muscle tissue? Where are they located?

____

____

____

Glenn continued, "There is **skeletal muscle** that covers our bones and allows us to move. It is under our voluntary control, of course. There is **smooth muscle** lining many of our body's passageways like the intestines. It is involuntary. And lastly, there is **cardiac muscle,** which is only found in the heart. It is also involuntary."

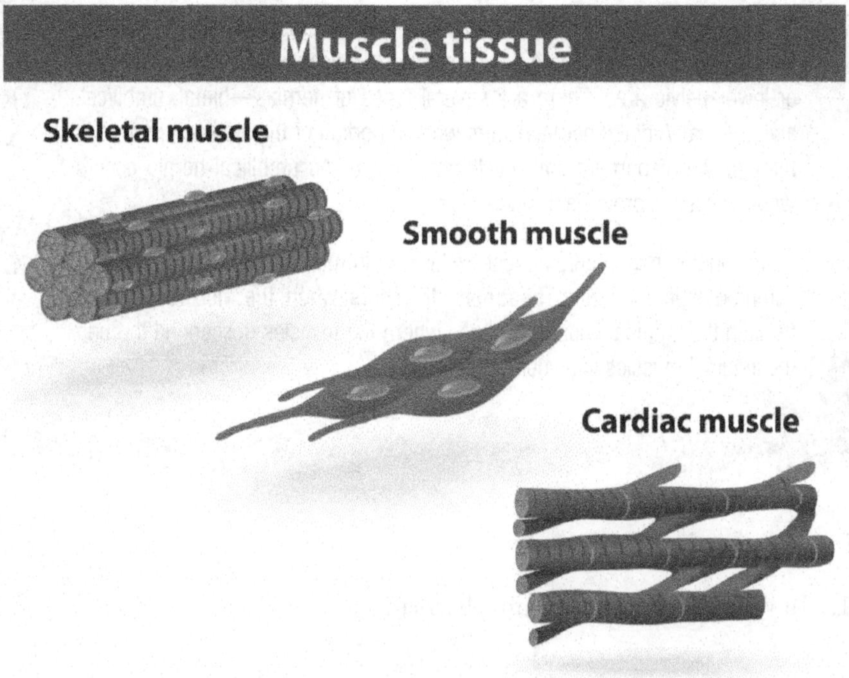

Muscle tissue

Skeletal muscle

Smooth muscle

Cardiac muscle

IMAGE 9.1 The Three Basic Muscle Types

STUDENT VIEW – WHY SHOULD I STUDY THIS?
HOW IS IT RELEVANT TO MY LIFE?

We use our skeletal muscles to move our bodies around. Sometimes we don't really stop to think about all that they do for us until we encounter a sore, painful muscle or have an injury or muscle strain. Because our bodies are literally flesh and blood, we will all have to deal with some type of noticeable muscle issue during our lifetimes.

Also, wouldn't it be nice to be able to know and actually name the major muscles of your body? Smarty pants!

CLINICAL APPLICATION: HERNIAS

You may have heard the term *hernia* or know of someone that has one. But do you actually know what a hernia is? Or what causes one?

It's a condition in which the viscera (in effect, internal soft organs like the intestine) protrude through a weak spot in the muscular wall of the abdomen or lower pelvic area. There are several types of hernias—hiatal, umbilical, and inguinal. A hiatal hernia occurs when a portion of the stomach protrudes through the diaphragm into the thoracic cavity. An umbilical hernia occurs when intestines protrude through the navel.

An inguinal hernia occurs almost exclusively in men and is usually the most common type to require treatment. It occurs when the viscera protrude through the inguinal canal—the area where the testicles descended through the inguinal muscles in a young male child.

Clinical Reflection Questions

1. In your own words, explain a hernia.

2. What is the most common type of hernia? Explain what it is.

SPECIAL CHARACTERISTICS OF MUSCLE TISSUE

Glenn stated, "All muscle cells have the following characteristics:

Responsiveness (excitability). Responsiveness is a property of all living cells, but muscle and nerve cells have developed this property to the highest degree.

When stimulated by chemical signals, stretch, and other stimuli, muscle cells respond with electrical changes across the plasma membrane.

"Conductivity. Stimulation of a muscle cell produces more than a local effect. The local electrical change triggers a wave of excitation that travels rapidly along the cell and initiates processes leading to contraction.

"Contractility. Muscle cells are unique in their ability to shorten substantially when stimulated. This enables them to pull on bones and other organs to create movement.

"Extensibility. In order to contract, a muscle cell must also be extensible—able to stretch again between contractions. Most cells rupture if they are stretched even a little, but skeletal muscle cells can stretch to as much as three times their contracted length.

"Elasticity. When a muscle cell is stretched and then released, it recoils to a shorter length. If it were not for this elastic recoil, resting muscles would be too slack."

FUNCTIONS OF MUSCLES

Glenn commented, "My specialty is skeletal muscle. Since there are over 600 muscles in the body, we will probably only get to survey a little less than one-third of them." He continued, "Muscles (all types) have different functions such as movement, stability, control of body passages and openings, heat production, and glycemic control of blood sugar."

"Hey, I've heard that term before—glycemic. Although, I'm not entirely sure what it means," commented Will.

Glenn replied, "Skeletal muscle uses a lot of glucose for fuel. The more muscle is exercised and built up, the more glucose it pulls from the blood. So, it can help to keep a person's sugar within normal range if you are doing weight-lifting, for example. If muscles become deconditioned and weakened from non-use (aka: couch potato syndrome), people can suffer from an increased risk of type 2 diabetes. Additionally, every carbohydrate you eat has a glycemic index. The **glycemic index** ranks carbohydrate foods on the basis of how they affect blood glucose."

SUMMARY OF THE FUNCTIONS OF MUSCLES

Producing Movement

Muscles produce movement by the action of muscles crossing joints between the bones of the skeleton, the muscles are connected to the joints/bones via tendons. For example when you extend your elbow the tendons in your elbow pull on the muscles to allow the movement to take place.

Maintaining Posture

The muscles define how well our bones and body are stabilized. For example if we train our trapezius muscles then our neck will have better support because the muscle is stronger and can take more force and weight resulting in a stronger neck.

Stabilizing Joints

Muscles play a role in the stabilization of the joints. The muscles limit movement in a joint or provide balance for a more stable joint.

Generating Heat

Muscles also produce heat within the body when they contract. This heat causes blood vessels in the skin to dilate, which will increase the blood flow to the skin. The heat that the muscles produce is energy with only around 20-25% of this energy being efficient mechanical energy. The other 75-80% of the energy is lost as heat through the skin. For example when an athlete starts to sweat that is the body releasing the excess energy as heat and the body releasing sweat to cool the skin down.

Assistance in Blood Circulation

When muscles contract they produce chemicals that act on the arterioles dilating them, regulating the blood flow for the required exercise being carried out. For example a power lifter will have to intake a lot of oxygen to provide to the muscles to lift such a large amount of static weight, so the blood vessels in the muscles allow this to happen by dilating and becoming bigger due to heat when it is required.

STUDY TIP!

The word *muscle* means "little mouse," which refers to the appearance of muscles rippling under the skin.

Glenn continued, "Skeletal muscle has a certain structure to it. If you have ever prepared a roast for cooking, you may have observed some similarities in what I am getting ready to show you. So, there are three levels of muscle tissue organization: epimysium, endomysium, and perimysium. The outside surface of an entire muscle is covered by a relatively thick and very tough connective tissue, the **epimysium**, which separates it from surrounding muscles. The collagen fibers of the epimysium are woven into particularly tight bundles that are wavy in appearance. These collagen bundles are connected to the perimysium. The **perimysium** divides the muscle into bundles typically containing about 100 to 150 muscle fibers, which form a fascicle. However, muscles that function in producing small or very fine movements have smaller fascicles containing relatively few fibers. Arteries and veins run through the inner layer or **endomysium,** which is a thin sleeve of loose connective tissue that surrounds each muscle fiber. Look at this poster; I keep it in my cubicle. It shows the layering or membranes that compose skeletal muscle tissue."

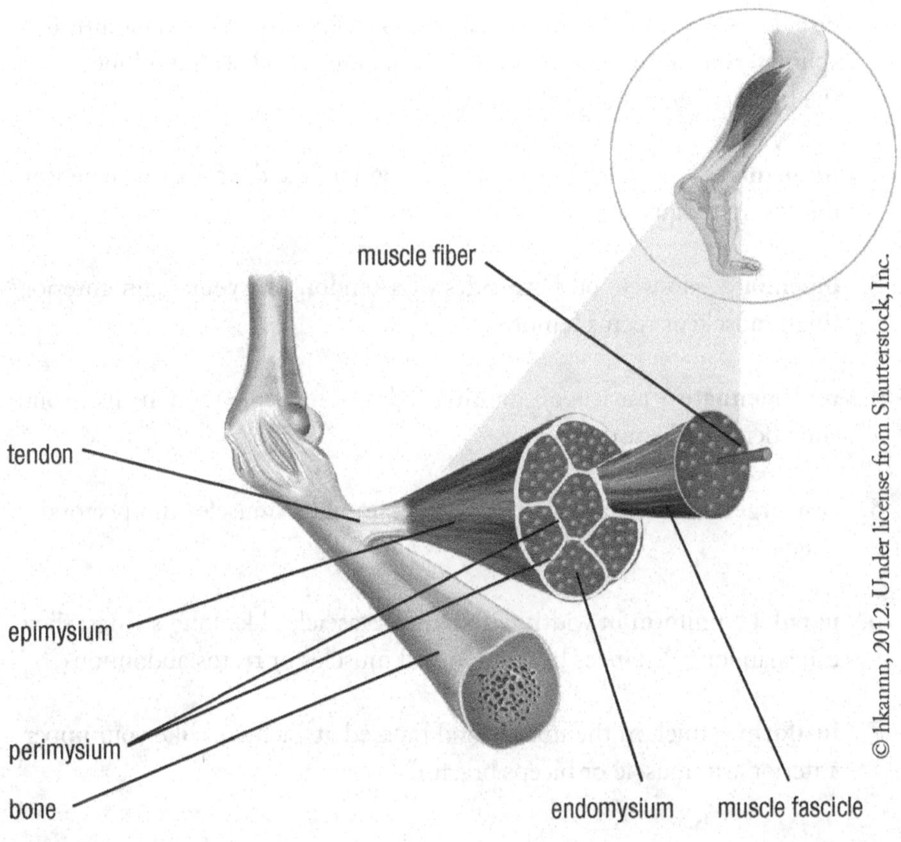

IMAGE 9.2 Glenn's Poster Showing the Structure of Skeletal Muscle

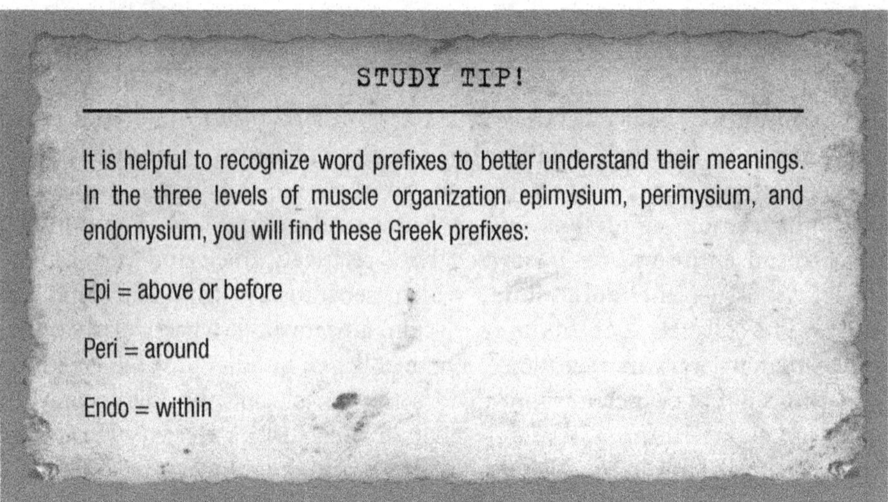

Glenn continued, "And muscles are actually classified according to their **fascicle** orientation (or shape). These same shapes occur over and over in human muscles but in different parts of the body. There are seven common fascicle arrangements or shapes:

1. **circular** – or round like the muscles surrounding the eyes and mouth; also sphincters forming rings around body openings such as the sphincters of the bladder controlling urine output

2. **unipennate** – muscle on one side of a tendon like the forearm's extensor digitorum longus

3. **bipennate** – muscle on both sides of a tendon like your main anterior thigh muscle or rectus femoris

4. **multipennate** – muscle on multiple sides of multiple tendons like your shoulder muscle or deltoid

5. **convergent** – or triangular like the main chest muscle: the pectoralis major

6. **parallel** – uniform in width with parallel fascicles like long strings; they can span long distances like abdominal muscles or rectus abdominus

7. **fusiform** – thick in the middle and tapered at each end like your upper anterior arm muscle or biceps brachii"

TABLE 9.1 Classification of Muscles According to Fascicle Orientation

Fascicle Type	Example of Shape
circular	
unipennate	
bipennate	

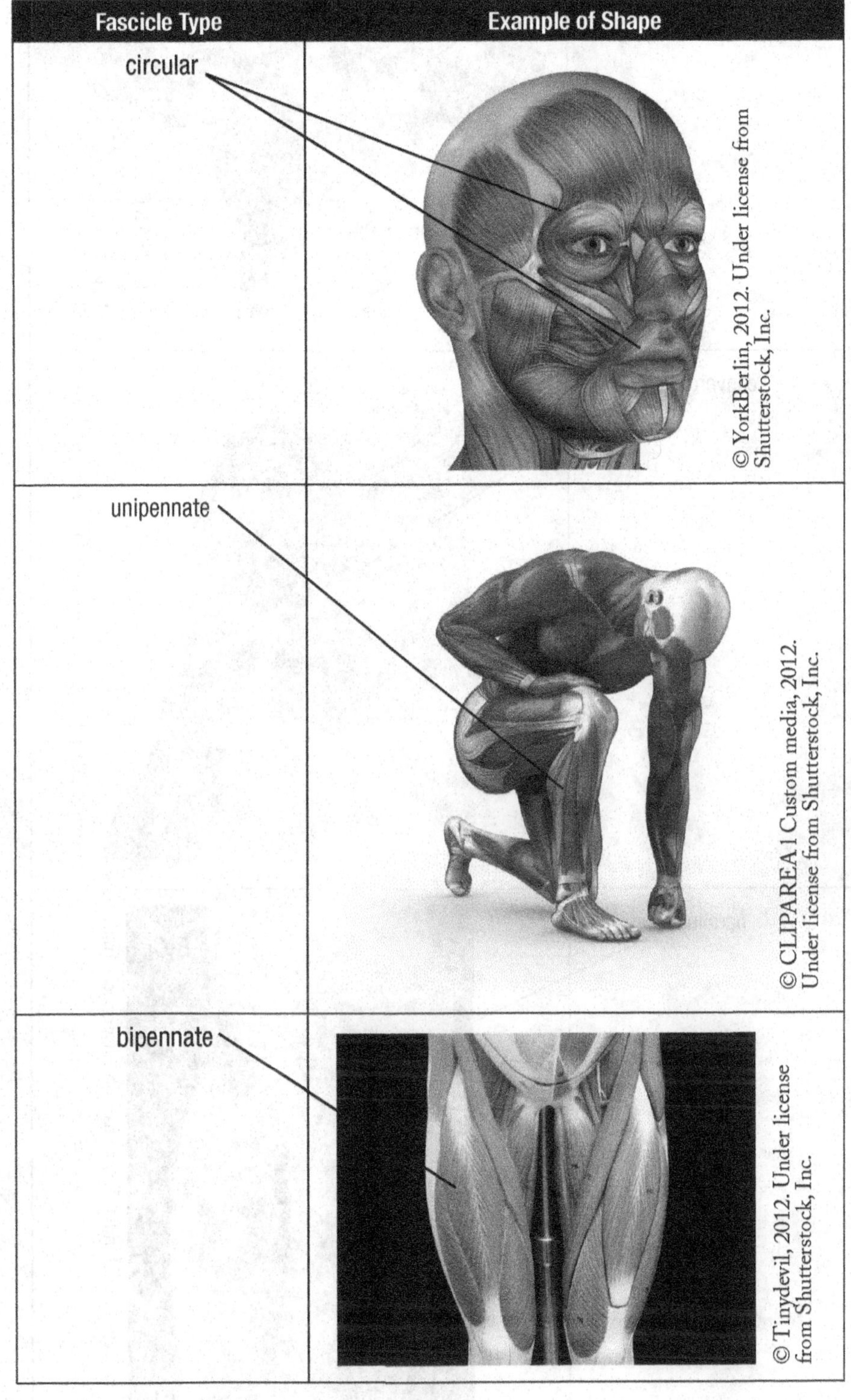

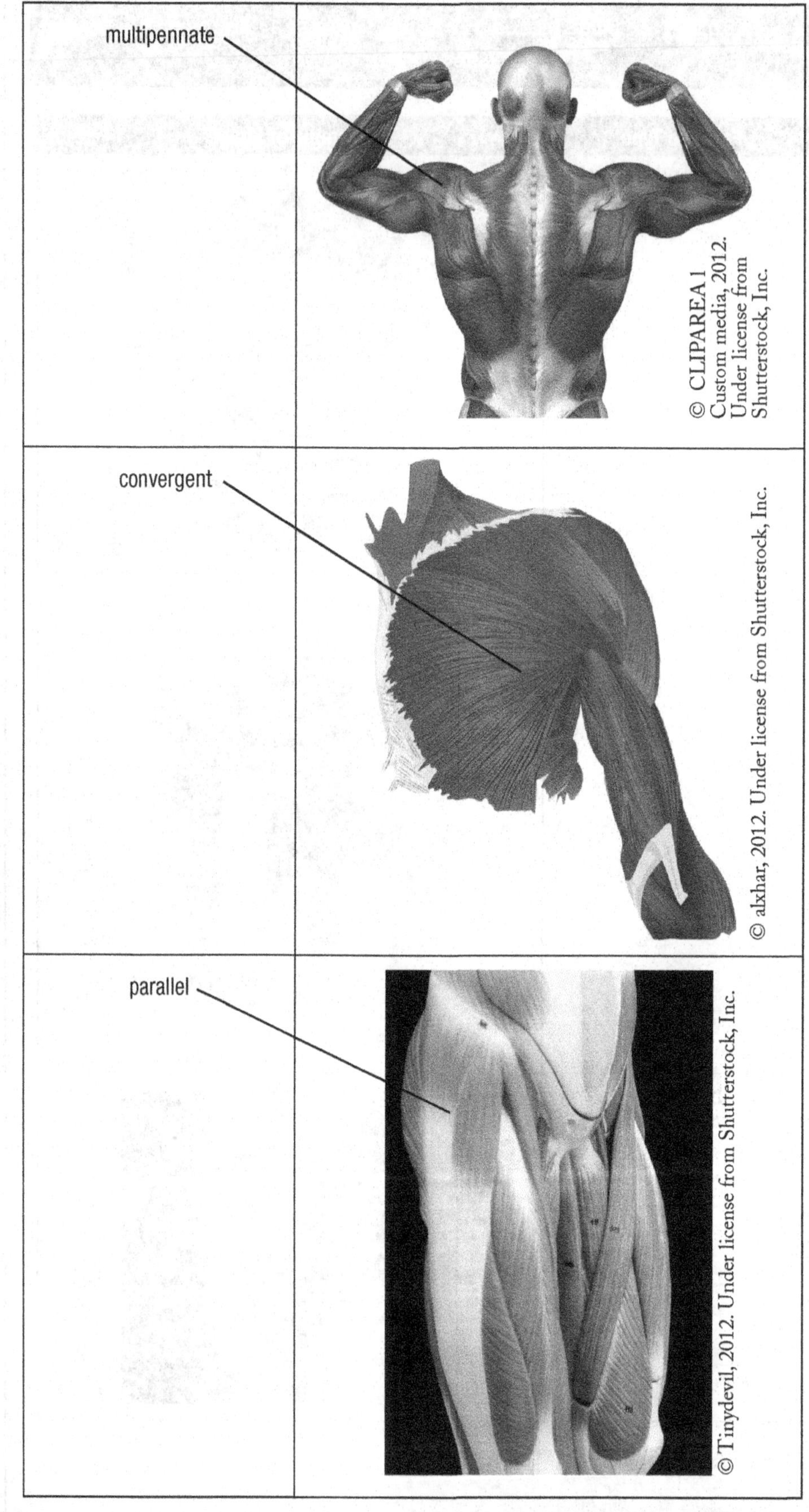
multipennate
convergent
parallel
© CLIPAREA l Custom media, 2012. Under license from Shutterstock, Inc.
© alxhar, 2012. Under license from Shutterstock, Inc.
© Tinydevil, 2012. Under license from Shutterstock, Inc.

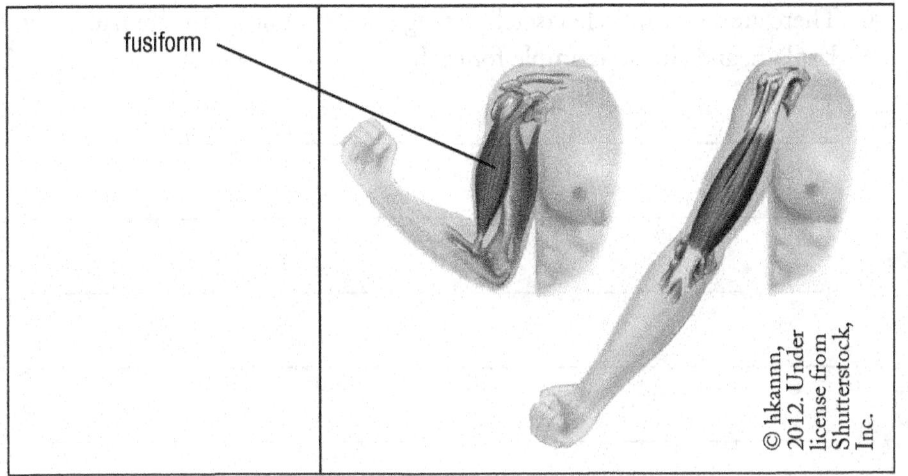

© hkannn, 2012. Under license from Shutterstock, Inc.

Self-Check – Continued

2. Label the epimysium, endomysium, and perimysium on the diagram.

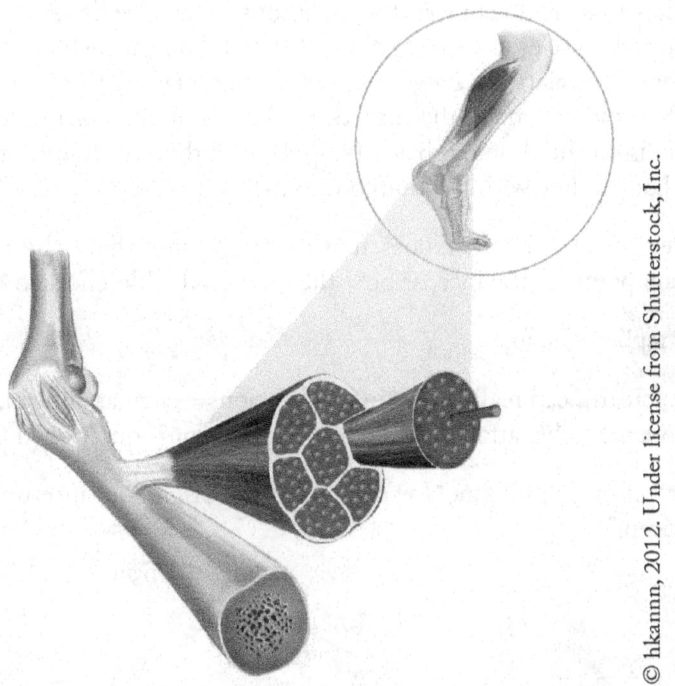

© hkannn, 2012. Under license from Shutterstock, Inc.

3. In your own words, explain the glycemic index. How does weight lifting help with excess glucose?

____

____

____

4. There are seven muscle fascicle arrangements found in the human body. Explain, and site an example for each.

____

____

____

____

____

MUSCLE ORIGINS AND ATTACHMENTS

Glenn continued, "As you know, skeletal muscle moves bone. It can move bone because it is attached to bone via tough, fibrous connective tissue called **tendon**. Most skeletal muscles are attached to a different bone at each end. The site of attachment to a relatively stable end is called an **origin**. The attachment site to the more mobile end is the **insertion**. For the biceps brachii, for example, the origin is on the shoulder blade (scapula) and the insertion is on the lower forearm bone in line with the thumb (radius)."

"So... because my shoulder does move a lot (some), that's the more stable connection point for the biceps? So—the origin?" double-checked Will.

"Right," replied Glenn.

"But the forearm can really move in all directions—circular, hinged, etc.—that is the more moveable attachment? So—the insertion?" questioned Grace.

"Correct," answered Glenn. "Let's see what the origin and insertion look like for the bicep."

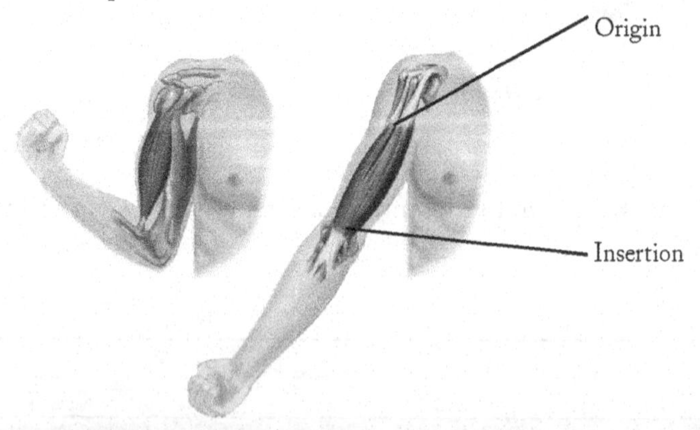

IMAGE 9.3 Origin and Insertion Attachment Points for Tendons of the Biceps Brachii

STUDY TIP!

How Muscles are Named

Location of muscle – bone or body region associated with the muscle

Shape of muscle – e.g., the deltoid muscle (deltoid = triangle)

Relative size – e.g., maximus (largest), minimus (smallest), longus (long)

Direction of fibers – e.g., rectus (fibers run straight), transversus, and oblique (fibers run at angles to an imaginary defined axis)

Number of origins – e.g., biceps (two origins) and triceps (three origins)

Location of attachments – named according to point of origin or insertion

Action – e.g., flexor or extensor, as in the names of muscles that flex or extend, respectively

MICROSCOPIC ANATOMY OF A SKELETAL MUSCLE FIBER

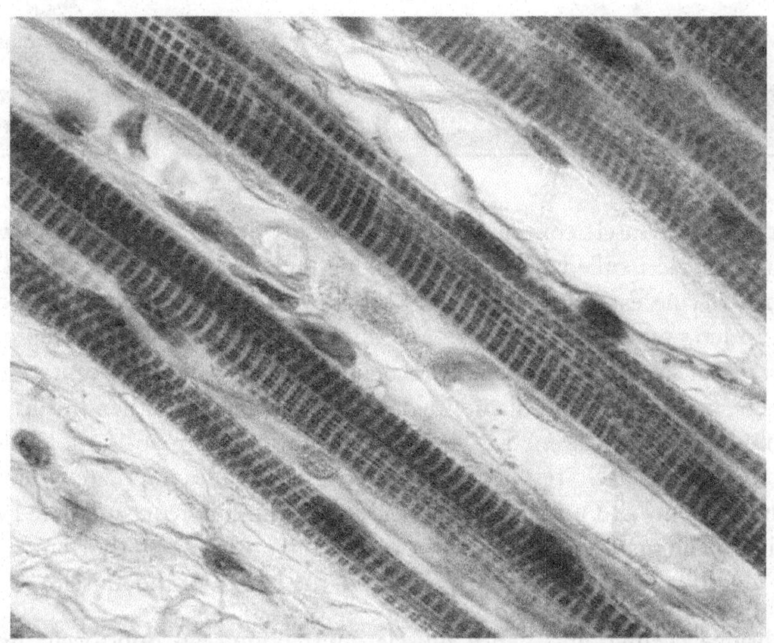

©Jose Luis Calvo, 2014. Under license from Shutterstock, Inc.

IMAGE 9.4 Skeletal Muscle Fibers Showing Striated Myofibrils 40x

Glenn continued, "Skeletal muscle fibers are very large, elongated cells. Roughly 80% of the content of each muscle fiber consists of long bundles of protein called **myofibrils**. The myofibrils, in turn, consist of two types of myofilament. One type of myofilament, called the thick filament, is composed of hundreds of molecules of a protein called **myosin**. The other type of myofilament, the thin filament, contains three different proteins: a structural protein called **actin** that can form bonds with myosin, a protein called **tropomyosin** that regulates binding between myosin and actin, and the calcium-binding troponin which regulates the position of tropomyosin. The two myofilaments are arranged in the myofibrils in distinctive repeated structures called **sarcomeres**. Each sarcomere contains a series of thin filaments at either end that partially overlap with thick filaments found in the center."

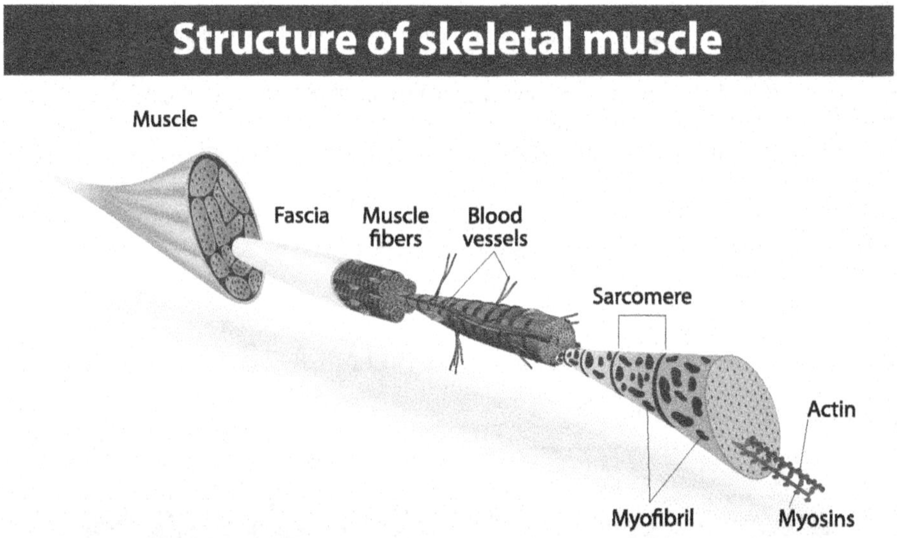

IMAGE 9.5 The Microanatomy of Skeletal Muscle

©Designua, 2014. Under license from Shutterstock, Inc.

He continued, "Muscle contracts through an ATP-driven interaction between actin and myosin called crossbridge cycling. First, the globular head of a myosin molecule extends laterally and binds with a complementary binding site on an actin molecule to form a bond called a crossbridge. Then, in a process called a power stroke, the globular head bends inward towards the center of the sarcomere, pulling the thin filament with it. The crossbridge then breaks, and the globular head of the myosin unbends, preparing the myosin molecule to repeat the process. As a result of many myosin molecules alternately binding the thin filaments and pulling them inward, the thin filaments are pulled over the thick filaments toward the center of the sarcomere, thus shortening the overall length of each sarcomere and, in turn, the length of the muscle.

"Crossbridge cycling, and hence muscle contraction, can only occur under specific conditions. This is because normally troponin positions tropomyosin on top of myosin-binding sites on the actin molecule. This blocks crossbridges from forming, and hence no contraction can take place. Only if troponin repositions tropomyosin to expose the myosin binding sites can crossbridge cycling occur. This occurs in response to an **action potential** being triggered in the skeletal muscle fiber, which leads to a series of events collectively called **excitation-contraction coupling**."

Physiology of Skeletal Muscle Fibers

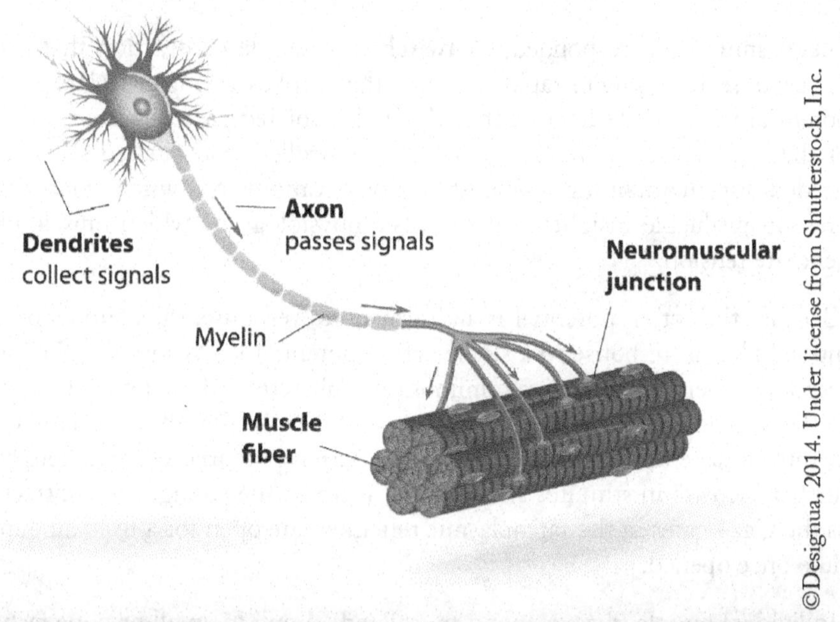

IMAGE 9.6 Motor Neurons Control Skeletal Muscle Movement

Grace asked, "How do we control our skeletal muscles?"

Glenn replied, "Nerve impulses that originate in the central nervous system cause muscles to contract. Both neurons and muscle tissue conduct electrical current by moving ions across cellular membranes. A motor neuron ends in a synapse with a muscle fiber. The neuron releases acetylcholine and transfers the action potential to the muscle tissue. The signal will travel through the tissue and trigger the contraction of individual sarcomeres. One synapse generally controls an entire muscle fiber. One motor neuron usually controls several adjacent muscle fibers. A group of fibers under the control of a single motor neuron is known as a **motor unit**.

"The contraction of a skeletal muscle fiber is triggered by an action potential occurring in the sarcolemma (plasma membrane) of that muscle fiber. The action potential propagates down the sarcolemma and is conducted down transverse tubules into the interior of the cell. This, in turn triggers the release of $Ca2+$ from the sarcoplasmic reticulum (a modified endoplasmic reticulum) into the cytosol. The $Ca2+$ binds to troponin on the thin filament and causes it to undergo a conformational change. This change in the shape of troponin shifts the position of tropomyosin on the thin filaments, exposing binding sites for myosin on the underlying actin and enabling crossbridge formation (the bonding of myosin on the thick filaments to actin on the thin filaments) to commence."

Muscle Twitch Parameters

Will asked, "What happens when my eyebrow unexpectedly twitches and I can't control it or make it stop?"

Glenn smiled and responded, "A **twitch** is a muscle contraction that occurs in response to a single, rapid stimulus that evokes a single, isolated action potential in a muscle fiber. Although single, isolated twitches are not in and of themselves very useful for generating controlled, coordinated movements needed for maintaining homeostasis, observations of twitch contractions present invaluable insights into the basic physiology by which muscle fibers generate tension.

"Because the action potential is an 'all or none' response, the contraction of a muscle fiber in response to a single action potential is likewise an all or none response. Therefore, there is a minimum stimulus strength that must be applied to the muscle fiber in order to reach threshold, evoke the action potential and, in turn, induce the contraction. Once the action potential occurs, though, no further increase in stimulus strength will increase the strength of contraction, as the Ca2+ gates in the sarcoplasmic reticulum are open for a fixed amount of time once opened.

"Individual muscle fibers respond to isolated stimuli in an all or none fashion. However, a muscle organ, such as the gastrocnemius muscle, is composed of many individual muscle fibers. By varying the number of motor units (groups of muscle fibers innervated by a singe somatic motor neuron) contracting at a given time, the amount of tension generated by the whole muscle can vary. Threshold is considered to be the level of stimulation required to trigger the smallest measurable contraction resulting from the excitation and contraction of the first few muscle fibers. If stimulus is increased above threshold into a range of stimulus intensities called submaximal stimuli, contraction strength will increase with stimulus intensity as progressively more and more muscle fibers in the muscle undergo contraction. Finally, when stimulus strength is increased above a certain level (maximal) no further increase in tension occurs, as all muscle fibers in the muscle are contracting."

Glenn continued, "A rather complex series of events occurs within the time course of a single twitch. The action potential is evoked upon application of the stimulus. That action potential, in turn, propagates down the length of the muscle fibers and triggers the excitation-contraction coupling process (release of Ca2+ from the sarcoplasmic reticulum, binding of Ca2+ to troponin, etc.). Once crossbridge cycling ensues, the muscle fibers contract, generating tension. Tension peaks, but then decreases as the activity of Ca2+ pumps in the sarcoplasmic reticulum reuptake Ca2+ from the cytosol, lowering the ability of actin and myosin to form crossbridges, and reducing tension generation as the fibers stretch back to their original length. These three basic stages (excitation-contraction coupling, tension generation, and relaxation) correlate with three different time phases during the twitch. During the latent period (the time between the application of the stimulus and the onset of contraction), excitation-contraction coupling takes place. During the contraction time (the time from the onset of contraction to peak tension), crossbridge cycling occurs at a high enough rate that the muscle fibers shorten. During the relaxation time (from peak tension to the point when tension returns to baseline), Ca2+ is being pumped back into the sarcoplasmic reticulum, and the muscle is stretching back to its original length."

Tetanus

Glenn continued, "Observations of twitch contractions within single muscle fibers or within whole muscle organs can yield important insights into the basic cellular processes involved in converting an electrical signal into a mechanical response by the muscle fibers. However, with a few exceptions, twitch-types of contractions are not the typical type of contraction that skeletal muscles inside the human body produce. That is because in order to enable coordinated body movement and the maintenance of balance and posture (the primary function of most skeletal muscles), tension must be sustained beyond the fraction of a second generated by a twitch. Therefore, most skeletal muscle contractions in the body are tetanic contractions.

"The basis of 'tetany,' or **'tetanus'** (not to be confused with the disease commonly called "lock jaw" caused by the bacterium **Clostridium tetani**) within a skeletal muscle organ can be somewhat confusing. Isolated individual muscle fibers are able to generate sustained levels of tension with high frequency stimulation. Thus, many action potentials can occur in the amount of time needed for a single twitch. If a muscle that is relaxing from a contraction is stimulated before it fully relaxes, the sarcoplasmic reticulum will release more Ca2+, and the cell will begin to contract again without fully relaxing. In effect, then, the twitches partially fused together. If stimulated at progressively higher frequency, the amount of relaxation that occurs in between each 'twitch' is progressively reduced, until a steady state of tension (tetanus, or tetany) is generated.

"In most of the tetanic contractions in the body, complete tetanus (contraction without any relaxation) is not common. Most sustained contractions are generated by a combination of twitches and partial-tetanic contractions by different motor units whose motor neurons are stimulating the fibers at different intervals and at different frequencies.

"Interestingly, the amount of tension generated during a tetanic contraction is often substantially higher than that of a maximal twitch. There are several reasons for this. First, when a muscle begins to contract, some of the tension generated by the muscle is absorbed by stretching elastic elements within the muscle's attachments. This can reduce the total tension generated on the attachments in a twitch contraction whereas in tetany, these elastic elements are fully stretched and more tension is exerted directly on the attachments.

"Secondly, recall that each time the muscle fibers undergo action potentials Ca2+ is released from the sarcoplasmic reticulum. The sarcoplasmic reticulum begins to reabsorb this Ca2+ almost as soon as it is released, but it does take time to totally recover all of the Ca2+. If the sarcoplasmic reticulum is induced by another action potential to release Ca2+ before it has fully recovered all of the Ca2+ previously released, then there will be overall more Ca2+ in the cytosol during the second contraction, more interaction between actin and myosin, and a stronger resultant contraction. Thus action potentials generated in rapid succession can have a summation effect on the strength of the contraction."

Functional Contraction Types

Glenn expounded, "The contractions generated by skeletal muscles are used for two basic functions: movement of the body and maintaining position and orientation of the body. **Isotonic** contractions are those that result in the muscle shortening in length, generating movement of a body part. In order for an isotonic contraction to occur, the muscle must contract with enough force to overcome the load applied to the muscle. **Isometric** contractions, in contrast, are contractions where the muscle is contracting and generating tension, but the muscle does not shorten in length as the force generated by the muscle is equal to the load placed on the muscle. Muscles that allow you maintain posture generate isometric contractions to counteract the force of gravity."

Electromyograms

Glenn continued, "The action potentials generated by contracting muscle alter the electrical charge in the surrounding extracellular fluid. These electrical changes are conducted through body fluids, and can be detected from the surface of the skin using electrodes applied to the skin. A variety of instruments can detect the differences in charge between the electrodes, amplify them, and generate recordings of these electrical changes called electromyograms (EMGs). EMGs are used diagnostically to detect damage to muscle or to the neural pathways responsible for triggering muscle contractions."

MAJOR MUSCLES OF HEAD AND NECK

Glenn explained, "At the top of the head is the scalp, also known as the **epicranius,** a general term that actually includes three components: the **frontalis** or frontal belly, which raises the eyebrows and draws the forehead downward; the **occipitalis** (muscle at back of head) which retracts the scalp; and **galea aponeurotica,** which is a tendon holding the frontalis to the occipitalis."

"So, the epicranius is a general term referring to the frontalis, galea aponeurotica, and the occipitalis?" double-checked Will.

"Yes," Glenn replied and then continued, "Superior to the temporal bone is the **temporalis** muscle that is capable of retracting and elevating, as well as lateral and medial motion of the mandible. On the side of the face in line with the sideburn area near the mandible, is the **masseter** that elevates the mandible as well as protrudes, retracts, and gives lateral and medial motion to the jaw. If you place your index finger vertically against the mandible in the sideburn area and clinch your back teeth, you will feel it move. In an indented area just under the zygomaticus major, is a muscle known as the **buccinator** that compresses the cheek against the teeth, directs food against back molars, and keeps the cheek away from the teeth when your mouth is closing. Surrounding the eyes is the circular **orbicularis oculi** which closes the eye during blinking, squinting, and sleep. The mouth is also surrounded by a circular muscle that is called the **obicularis oris** that closes or protrudes the lips. Immediately under the eye and closest to the nose is the **levator labii superioris** which elevates and everts the upper lip for sad, sneering, or serious expressions."

"Like this?" Grace replied as she snarled and curled her lip. The guys laughed.

"Then we have the two strips here on the cheek. They are both zygomaticus muscles. The smaller, more superior one is the **zygomaticus minor**—it elevates the upper lip as well as exposes the upper teeth during smiling or sneering. The inferior one is the **zygomaticus major** which draws the mouth upward and laterally when you laugh."

Glenn continued explaining, "Your lower chin and anterior neck is covered by a sheet of muscle called the **platysma** that draws the lower lip and angle of the mouth downward in expressions of horror or surprise. If you peel up the platysma, underneath you will see the **sternohyoid** muscle, which depresses the hyoid bone. And a much larger muscle called the **sternocleidomastoid** which rotates the head left and right, draws the head straight forward and down, and provides unilateral motion by allowing the head to tilt upward and toward the opposite side."

"Wow! That's a lot of muscles in just the head and neck," exclaimed Will.

"Just remember, I am sharing only the major muscles with you. There are many muscles that we are purposefully omitting. Hey, I have some posters showing the major muscles of the head and neck. Lets take a look. I think it will help you understand better," said Glenn.

Self-Check – Continued

5. Explain origin and insertion.

____

____

____

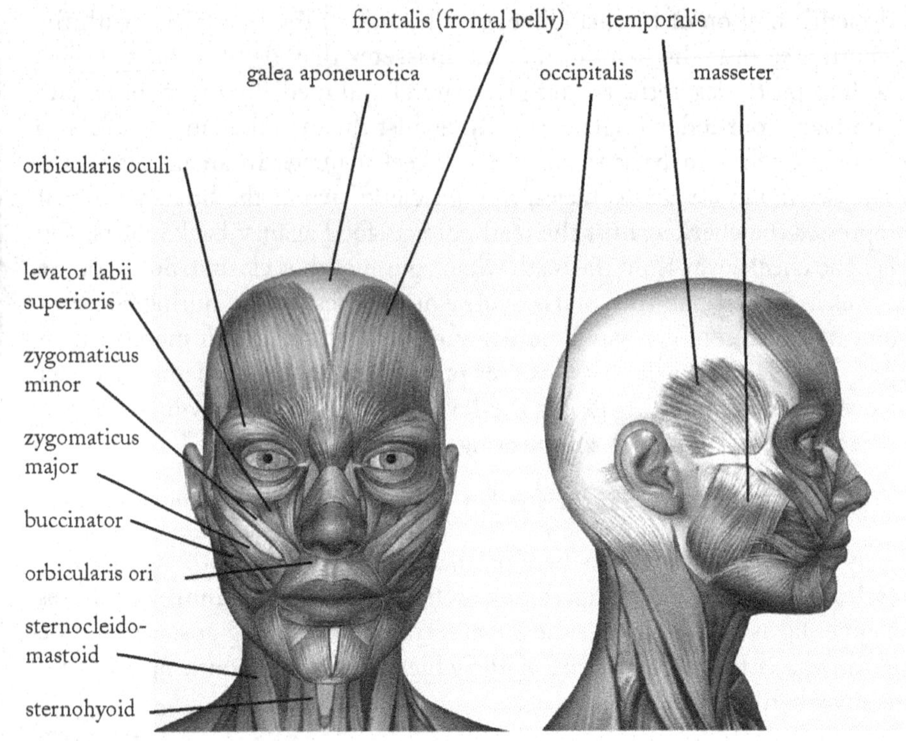

IMAGE 9.7 Major Muscles of the Head and Neck (Anterior and Lateral Aspects)

MAJOR MUSCLES OF THE TORSO, ARMS AND LEGS

Glenn asked, "Now we're going to look at the major muscles of the torso and their functions. Do you know the muscles located at the top of the shoulders?"

"Isn't it the trapezius?" replied Grace.

Glenn answered, "It is! The main thing the **trapezius** does is extends and laterally flexes the neck." Glenn pointed to his side muscles and challenged, "And these?"

"I believe those are the external obliques," commented Will.

"Right again. And the **external obliques** stabilize the vertebral column during heavy lifting, maintain posture, aid in the forceful expulsion of various products from the body such as breath, urine, feces, and vomit," added Glenn.

"Let's look at this poster (Table 9.2) to help us learn the remainder of the muscles. Check it out!"

TABLE 9.2 Major Muscles of the Torso, Arms, and Legs and Their Functions

Muscle	Function
Rectus abdominis	Flexes lumbar column allowing for forward bending at the waist
Pectoralis major	Flexes, adducts, and plays a part in the rotation of the humerus as in hugging; aids in deep inhalation
Latissimus dorsi	Adducts and medially rotates humerus; extends the shoulder joint in a motion similar to rowing a boat; creates backward swing of arm as in bowling; allows things to be pulled down that are over your head; pulls body forward and upward; and sustained forceful exhalation as in playing a musical instrument
Deltoid	Flex and medially as well as laterally rotate arm; arm abduction; swinging motion of arm
Teres minor	Helps control the action of the deltoid; prevents head of humerus from sliding upward as arms are abducted; rotates humerus laterally
Teres major	Extends and medially rotates humerus; helps arms swing
Infraspinatus	Similar to teres minor—helps control action of deltoid; prevents head of humerus from sliding upward; rotates humerus laterally
Brachialis	Creates elbow flexion
Biceps brachii	Quick supination of forearm; helps in elbow flexion; helps with slight shoulder flexion; origin tendon of this muscle helps stabilize the shoulders by holding the head of the humerus against the glenoid fossa of the scapula
Triceps brachii	Extends elbow; long head extends and adducts the humerus
Brachioradialis	Flexes elbow
Flexor carpi radialis	Flexes wrist anteriorly; aids in radial flexion of the wrist
Flexor carpi ulnaris	Flexes wrist anteriorly; aids in ulnar flexion of the wrist
Extensor carpi radialis longus	Extends wrist; aids in radial flexion of the wrist
Gluteus minimus	Abduct and medially rotate thigh; shift weight of trunk toward limb with foot on ground as other foot is lifted
Gluteus maximus	Extends thigh at hip (as in stair climbing); abducts thigh; elevates trunk after stooping; prevents trunk from pitching forward during walking and running; helps stabilize femur on tibia
Gracilis	Flexes and medially rotates tibia at knee
Pectineus	Flexes and adducts thigh
Rectus femoris	Extends knee; flexes thigh at hip

TABLE 9.2 Major Muscles of the Torso, Arms, and Legs and Their Functions

Vastus lateralis	Extends knee; keeps patella in grove on femur during knee movements
Vastus medialis	Same as vastus lateralis
Sartorius	Helps in knee and hip flexion (as in sitting or climbing); abducts and laterally rotates thigh
Biceps femoris	Flexes knee; extends hip; elevates trunk from stooping posture; laterally rotates tibia on femur when knee is flexed; laterally rotates femur when hip is extended; counteracts forward bending at hips
Semitendinosus	Flexes knee; medially rotates tibia on femur when knee is flexed; medially rotates femur when hip is extended; counteracts forward bending at hips
Semimembranosus	Same as semitendinosus
Tibialis anterior	Dorsiflexes and inverts foot; resists backward tipping of body
Gastrocnemius	Plantar flexes foot, flexes knee; active in walking, running and jumping
Soleus	Plantar flexion of foot; keeps leg steady on ankle during standing

NOTE: face not included on this muscle list

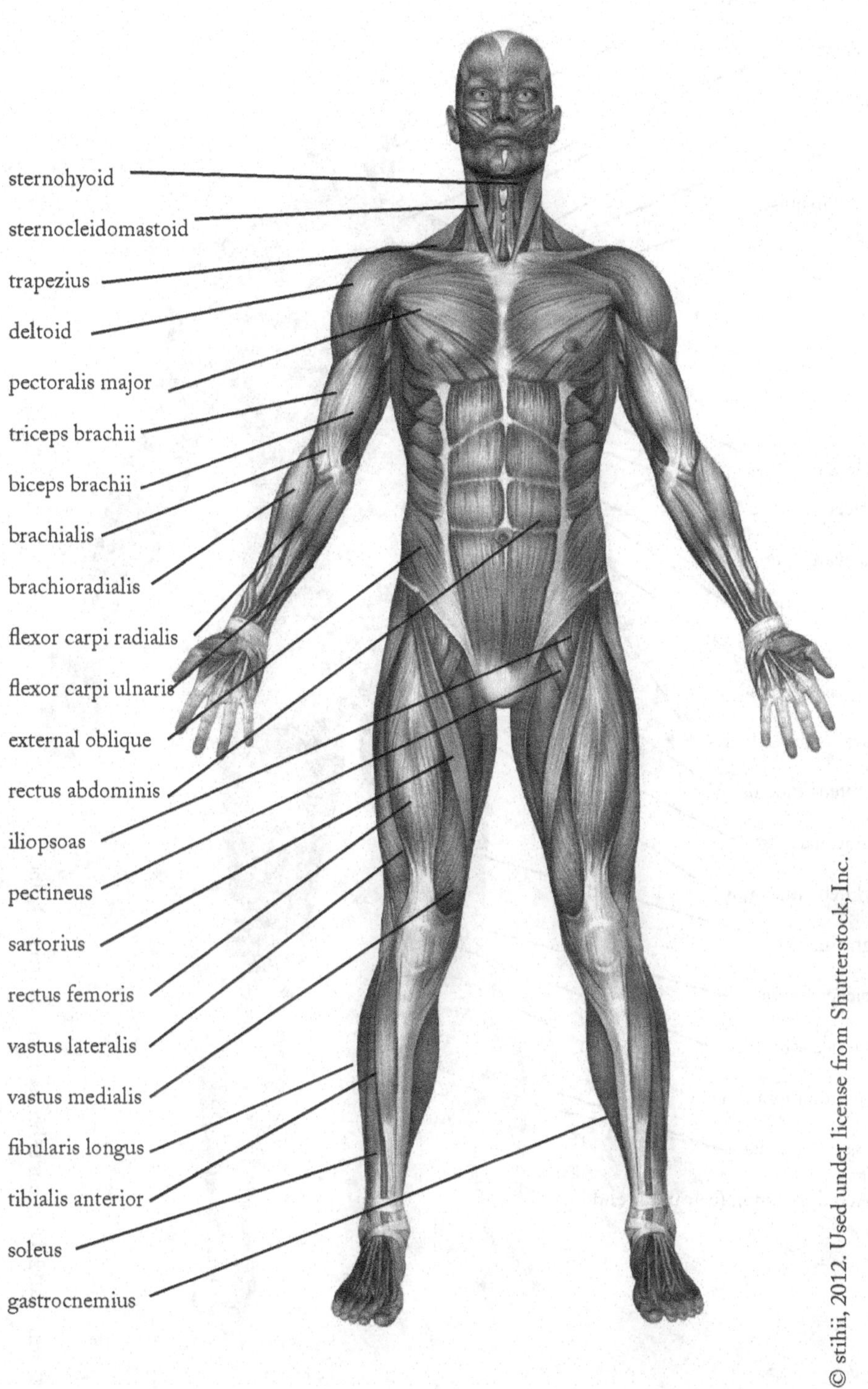

IMAGE 9.8 The Muscular System — Selected Major Muscles Only (anterior aspect)

A Visit With an Athletic Trainer to Learn About Muscles

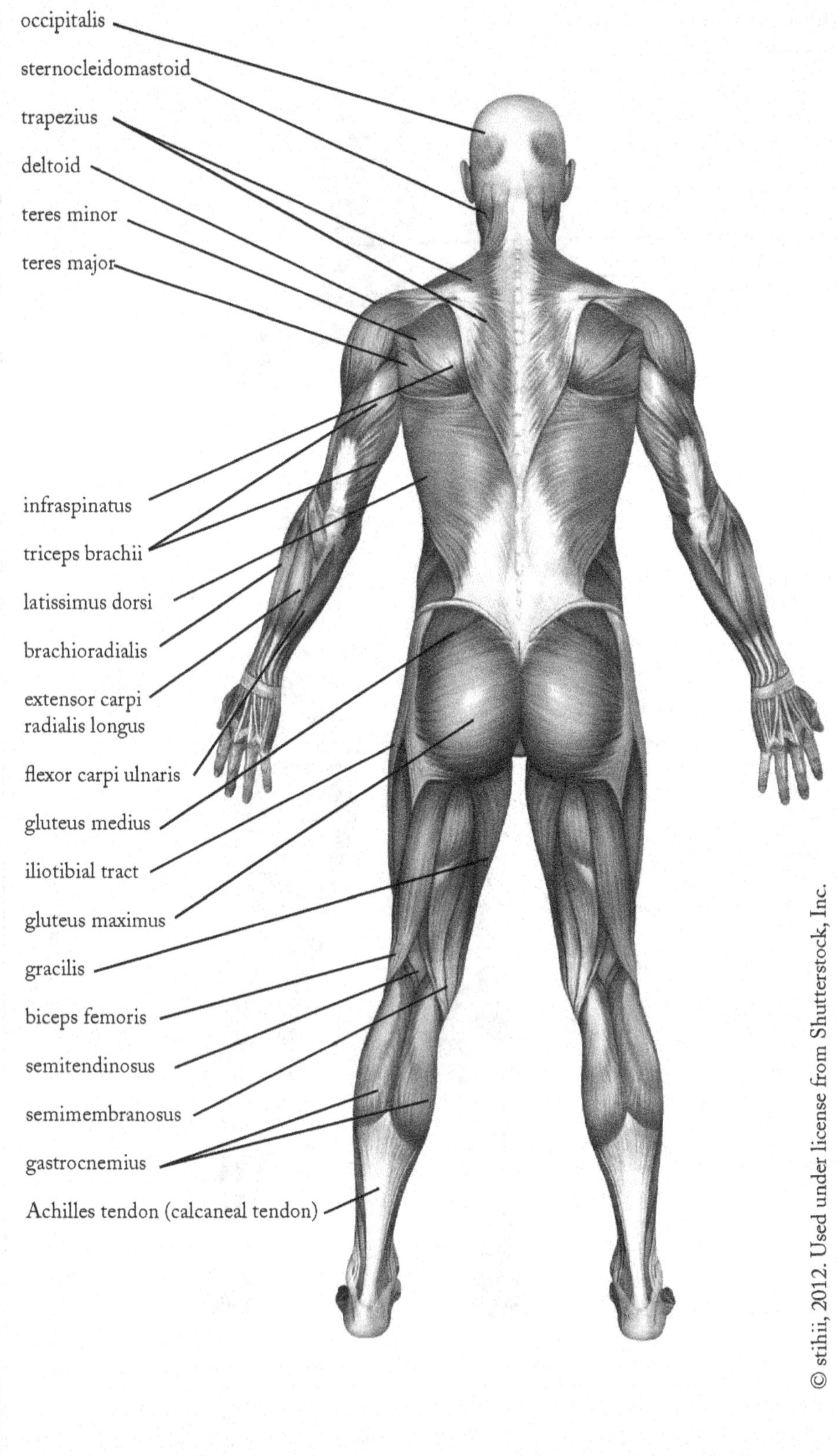

© stihii, 2012. Used under license from Shutterstock, Inc.

IMAGE 9.9 Selected Major Muscles Only (Posterior Aspect)

6. Practice labeling the major muscles of the head and neck.

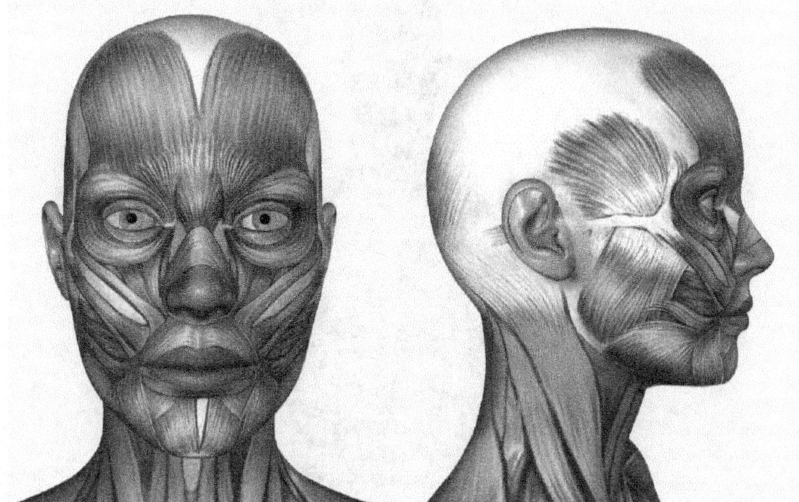

© stihii, 2012. Used under license from Shutterstock, Inc.

7. Practice labeling the major anterior muscles on the diagram below.

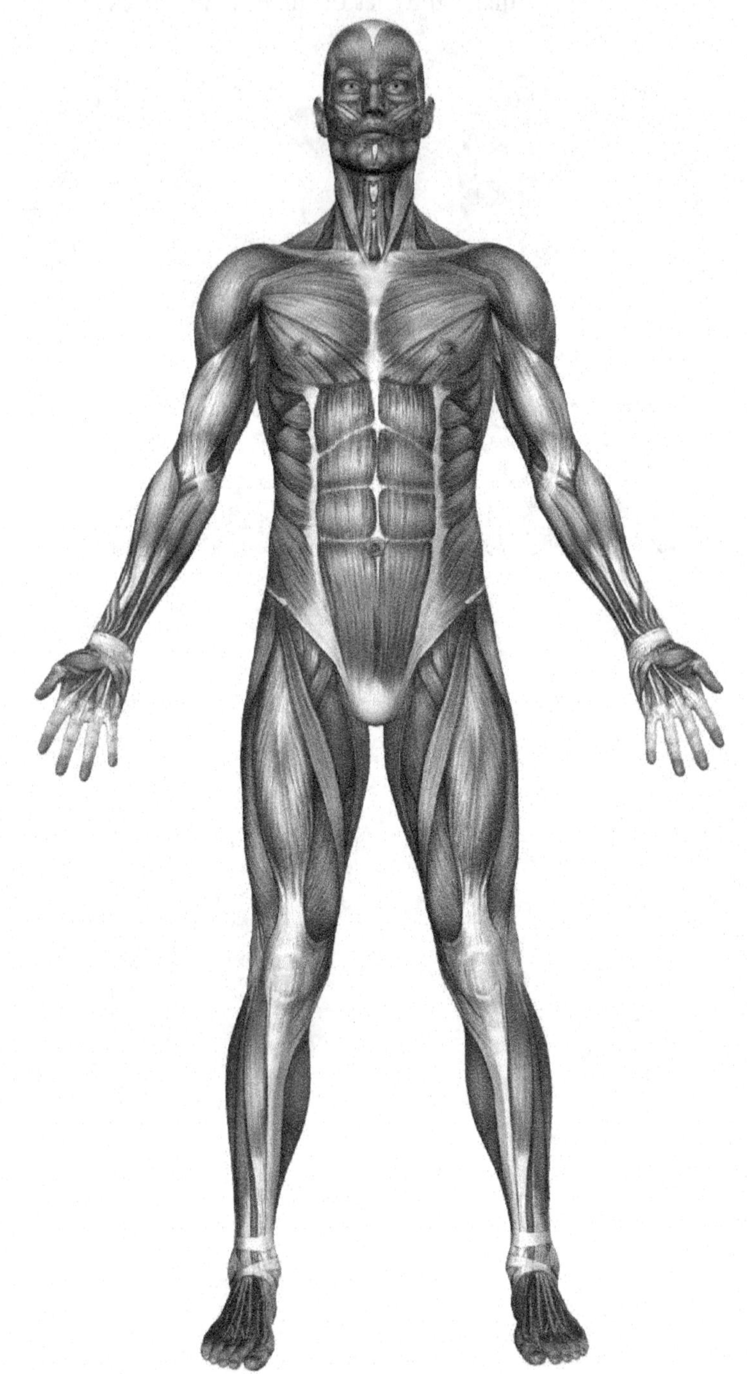

© stihii, 2012. Used under license from Shutterstock, Inc.

8. Practice labeling the major posterior muscles on the diagram below.

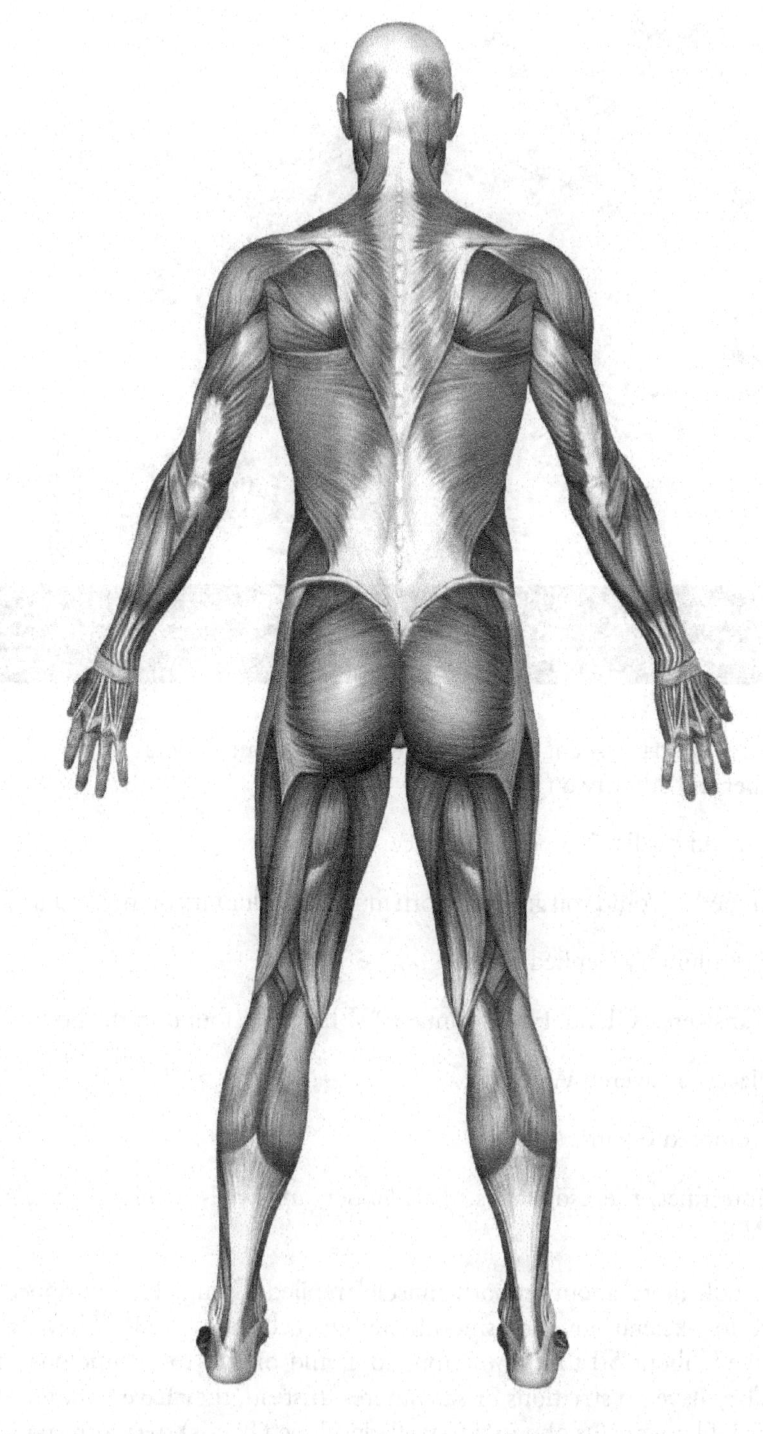

© stihii, 2012. Used under license from Shutterstock, Inc.

SMOOTH MUSCLE

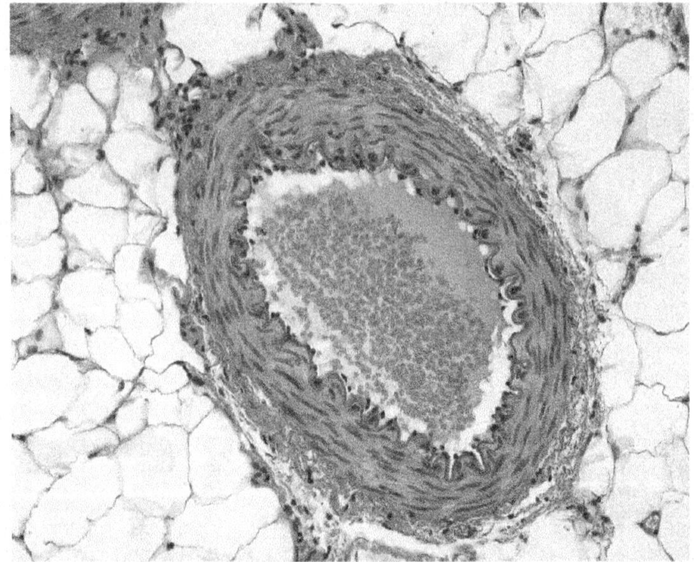

IMAGE 9.10 Cross Section of an Artery Wall Showing Smooth Muscle

Glenn asked, "We have spent a lot of time talking about skeletal muscle. Do you remember the other two types?"

"Yes, smooth and cardiac" responded Grace.

Glenn continued, "Would you guess smooth muscle is voluntary or involuntary?"

"Definitely involuntary" replied Grace.

"Ok, good" answered Glenn. He continued, "Where is it found in the body?"

"In many places" answered Will.

"Like...?" prompted Glenn.

"Like the intestines, the esophagus, the bladder, and walls of blood vessels" answered Will.

"Right, lets talk more about smooth muscle" replied Glenn. He continued, "Compared to skeletal muscle, smooth-muscle cells are small. They are spindle-shaped, about 50 to 200 microns long and only 2 to 10 microns in diameter. They have no striations or sarcomeres. Instead, they have bundles of thin and thick filaments (as opposed to well-developed bands) that correspond to myofibrils. In smooth-muscle cells, intermediate filaments are interlaced through the cell much like the threads in a pair of 'fish-net' stockings. The intermediate filaments anchor the thin filaments and correspond to the Z-disks of skeletal muscle. Unlike skeletal-muscle cells, smooth-muscle cells have no troponin, tropomyosin or organized sarcoplasmic reticulum."

Glenn added, "Most smooth muscle is organized into sheets of closely apposed fibers. These sheets occur in the walls of all but the smallest blood vessels and in the walls of hollow organs of the respiratory, digestive, urinary, and reproductive tracts. In most cases, two sheets of smooth muscle are present with their fibers oriented at right angles with one another (as in the intestine). In the longitudinal layer, the muscle fibers run parallel to the long axis of the organ. Consequently, when the muscle contracts, the organ dilates and shortens. In the circular layer, the fibers run around the circumference of the organ. Contraction of this layer constricts the lumen (cavity) of the organ and causes the organ to elongate. The alternating contraction and relaxation of the opposing layers mixes substances in the lumen and squeeze's them through the organ's internal pathway. This propulsive action is called peristalsis (as in the esophagus)."

Glenn continued, "As in skeletal-muscle cells, contraction in a smooth-muscle cell involves the forming of crossbridges and thin filaments sliding past thick filaments. However, because smooth muscle is not as organized as skeletal muscle, shortening occurs in all directions. During contraction, the smooth-muscle cell's intermediate filaments help to draw the cell up, like closing a drawstring purse."

He added, "Calcium ions regulate contraction in smooth muscle, but they do it in a slightly different way than in skeletal muscle. Calcium ions come from outside of the cell. Calcium ions bind to an enzyme complex on myosin, called calmodulin-**myosin light chain kinase**."

CARDIAC MUSCLE TISSUE

Glenn stated, "You will be covering heart tissue in detail in level 2 anatomy and physiology (Book 2) so our discussion on this muscle type will be cursory for now. But let me ask you... is cardiac muscle involuntary or voluntary?"

"It is involuntary" exclaimed Will.

Grace added, "I cannot fathom trying to sleep at night while commanding my heart to beat! I'd never get any sleep and if I did I would surely die!"

Glenn continued, "Cardiac muscles are the muscles of the heart. They are self-contracting, autonomically regulated and must continue to contract in rhythmic fashion for the whole life of the organism. Hence they have special features.

"The cells are Y shaped (branched) and are shorter and wider than skeletal muscle cells. They are predominately mononucleated. The arrangement of actin and myosin is similar to skeletal striated muscle."

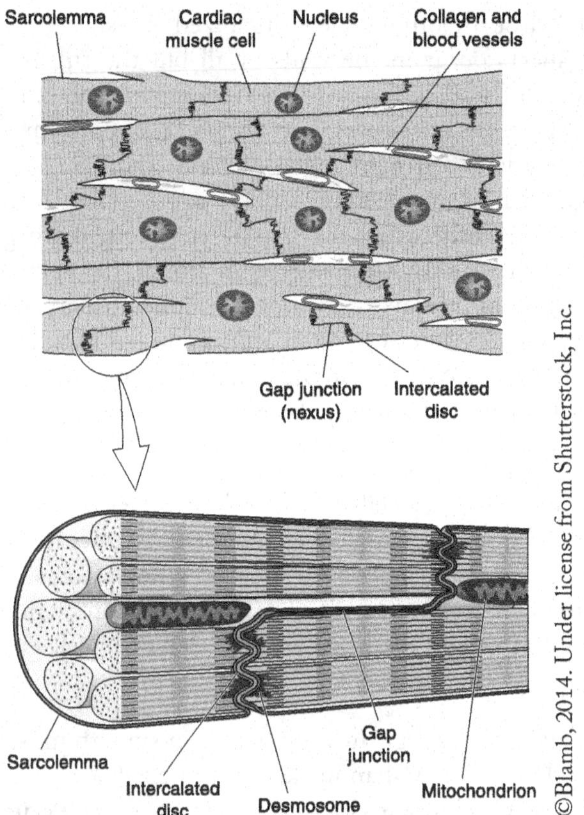

IMAGE 9.11 Cardiac Muscle Tissue

Glenn added, "Some of the cardiac muscle cells are auto-rhythmic, (i.e., they contract even in the absence of neuronal innervation known as pacemaker cells). Intercalated disks are located between cardiac muscles cells. These contain gap junctions which provide communicating channels between cells. The intercalated disks allow waves of depolarizations to sweep across the cells thus synchronizing muscle contraction.

"Repolarization takes much longer to occur and thus cells cannot be stimulated at high frequency. The advantage is that cardiac muscle is prevented from going into tetanus."

WRAP-UP

Grace looked at her watch mindful of Glenn's time and announced, "I know time is short. Thanks for helping us."

Glenn answered, "No problem. Happy to help out."

As Grace and Will left the building, they reflected on all that Glenn had helped them with including the structure of muscle tissue, muscle fascicles, origins and insertions, as well as all the major muscles.

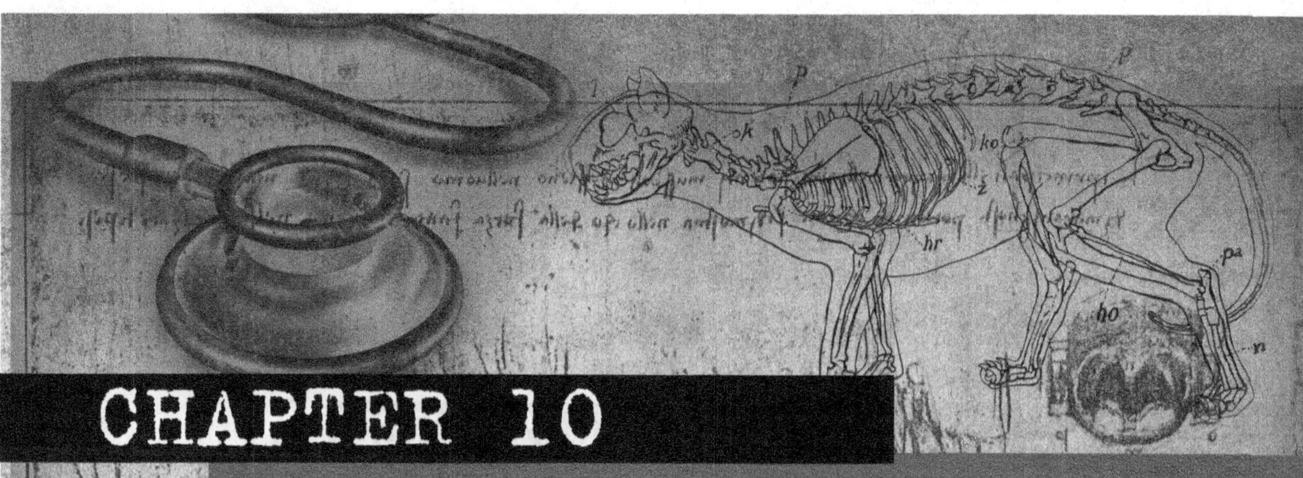

CHAPTER 10

An Interview with a Neurosurgeon: Learning the Fundamentals of Neurobiology

KEY WORDS

The study of Anatomy & Physiology involves many new vocabulary words. It is helpful to gain familiarity with these new key words, just as you would a foreign language.

action potential	excitability	oligodendrocytes
associative neurons	external auditory canal	optic nerve
astrocytes	fovea	otoliths
auditory tube	frontal lobe	parasympathetic division
axon	grey matter	parietal lobe
axon terminals	gyri	perilymph pineal gland
brain stem	hypothalamus	perineurium
cerebellum	incus	pinna
cerebral hemispheres	lateralization	pituitary gland
cerebrospinal fluid	lens	PNS
cerebrum	longitudinal fissure	pons
choroid	malleus	pupil
ciliary body	medulla oblongata	resting potential
CNS	microglia	retina
cochlea	motor cortex	rods
conductivity	motor (efferent) division	saccule
cones	motor neurons	satellite cells
cornea	myelin sheath	Schwann cells
corpus callosum	neurobiology	sclera
dendrites	neuroglia	secretion
electrochemical	neurons	semicircular canals
endoneurium	neurotransmitters	sensory (afferent) division
ependymal cells	nodes of Ranvier	sensory cortex
epineurium	occipital lobe	sensory neurons

somatic motor division	synapse	ventricles
somatic sensory division	temporal lobe	vestibule
stapes	thalamus	visceral motor division
sulci	tympanic membrane	visceral sensory division
sympathetic division	utricle	white matter

STUDENT STUDY GUIDE/OUTCOMES

It is helpful to have an idea of what you need to learn before proceeding. Here is a study guide to assist you:

1. Describe the function of the nervous system.
2. Describe the major anatomical and functional subdivisions.
3. Draw and label the parts of a neuron.
4. Explain the difference between a sensory neuron, a motor neuron, and an associative neuron.
5. Describe the various types of supporting cells in the nervous system and what they do.
6. Explain the basic parts of the brain and what they do.
7. Identify the basic parts of the brain on a diagram.
8. Be able to identify the structures of the eye and what they do.
9. Explain the physiology of how we see.
10. Be able to identify the structures of the ear and what they do.
11. Explain the physiology of how we hear.
12. Explain gustatory hairs and how you taste.

Grace and Will have been asked by their college Anatomy instructor to gather as much knowledge as they can about **neurobiology**—the study of the human nervous system. In order to better understand this complex system, they have enlisted the help of a neurosurgeon, Dr. Christine Schmidt, also known as "the nerve doc." Grace and Will have arranged to meet her at her office and later follow her on rounds at the hospital.

Grace and Will finally get to meet her. Out of the main office walked a tall, strikingly good-looking, middle-aged blonde, "Hi, I'm Dr. Christine Schmidt—welcome."

"Hi, I'm Will and… this is Grace," introduced Will. He continued, "Thanks for meeting with us today."

"What are you most interested in?" questioned Dr. Schmidt.

"While we really need a crash course in the nervous system, we're also fascinated by diseases such as Alzheimer's," replied Grace.

"Okay, we can start there. We have quite a bit of ongoing research happening right now at our facility," Dr. Schmidt explained.

CLINICAL APPLICATION: WHAT'S THE LATEST WITH ALZHEIMER RESEARCH?

Perhaps you are familiar with the classic symptoms of Alzheimer's including memory loss, progressive loss of ability to care for one's self, and disorientation? But do you know what is actually happening in the brain to cause these issues?

There is actual atrophy of the gyri (folds) in the cerebral cortex of the brain as well as the hippocampus (the main center of memory). Additionally, the nerve cells of the brain show neurofibrillary tangles or dense masses of broken and twisted cytoskeleton. In the intracellular spaces, there are plaques consisting of clumps of cells, altered nerve fibers, and B-amyloid protein (rarely seen in elderly people without Alzheimer's).

Research efforts are currently focused on further study of three chromosomes found to be involved in the disease—chromosome 1, 14, and 21. Of particular note is that people with Down's syndrome (trisomy 21 with three copies of chromosome 21 instead of two), tend to show early onset Alzheimer's disease. Regarding treatment, the focus is on trying to halt B-amyloid formation or to activate the body's own immune system to clear out B-amyloid from brain tissue. There have been some severe side effects in going this route and researchers are trying to figure out how to halt those side effects before they continue in this direction. It is well documented that Alzheimer's patients show deficient levels of acetylcholine and nerve growth factor. Research in those areas has only yielded modest results. So the search for the cure continues.

OVERVIEW OF THE NERVOUS SYSTEM

Dr. Schmidt joked, "Has anyone ever told you that you are red hot?"

Grace and Will looked stunned. Will added, "I don't think so."

Dr. Schmidt continued, "What I mean by that is your nervous system is wired to send messages quickly to cells both electrically and chemically, which can also be thought of as being 'wired' **electrochemically**. It functions and carries out its tasks through sense organs and simple sensory nerve endings that receive information about changes in the body and the external environment which transmits messages to the **central nervous system (CNS)**, the CNS determines if a response is required, and executes responses if needed."

"Ok, that makes sense," answered Grace.

Dr. Schmidt continued, "So, the nervous system actually has two major anatomical subdivisions—the CNS (that we have been talking about) which consists of the brain and spinal cord, and the **peripheral nervous system or PNS** which consists of nerves emerging from the spinal cord that extend through the body and limbs. The PNS is further divided into the **sensory (afferent) division** which carries signals from sense organs and nerve endings to the CNS, and the **motor (efferent) division** which carries signals from the CNS to glands and muscle cells to carry out the body's responses.

"And, the sensory division is further broken down into the **somatic sensory division** that carries signals from receptors in the skin, muscles, bones and joints; and the **visceral sensory division** which carries signals from the thoracic (chest area) and abdomen such as the heart, lungs, stomach, and bladder.

"The motor division is further broken down into the **somatic motor (or voluntary) division** that carries signals to the skeletal muscles; and the **visceral motor (or autonomic) division** also known as the **ANS** which carries signals to glands, the heart, and smooth muscle in the body.

"Additionally, the ANS is divided into the **sympathetic division,** that can be thought of as the 'fight or flight' division because it prepares you to arouse for action if needed, by accelerating the heartbeat and increasing respiration; and the **parasympathetic division** which tends to have a calming effect like slowing the heartbeat. Here's a pocket chart that you can take with you to study from."

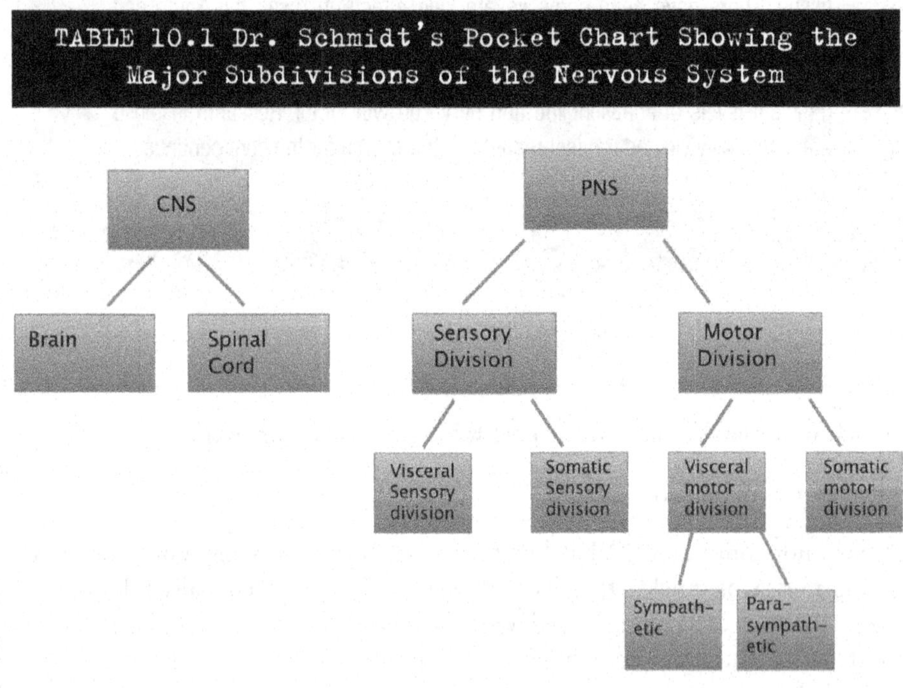

Self-Check – Answer these questions now before you move on:

1. Distinguish between the CNS and the PNS, and between visceral and somatic divisions of the sensory and motor divisions.

____

____

____

____

2. What is another name for the visceral motor nervous system? What are its two subdivisions? What are their functions?

____

____

____

____

____

> **STUDENT VIEW – WHY SHOULD I STUDY THIS?
> HOW IS IT RELEVANT TO MY LIFE?**
>
> Perhaps you or someone you know is afflicted by some type of disorder of the nervous system? You may recognize some of these disorders including, but not limited to Alzheimer's, Parkinson's, multiple sclerosis, and severe migraines. Learning how the nervous system functions may help you better understand how some of these disorders happen.

BASIC STRUCTURE OF A NEURON

"Let's look at the structure of a motor neuron," commented Dr. Schmidt.

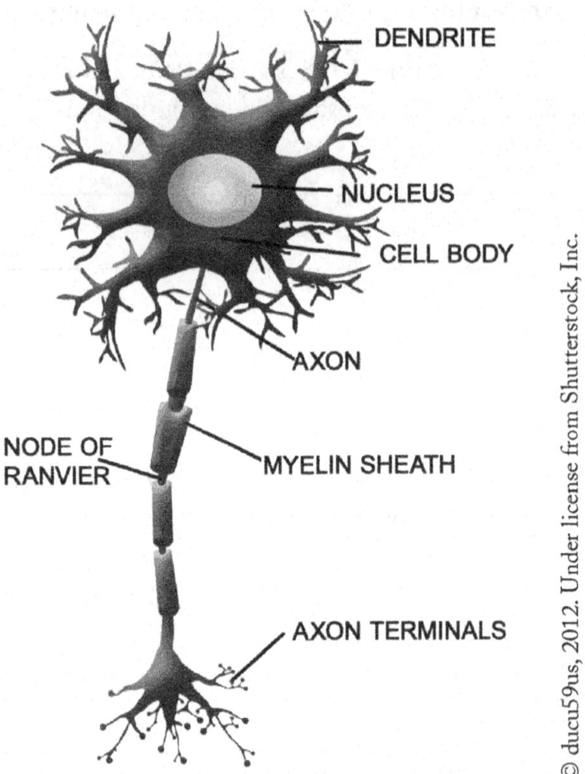

IMAGE 10.1 Structure of a Motor Neuron

"Why is it so funny looking? What are those tree branch-looking structures?" asked Grace.

"There is an old saying in biology that goes like this, 'Function follows structure and structure follows function.' What do you suppose I mean by that?" asked Dr. Schmidt.

"Maybe that things look like they do for a reason, that the way they look has something to do with their function…" replied Grace thoughtfully.

"Yes, absolutely. So, neurons like this are long and branched for fast communication," replied Dr. Schmidt. She continued, "The branches that receive the electrical signals are **dendrites**. The electrical signal then transmits to the cell body and nucleus and passes on toward the **axon** (or tail). The signal continues down the axon toward the **axon terminals** (also dendrites). In the same way that electrical wiring in your home is covered with insulation (i.e., copper wire wrapped in plastic), the axon is wrapped in a type of fatty insulation called the **myelin sheath**. Where you see spaces between the myelin sheath cells, there are the **nodes of Ranvier** (or gaps in the myelin sheath)." **NOTE:** Not all axons are myelinated.

Dr. Schmidt continued, "Neurons do not touch one another. There are junctions (gaps) between neurons called a **synapse**. There are electrical synapses and chemical synapses. An electrical synapse is a physical connection between two neurons with a pore allowing charged particles (ions) to pass from one neuron to the other. Unless there are a large number of pores, or the pores are very large, the connection is weak, and doesn't cause an **action potential** in one neuron to result in an action potential in the other. Since electrical synapses can cause either neuron to influence the other, these are bidirectional synapses.

"A chemical synapse is a junction between two neurons. The first (presynaptic) neuron releases a chemical (a neurotransmitter) which rapidly (in as little as 50 microseconds) crosses the small space across the synapse and binds to a receptor on the second (postsynaptic) neuron. This receptor may cause the direct or indirect opening of an ion-specific pore, allowing influx of sodium, calcium, or chloride or outflux of potassium ions. Since the presynaptic neuron can affect the postsynaptic neuron, these are unidirectional synapses.

"Originally synapses were thought to be largely electrical, but it turned out that the vast majority are chemical. A large number of neurotransmitter and receptors exist in the brain (the main transmitters being glutamate and GABA). Electrical synapses appear to be used to allow large networks of neurons to act in a more unified manner (for example, a single neuronal subtype in the cortex is wired to its identical neighbors). The synapse is filled with a liquid chemical called a **neurotransmitter**. There are many types of neurotransmitters but some you may recognize are substances like dopamine, acetylcholine, epinephrine, and norepinephrine. Neurotransmitters are the chemical component of the nervous system. They are excellent conductors of electricity. In some disorders like Alzheimer's there is a shortage of acetylcholine. Yet other disorders like Parkinson's appear to result from a shortage of dopamine. Manic depression (also known as bipolar) is associated with a lack of norepinephrine."

Will asked, "Where are neurotransmitters made?"

"Great question Will" replied Dr. Schmidt. She continued, "Neurotransmitters are made in the cell body of the neuron and then transported down the axon to the axon terminal. Molecules of neurotransmitters are stored in small "packages" called vesicles (see the picture on the right). Neurotransmitters are released from the axon terminal when their vesicles "fuse" with the membrane of the axon terminal, spilling the neurotransmitter into the synaptic cleft. Neurotransmitters are made in the cell body of the neuron and then transported down the axon to the axon terminal. Molecules of neurotransmitters are stored in small "packages" called vesicles (see Image 10.3). Neurotransmitters are released from the axon terminal when their vesicles "fuse" with the membrane of the axon terminal, spilling the neurotransmitter into the synaptic cleft."

Grace asked, "How do the neurotransmitters know how to work?"

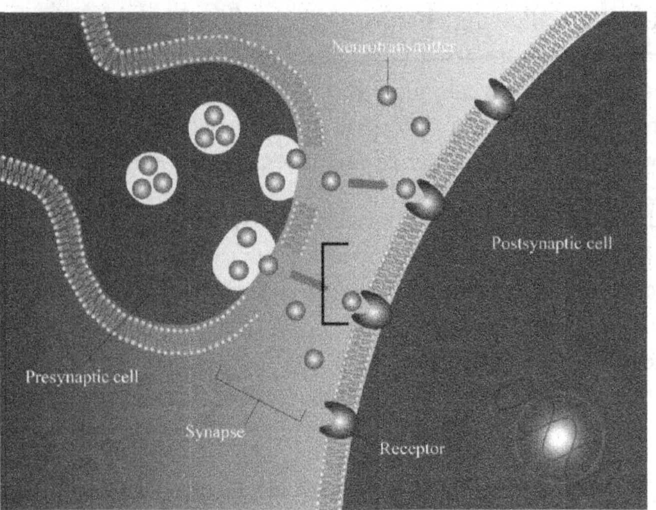

IMAGE 10.2 Neurotransmitter Physiology

Dr. Schmidt answered, "Neurotransmitters will bind only to specific receptors on the postsynaptic membrane that recognize them."

Will questioned, "Is there anything that stops or interrupts them?"

Dr. Schmidt replied, "Ah-h-h yes Will." She continued, "The action of neurotransmitters can be stopped by four different mechanisms:

1. Diffusion: the neurotransmitter drifts away, out of the synaptic cleft where it can no longer act on a receptor.
2. Enzymatic degradation (deactivation): a specific enzyme changes the structure of the neurotransmitter so it is not recognized by the receptor. For example, acetylcholinesterase is the enzyme that breaks acetylcholine into choline and acetate.
3. Glial cells: astrocytes remove neurotransmitters from the synaptic cleft.
4. Reuptake: the whole neurotransmitter molecule is taken back into the axon terminal that released it. This is a common way the action of norepinephrine, dopamine and serotonin is stopped...these neurotransmitters are removed from the synaptic cleft so they cannot bind to receptors."

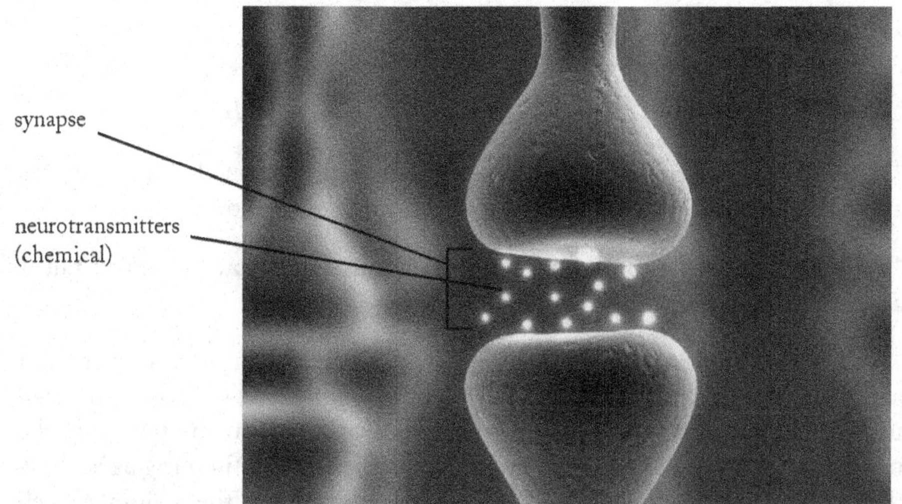

IMAGE 10.3 Synapse Between Two Neurons with Neurotransmitter

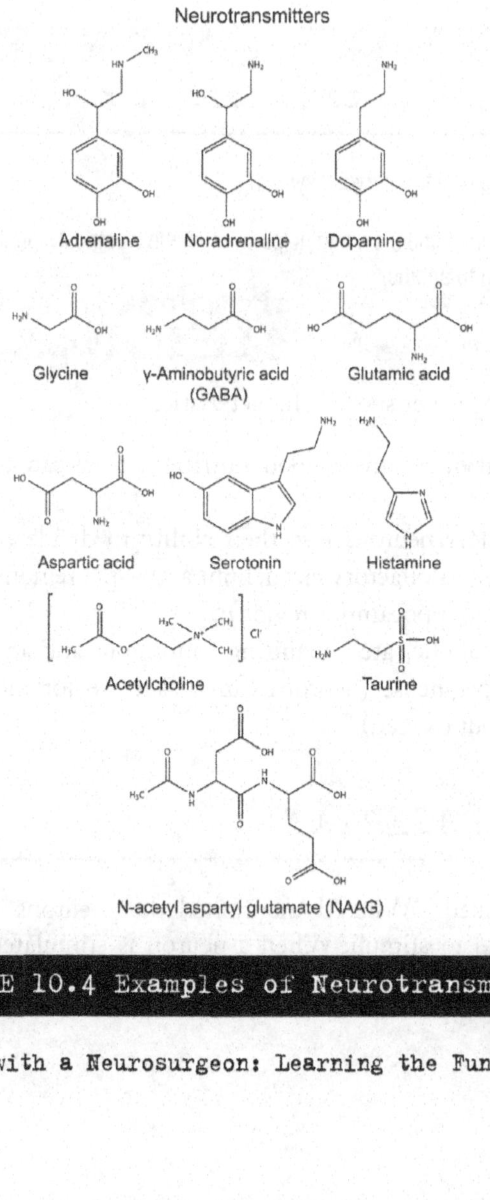

IMAGE 10.4 Examples of Neurotransmitters

NEURONS

"So, when do we get to talk about neurons?" questioned Will.

"Now is a great time," answered Dr. Schmidt. She continued, "What do you already know about them? And what would you like to know?"

"Well, I know they are the main type of cell within the nervous system overall—also known as nerve cells," replied Will.

"Yeah, that's right. **Neurons** are the main structural units that carry out communication," replied Dr. Schmidt. She continued, "There are estimated to be about a trillion neurons in the nervous system—there are ten times the number of neurons in your body than the number of stars in our galaxy. And, they exhibit some interesting characteristics that make them unique such as **excitability**—they respond to environmental changes and stimuli; they are **conductive**—they produce electrical signals that are quickly passed to other cells at distant locations; and they **secrete** chemical substances called **neurotransmitters** that bridge the gaps between neurons."

STUDY TIP!

neuro = pertaining to the nervous system

transmitter = a brain chemical that helps conduct the electrical signal traveling from one neuron to another

"They also have some other special characteristics:

1. Extreme longevity - given good nutrition, they can function over a lifetime.
2. Amniotic - most neurons lose their ability to divide (there are some exceptions such as olfactory epithelium and some regions involved with memory in the hippocampus region).
3. Very high metabolic rate - requiring continuous and adequate supplies of oxygen and glucose (neurons cannot survive for more than a few minutes without oxygen)."

MEMBRANE POTENTIALS

Dr. Schmidt continued, "When I mentioned that neurons were excitable, I mean they respond to stimuli. When a neuron is stimulated, an electrical impulse is generated and conducted along the length of its axon (tail). This response is called **action potential** (or nerve impulse). Curiously, this impulse is always the same regardless of the source or type. Plasma membranes, surrounding cells, have a variety of membrane proteins that act as ion channels (selectively allowing ions to pass through them). For example, a potassium (K+) ion channel only allows potassium to pass through it."

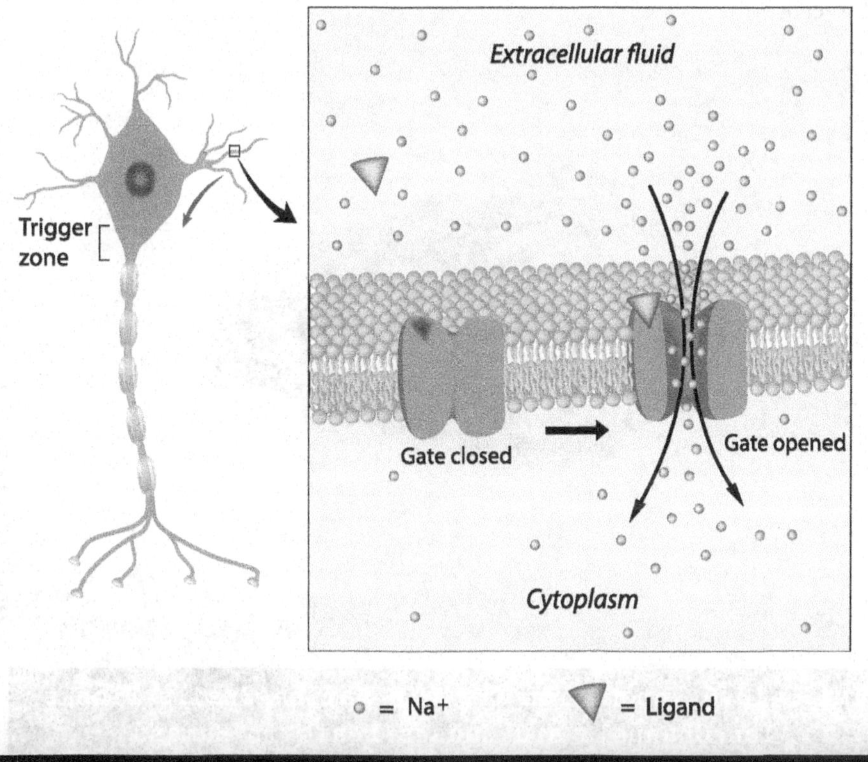

IMAGE 10.5 Neuron and Local Potential

Dr. Schmidt continued, "There are different types of ion channels. The three main groups of ion channels are:

1) the voltage-gated channels such as the sodium and potassium channels of the nerve axons and nerve terminals;
2) the extracellular ligand-activated channels which include channels such as GABA and glycine receptor channels, most of which are regulated by ligands that are "neurotransmitters." These channels are often named according to the ligand they bind to; and
3) intracellular ligand-gated ion channels. These include ion channels involved in sense perception."

She continued, "There are also two miscellaneous ion channel groups: The mechanosensory and volume-regulated channels have their own grouping, but they are still in the process of being classified. We have made a fifth 'catch-all' group, miscellaneous 2, which includes any ion channels not included in the other groups. This group includes the GAP junctions, peptide ion channels like Gramicidin, and various venomous insect toxins like the conus toxins from cone shells. Researchers have shown that the different venom components have specific targets in the nervous system — they blocked different ion channels in cell membranes, which control the transmission of impulses by nerves, or specific receptors for the neurotransmitter chemicals that transmit signals from cell to cell."

©Weblogiq, 2014. Under license from Shutterstock, Inc.

IMAGE 10.6 Venom from Certain Marine Snails Blocks Ion Channels and Causes Death in Humans

RESTING MEMBRANE POTENTIALS

Dr. Schmidt added, "When a neuron is not sending a signal, it is 'at rest.' When a neuron is at rest, the inside of the neuron is negative relative to the outside. Although the concentrations of the different ions attempt to balance out on both sides of the membrane, they cannot because the cell membrane allows only some ions to pass through channels (ion channels). At rest, potassium ions (K+) can cross through the membrane easily. Also at rest, chloride ions (Cl-) and sodium ions (Na+) have a more difficult time crossing. The negatively charged protein molecules (A-) inside the neuron cannot cross the membrane. In addition to these selective ion channels, there is a pump that uses energy to move three sodium ions out of the neuron for every two potassium ions it puts in. Finally, when all these forces balance out, and the difference in the voltage between the inside and outside of the neuron is measured, you have the resting potential. The resting membrane potential of a neuron is about -70 mV (mV = millivolt), which means that the inside of the neuron is 70 mV less than the outside. At rest, there are relatively more sodium ions outside the neuron and more potassium ions inside that neuron."

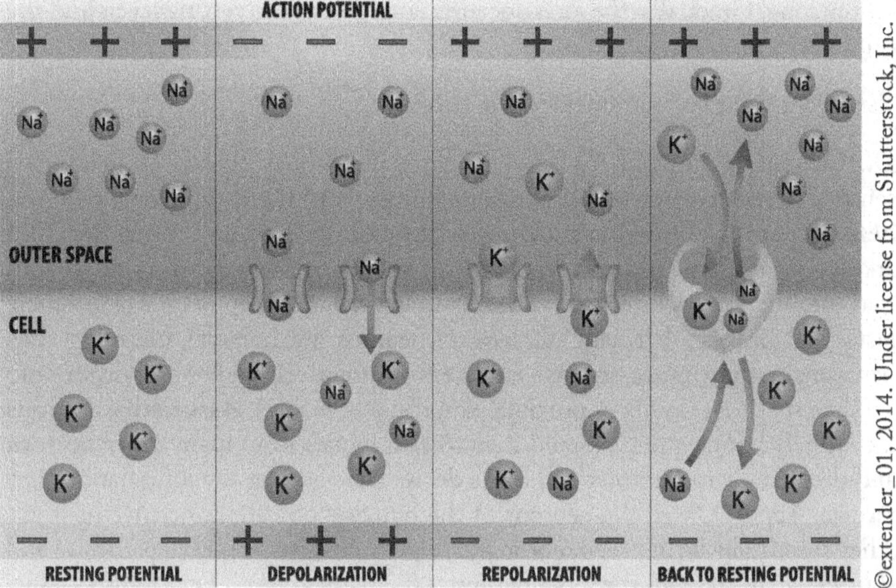

IMAGE 10.7 Nerve Impulse Action Potential in a Neuron

More About Action Potential

Will explained, "From what I understand, the resting potential tells about what happens when a neuron is at rest. An action potential occurs when a neuron sends information down an axon, away from the cell body. Neuroscientists use other words, such as a 'spike' or an 'impulse' for the action potential. The action potential is an explosion of electrical activity that is created by a depolarizing current. This means that some event (a stimulus) causes the resting potential to move toward 0 mV. When the depolarization reaches about -55 mV a neuron will fire an action potential. This is the threshold. If the neuron does not reach this critical threshold level, then no action potential will fire. Also, when the threshold level is reached, an action potential of a fixed sized will always fire... for any given neuron, the size of the action potential is always the same. There are no big or small action potentials in one nerve cell - all action potentials are the same size. Therefore, the neuron either does not reach the threshold or a full action potential is fired - this is the 'ALL OR NONE' principle."

Grace added, "Action potentials are caused when different ions cross the neuron membrane. A stimulus first causes sodium channels to open. Because there are many more sodium ions on the outside, and the inside of the neuron is negative relative to the outside, sodium ions rush into the neuron. Remember, sodium has a positive charge, so the neuron becomes more positive and becomes depolarized. It takes longer for potassium channels to open. When they do open, potassium rushes out of the cell, reversing the depolarization. Also at about this time, sodium channels start to close. This causes the action potential to go back toward -70 mV (a repolarization). The action potential actually goes past -70 mV (a hyperpolarization) because the potassium channels stay open a bit too long. Gradually, the ion concentrations go back to resting levels and the cell returns to -70 mV."

"Excellent you two!" commented Dr. Schmidt.

Dr. Schmidt continued, "There are three functional classes of neurons—sensory (afferent) neurons, association neurons, and motor (efferent) neurons. **Sensory neurons** begin in almost every organ of the body and end in the CNS and are specialized to detect heat, light, pressure, and chemicals and transmit that information back to the CNS. Examples include receptors for pain, smell, taste, and hearing. Virtually all sensory neurons are unipolar (meaning they have single, short processes that emerge from their cell bodies and divide into a T-like split that results in proximal and distal branches). **Associative neurons** are entirely within the CNS and they receive signals from many other neurons and decide how to interpret and make decisions regarding the information they receive. About 90% of the neurons of the human body are associative neurons. They specifically lie between motor and sensory neurons in special pathways and shuttle signals through the CNS pathways (help integrate). They are also known as interneurons and are almost always multipolar (meaning they have three or more processes - one axon and the rest dendrites; they are the major neuron type in the CNS). **Motor neurons,** on the other hand, send signals predominantly to muscle and gland cells. They conduct messages away from the CNS. Motor neurons are also known as efferent neurons. Motor neurons are multipolar. Except for some neurons in the ANS, their cell bodies are located in the CNS."

Basic Principles of Electricity

The human body is electrically neutral - meaning it has the same number of positive and negative charges. However, there are areas where one type - let's say positive - predominates making an entire region positively charged. Because opposite charges attract each other, energy must be used (work must be done) to separate them.

Helpful Definitions

voltage (V) - the measure of potential energy generated by separate charges; always measured between two points (known as potential difference)

current (I) - the flow of electrical charge from one point to another

Types of Neurons

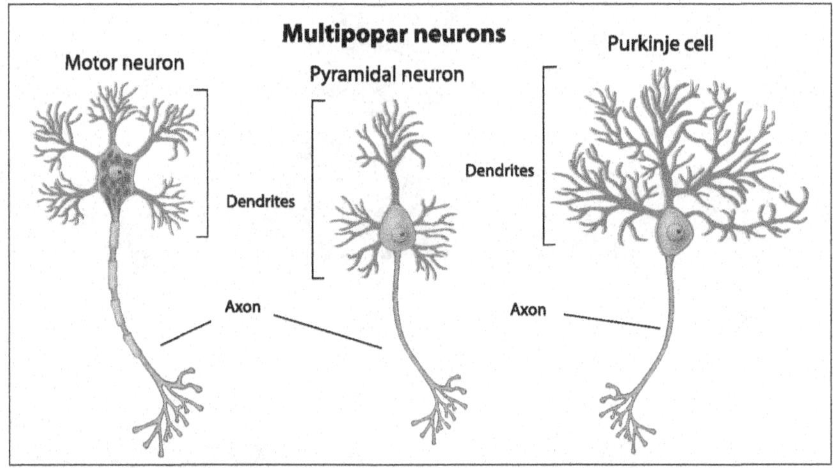

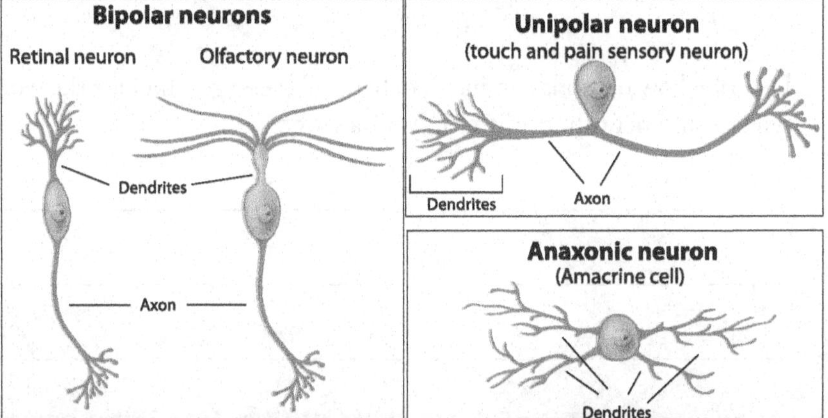

IMAGE 10.8. Different Types of Neurons

Self-Check – Answer these questions before you continue:

3. Describe three characteristics of neurons.

____

____

____

4. Explain the three functional classes of neurons and what they do.

____

____

____

5. Draw a motor neuron and label its main structures.

6. Describe how neurons conduct biochemical messages. Include the words synapse and neurotransmitter in your answer.

____

____

____

SUPPORTIVE CELLS – NEUROGLIA (GLIAL CELLS)

"What most people don't know is that neurons are supported by special cells called **neuroglia** (or glial cells). And these neuroglia outnumber neurons 10 to 1. It has been estimated that Einstein's brain had 73% more neuroglia than most adult males. This is significant because neuroglia play a supporting and nourishing role for neurons. Glial cells are non-neural cells that also perform "housekeeper" functions such as clearing out debris and excess materials. There are several different types of glial cells: astrocytes, oligodendrocytes, microglia, ependymal cells, satellite cells and Schwann cells (see Table 10.2 for more about neuroglia). Astrocytes are found in the brain's capillaries and form the blood-brain barrier that restricts what substances can enter the brain. They are the most abundant of the glial cells. Microglia are extremely small cells of the central nervous system that remove cellular waste and protect against microorganisms. Ependymal cells line the central cavities of the brain and spinal cord where they form a fairly permeable barrier between the cerebrospinal fluid that fills those cavities and the tissue fluid bathing the cells of the CNS. Oligodendrocytes are central nervous system structures that wrap some neuronal axons to form an insulating coat known as the myelin sheath.

Schwann cells are peripheral nervous system structures that wrap some neuronal axons to form an insulating coat known as the myelin sheath," added Dr. Schmidt.

"Wow! That's pretty interesting! Is there anything else about Einstein's brain that is significant?" asked Will.

"Actually, it was also about 15% larger than the average male brain," expounded Dr. Schmidt. "There are six basic types of neuroglia. Check out this poster in my office."

TABLE 10.2 Dr. Schmidt's Poster Showing the Six Types of Neuroglia and Their Functions	
Neuroglia of the CNS	
Type	**Function**
oligodendrocytes	Form myelin in brain and spinal cord
ependymal cells	Line cavities of brain and spinal cord; secrete and help circulate cerebrospinal fluid
microglia	Phagocytize (destroy) bacteria that cause infection
astrocytes	Most abundant type; cover brain surface and non-synaptic portion of neurons; produce growth factors that stimulate neurons; nourish neurons
Neuroglia of the PNS	
Schwann cells	Form myelin around PNS nerve fibers
satellite cells	Provide electrical insulation and help regulate the chemical environment of neurons

continued on next page...

CNS – THE BRAIN

"So, we've talked about neurons and neuroglia… when do we get to talk about the brain?" asked Grace enthusiastically.

"Now is a great time!" replied Dr. Schmidt. She added, "The brain in a female weighs approximately 3.2 lbs. In males about 3.5 lbs although in terms of brain mass versus body mass they are the same relative mass for men and women. The brain is composed of white matter and grey matter. **Grey matter**, which has a pinkish-grey color in the living brain, contains the cell bodies, dendrites

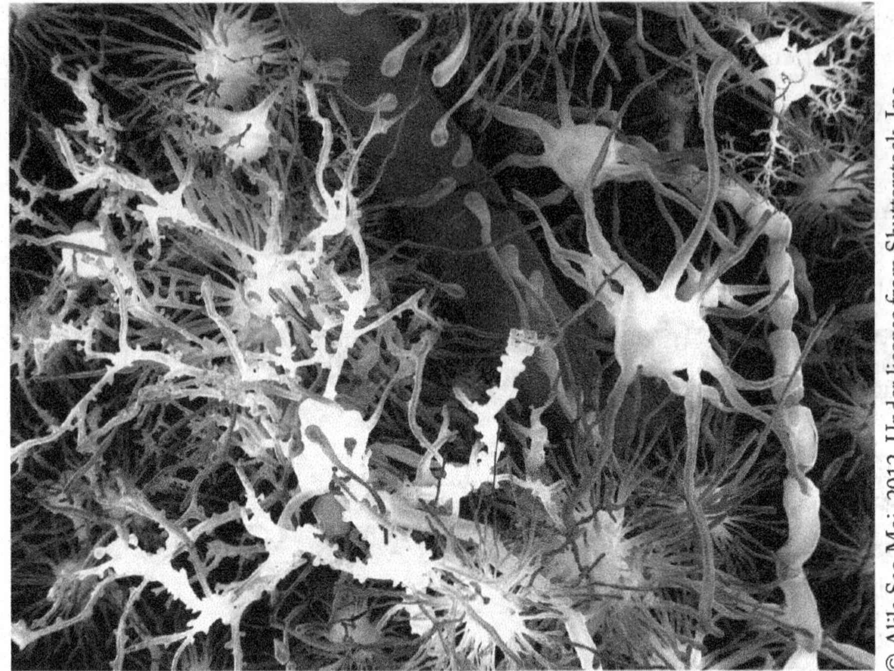

IMAGE 10.9. Three of the Six Types of Glial Cells Featured Here — Microglia, Oligodendrocytes and Astrocytes

and axon terminals of neurons, so it is where all synapses are. **White matter** is made of axons connecting different parts of grey matter to each other. White matter is white because of myelin (a mixture of proteins and phospholipids forming a whitish insulating sheath around many nerve fibers, increasing the speed at which impulses are conducted)." She continued, "We can think of the brain being composed overall of three main parts when we first view it, as in this specimen where we can see the **cerebrum** (two caps on top), the **cerebellum** (posterior and inferior), and the **brain stem** that includes the midbrain, pons, and medulla oblongata. The cerebrum is involved in several functions of the body including determining intelligence, determining personality, thinking, perceiving, producing/understanding language, interpretation of sensory impulses, motor function, and planning and organization of touch sensation. The cerebrum is approximately 83% of the brain's volume and consists of two half globes called **cerebral hemispheres**. Both cerebral hemispheres and the cerebellum have an outer layer or cortex which represents grey matter. With regard to white and grey matter, this pattern changes as you descend through the brain stem - the cortex disappears but scattered grey matter regions are seen within the white matter. Further, each cerebral hemisphere is chiefly concerned with the sensory and motor functions of opposite sides of the body. This means the left side of the cerebral hemisphere controls the right side of the body and vice versa."

Dr. Schmidt continued, "Going back to the cerebral cortex for a moment - do you know that it contains three kinds of functional areas? That would be motor areas, sensory areas and association areas."

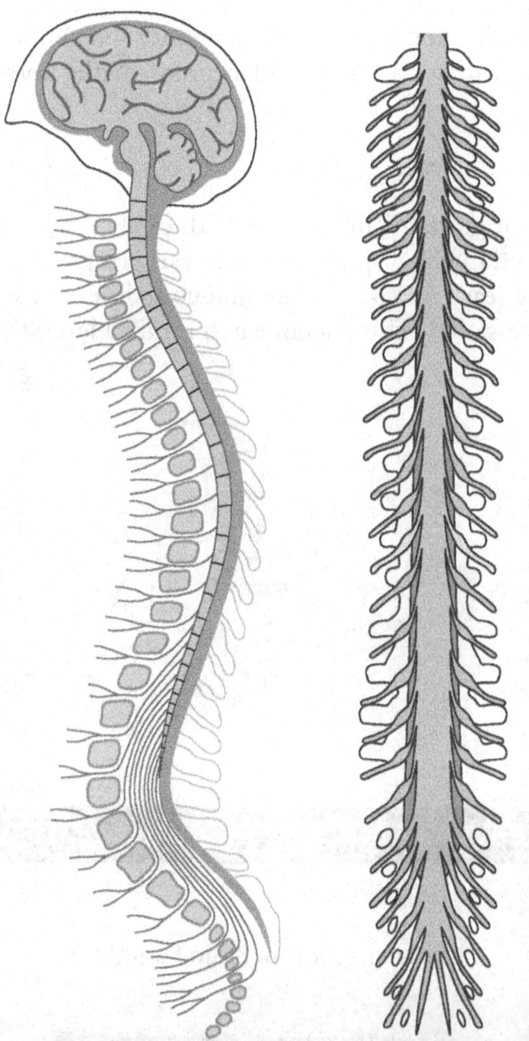

IMAGE 10.10. The Brain and Spinal Cord Comprise the CNS

STUDY TIP!

In the CNS:

Grey matter = soma (also called cell bodies)

White matter = axons (the myelin sheaths are white)

Will warned, "Yeah, but don't confuse the sensory and motor areas of the cortex with sensory and motor neurons. All neurons in the cortex are associative neurons."

'Good point" added Grace.

Dr. Schmidt continued, "While we use both cerebral hemispheres for almost every activity, and the hemispheres appear nearly identical - there is actually a division of labor (each hemisphere has unique abilities not shared by the other hemisphere necessarily). This phenomenon is called **lateralization**.

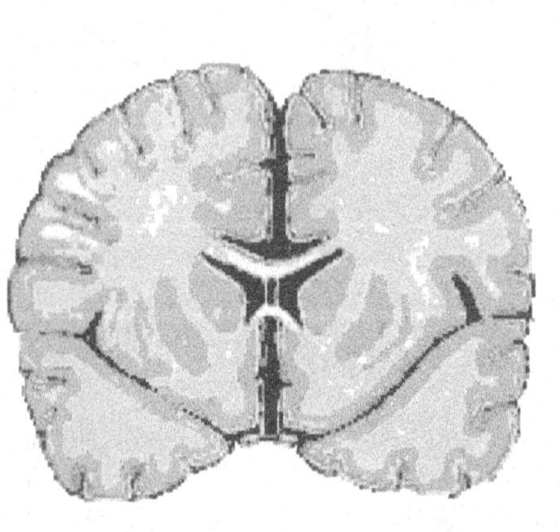

© Laschon Robert Paul, 2014. Under license from Shutterstock, Inc.

IMAGE 10.11 White and Grey Matter in the Brain

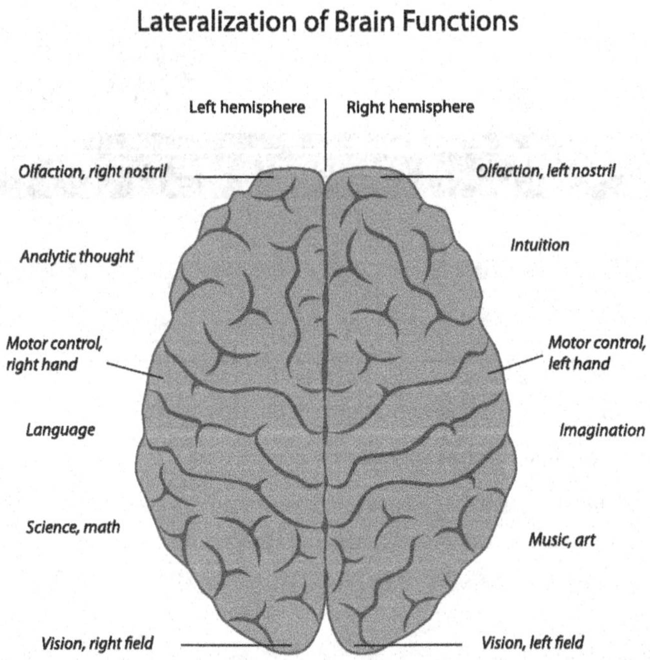

Lateralization of Brain Functions

© Alila Medical Media, 2014. Under license from Shutterstock, Inc.

IMAGE 10.12 Lateralization

"Each cerebral hemisphere has heavy folds that are raised slightly called **gyri** (pronounced JY-rie) and shallow depressions called **sulci** (pronounced SUL-see). Further, the two globes have a deep fissure that separates them called the **longitudinal fissure**. The cerebral hemispheres connect to the rest of the brain via a network of nerves in an area called the **corpus callosum**. The cerebellum is the second largest region of the brain making up about 10% of its volume but containing 50% of the brain's neurons. The cerebellum is located at the base of the brain, just above the brain stem, where the spinal cord meets the brain, and is made of two hemispheres. The cerebellum receives information from the sensory systems, the spinal cord, and other parts of the brain and then regulates motor movements. The cerebellum coordinates voluntary movements such as posture, balance, coordination, and speech, resulting in smooth, balanced muscular activity. It is also important for learning motor behaviors. It is a relatively small portion of the brain -- about 10% of the total weight, but it contains half the brain's neurons. The cerebellum is located at the base of the brain, just above the brain stem, where the spinal cord meets the brain, and is made of two hemispheres. The cerebellum receives information from the sensory systems, the spinal cord, and other parts of the brain and then regulates motor movements. The cerebellum coordinates voluntary movements such as posture, balance, coordination, and speech, resulting in smooth, balanced muscular activity. It is also important for learning motor behaviors. It is a relatively small portion of the brain -- about 10% of the total weight."

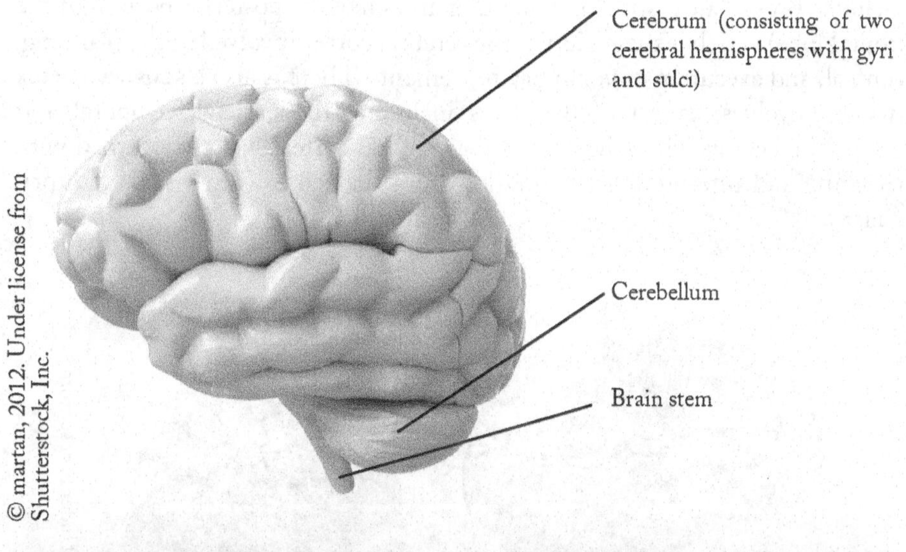

IMAGE 10.13 External Brain

Dr. Schmidt continued, "Additionally, the cerebral cortex has lobes—the parietal lobe, temporal lobe, occipital lobe, and frontal lobe featured below. The **parietal lobe** is involved with cognition, information processing, pain and touch sensation, spatial orientation, speech, and visual perception. The **frontal lobe** handles movement, decision-making, problem solving, and planning. The **temporal lobe** plays an important role in organizing sensory input, auditory perception, language and speech production, as well as memory association and formation. The **occipital lobe** deals with visual perception and color recognition."

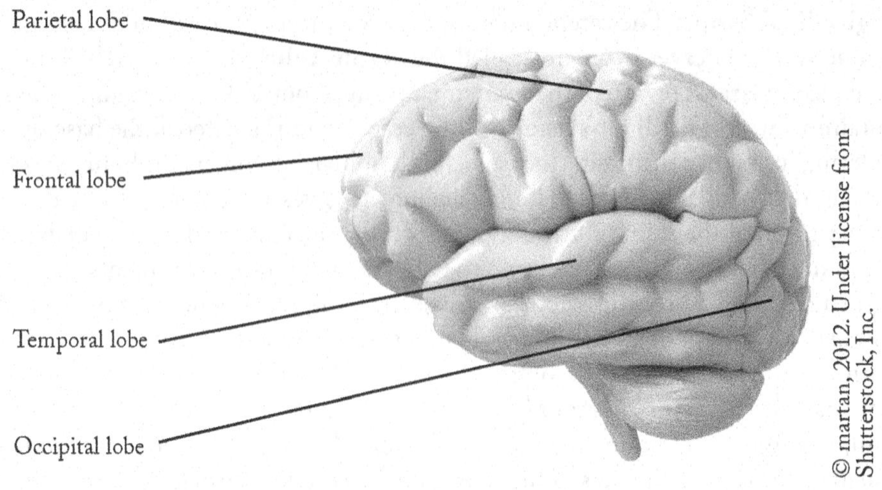

IMAGE 10.14 Brain's Basic Lobe Areas

Dr. Schmidt continued, "I would also like to introduce the **motor cortex** (which includes Broca's area (speech); and other areas like the posterior portion of the frontal lobe) which is the region of the cerebral cortex involved in the planning, control, and execution of voluntary movements. There is also a **sensory cortex** (which involves several of the lobes including the parietal, temporal and occipital lobes) which is the part of the cerebral cortex that is concerned with receiving and interpreting information from the senses (i.e., visual, auditory, olfactory)."

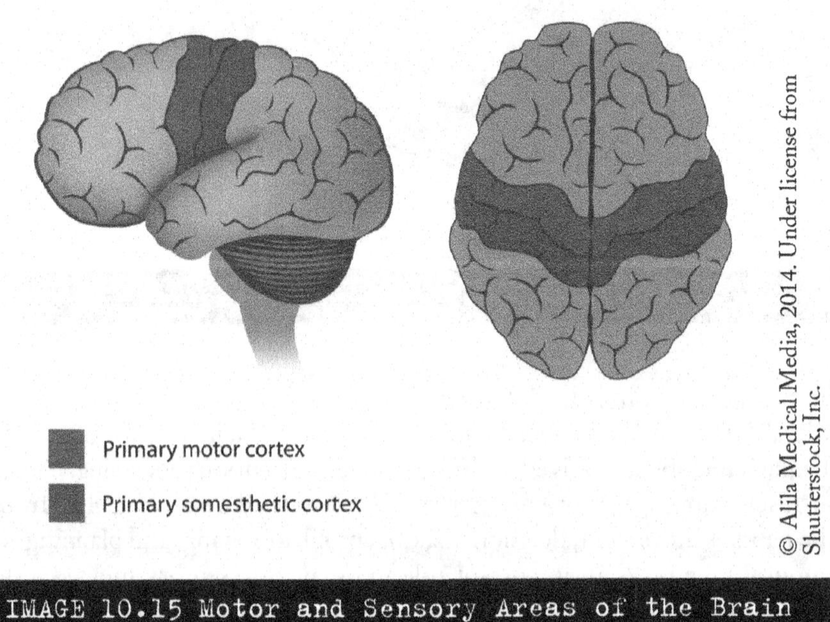

IMAGE 10.15 Motor and Sensory Areas of the Brain

Dr. Schmidt added, "You can better see the corpus callosum as well as the brain stem and various other structures on this internal diagram."

Dr. Schmidt paused and asked, "Are there any questions?"

Will asked, "I know we talked about the corpus callosum earlier but what does it do exactly?"

Dr. Schmidt replied, "The corpus callosum is involved in several functions of the body including communication between brain hemispheres, eye movement, maintaining the balance of arousal, and attention tactile localization."

Self-Check – Answer these questions before you continue:

7. Describe the parts of the brain stem and their functions.

____

____

____

8. Describe the functions of these parts of the brain: hypothalamus, pituitary gland, corpus callosum, thalamus, and midbrain.

____

____

____

Dr. Schmidt continued, "Lets talk about the diencephalon next."

Grace mentioned, "The diencephalon is the central core of the forebrain and consists of the thalamus, hypothalamus and pineal gland."

"Yes" replied Dr. Schmidt. She continued, "The thalamus is a large, dual lobed mass of grey matter buried under the cerebral cortex. It is involved in sensory perception and regulation of motor functions. The thalamus is a limbic system structure and it connects areas of the cerebral cortex that are involved in sensory perception and movement with other parts of the brain and spinal cord that also have a role in sensation and movement. As a regulator of sensory information, the thalamus also controls sleep and awake states of consciousness.

"The thalamus is involved in several functions of the body including motor control, receiving auditory, somatosensory and visual sensory signals, relaying sensory signals to the cerebral cortex, and controlling sleep and awake states."

HUMAN BRAIN

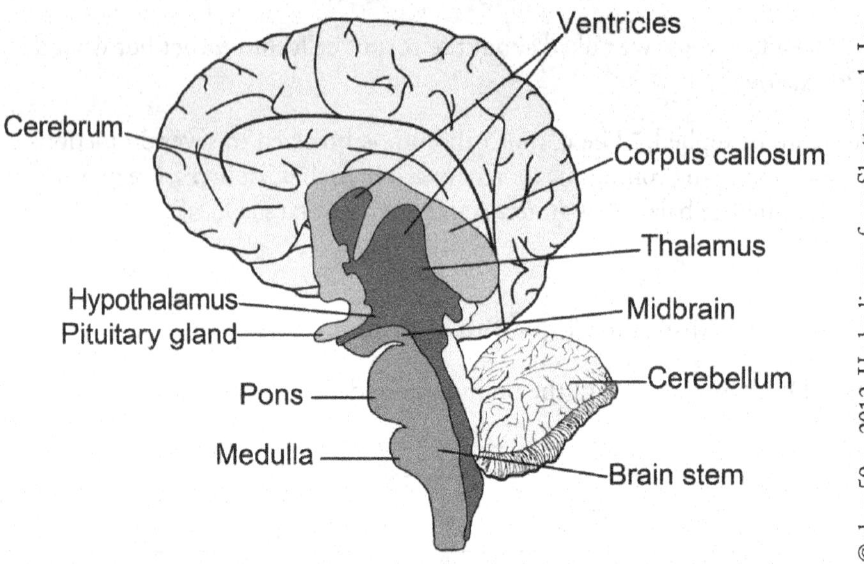

IMAGE 10.16 Internal Brain

She continued, "The hypothalamus is about the size of a pearl, the hypothalamus directs a multitude of important functions in the body. It is the control center for many autonomic functions of the peripheral nervous system. Connections with structures of the endocrine and nervous systems enable the hypothalamus to play a vital role in maintaining homeostasis. For example, blood vessel connections between the hypothalamus and pituitary gland allow hypothalamic hormones to control pituitary hormone secretion. As a limbic system structure, the hypothalamus also influences various emotional responses. The hypothalamus is involved in several functions of the body including autonomic function control, endocrine function control, homeostasis, motor function control, food and water intake regulation, and sleep-wake cycle regulation."

Dr. Schmidt concluded, "The pineal gland (epithalamus) is a pine cone shaped gland of the endocrine system. A structure of the diencephalon of the brain, the pineal gland produces several important hormones including melatonin. Melatonin influences sexual development and sleep-wake cycles. The pineal gland is composed of cells called pinealocytes and cells of the nervous system called glial cells. The pineal gland connects the endocrine system with the nervous system in that it converts nerve signals from the sympathetic system of the peripheral nervous system into hormone signals. The pineal gland is involved in several functions of the body including secretion of the hormone melatonin, regulation of endocrine functions, conversion of nervous system signals to endocrine signals, causes feeling of sleepiness, and influences sexual development."

"I've always been fascinated by the different areas of the brain and what they do," commented Will.

"Here's a chart showing the basic functions," replied Dr. Schmidt.

TABLE 10.3 Dr. Schmidt's Chart Showing the Main Areas of the Brain and Their Functions

Area	Function
cerebrum	Determines intelligence and personality, thinking, perceiving, producing and understanding language, interpretation of sensory impulses, motor function, planning and organization, touch sensation
corpus callosum	Allows communication between both hemispheres
hypothalamus	Autonomic function control, endocrine function control, homeostasis, motor function control, food and water intake regulation, sleep-wake cycle regulation
thalamus	Relay center for sensory impulses to the cerebral cortex
ventricles	Four communicating cavities within the cerebral hemispheres; filled with cerebrospinal fluid
pituitary gland	The master gland of the body; controls the thyroid, adrenals, and gonads
midbrain	Relay station for auditory and visual information; smallest part of the brain
pons	Connects the cerebral cortex to the medulla oblongata; serves as a communications and coordination center between the two hemispheres of the brain; helps in the transferring of messages between various parts of the brain and the spinal cord
medulla oblongata	Lowest portion of the brain stem; controls breathing, digestion, and heart rate
cerebellum	Controls motor movement coordination, balance, equilibrium, and muscle tone
pineal gland	Secretes melatonin; responsible for sleep and wake cycles

MORE ABOUT THE VENTRICLES

Dr. Schmidt continued, "Within the cerebral hemispheres are four cavities or **ventricles** that have the ability to communicate and are filled with cerebrospinal fluid. **Cerebrospinal fluid** is a clear, colorless fluid that bathes the entire CNS. It is produced by a structure called the *choroid plexus* in the lateral, third, and fourth ventricles. The cerebrospinal fluid can be tapped into as in a spinal tap to analyze it under the microscope if disease is present. This is a common procedure when meningitis is suspected, for example. The functions of the four ventricles are to absorb physical shocks to the brain, distribute nutritive materials to the nervous system/remove wastes and to provide a stable chemical environment."

THE BRAIN STEM

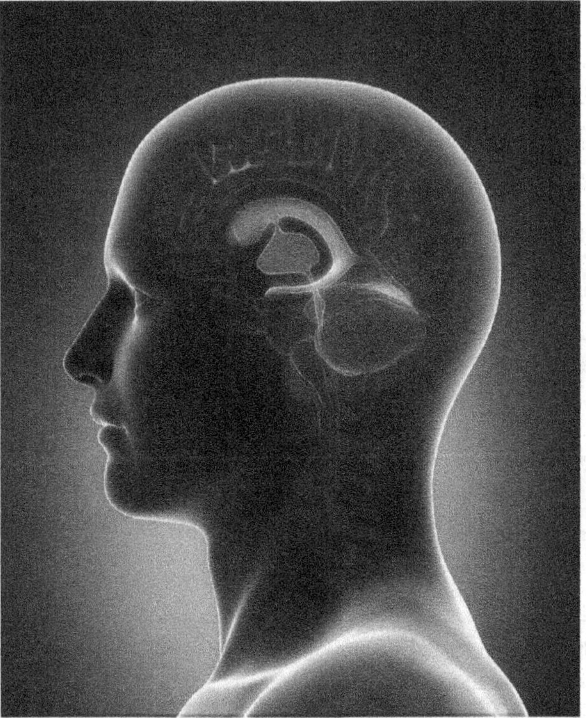

© CLIPAREA | Custom media, 2014. Under license from Shutterstock, Inc.

IMAGE 10.17 Ventricles in the Brain

THE SPINAL CORD

Dr. Schmidt reminded Grace and Will, "No discussion about the CNS is complete without discussing the spinal cord. The spinal cord is the most important structure between the body and the brain. The spinal cord extends from the magnum foramen where it is continuous with the medulla to the level

VERTEBRAL COLUMN

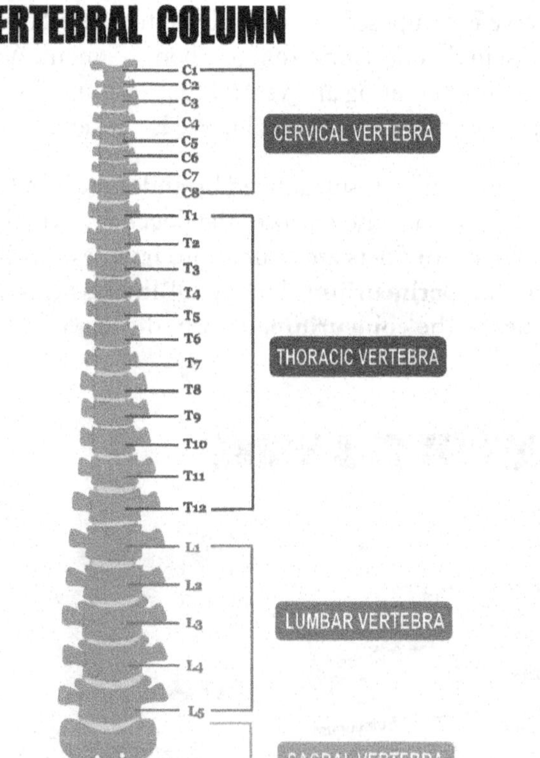

IMAGE 10.18 Spinal Cord Regions

of the first or second lumbar vertebrae. It is a vital link between the brain and the body, and from the body to the brain. The spinal cord is 40 to 50 cm long and 1 cm to 1.5 cm in diameter. Two consecutive rows of nerve roots emerge on each of its sides. These nerve roots join distally to form 31 pairs of spinal nerves. These nerves are divided into 8 cervical, 12 thoracic, 5 lumbar, 5 sacral, and 1 coccygeal nerve. The spinal cord is a cylindrical structure of nervous tissue composed of white and gray matter, is uniformly organized and is divided into four regions: cervical (C), thoracic (T), lumbar (L) and sacral (S), each of which is comprised of several segments. The spinal cord contains motor and sensory nerve fibers to and from all parts of the body. Each spinal cord segment innervates a dermatome (an area of skin that is mainly supplied by a single spinal nerve)."

Dr. Schmidt continued, "All spinal nerves, except the first, exit below their corresponding vertebrae. In the cervical segments, there are 7 cervical vertebrae and 8 cervical nerves. C1-C7 nerves exit above their vertebrae whereas the C8 nerve exits below the C7 vertebra. It leaves between the C7 vertebra and the first thoracic vertebra. Therefore, each subsequent nerve leaves the cord below the corresponding vertebra. In the thoracic and upper lumbar regions, the difference between the vertebrae and cord level is three segments. Therefore, the root filaments of spinal cord segments have to travel longer distances to reach the corresponding intervertebral foramen from which the spinal nerves emerge. The lumbosacral roots are known as the cauda equina.

"Each spinal nerve is composed of nerve fibers that are related to the region of the muscles and skin that develops from one body segment. A spinal segment is defined by dorsal roots entering and ventral roots exiting the cord, (i.e., a spinal cord section that gives rise to one spinal nerve is considered as a segment).

"Within a nerve, each axon is surrounded by **endoneurium** (a delicate layer of loose connective tissue that also encloses the fiber's associated myelin sheath or neurolemma). Groups of fibers are bound into bundles (fascicles) by a coarser tissue wrapping, the **perineurium**. Finally, all the fascicles are enclosed by a tough fibrous sheath, the **epineurium**, to form the nerve."

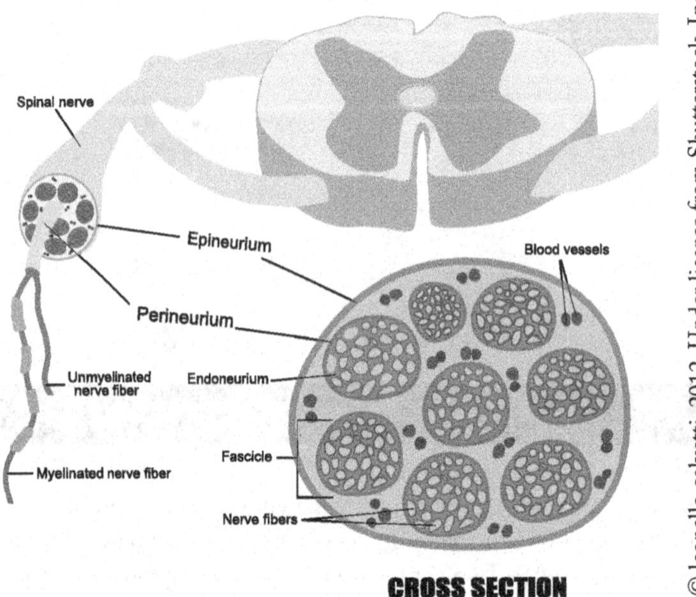

IMAGE 10.19 Nerve Anatomy

CRANIAL NERVES

Dr. Schmidt continued, "Twelve pairs of cranial nerves are associated with the brain. The first two pairs attach to the forebrain, and the rest are associated with the brain stem. Other than the vagus nerves, which extend into the abdomen, cranial nerves serve only head and neck structures. In most cases, the names of cranial nerves reveal either the structures they serve or their functions. The nerves are usually numbered with Roman numerals from the most roastral to to the most caudal. This is a mini-introduction to the nerves, their origins and what they control:

I Olfactory nerve - transmits the sense of smell; Located in olfactory foramina of ethmoid

II Optic nerve - transmits visual information to the brain; Located in optic canal

III Oculomotor nerve - innervates levator palpebrae superioris, superior rectus, medial rectus, inferior rectus, and inferior oblique, which collectively perform most eye movements; Located in superior orbital fissure

IV Trochlear nerve - innervates the superior oblique muscle, which depresses, pulls laterally, and intorts the eyeball; Located in superior orbital fissure

V Trigeminal nerve - receives sensation from the face and innervates the muscles of mastication; Located in superior orbital fissure (ophthalmic branch), foramen rotundum (maxillary branch), and foramen ovale (mandibular branch)

VI Abducens nerve - innervates the lateral rectus, which abducts the eye; Located in superior orbital fissure

VII Facial nerve - provides motor innervation to the muscles of facial expression and stapedius, receives the special sense of taste from the anterior 2/3 of the tongue, and provides secretomotor innervation to the salivary glands (except parotid) and the lacrimal gland; Located and runs through internal acoustic canal to facial canal and exits at stylomastoid foramen

VIII Vestibulocochlear nerve (or auditory-vestibular nerve) - senses sound, rotation and gravity (essential for balance & movement; Located in internal acoustic canal

IX Glossopharyngeal nerve- receives taste from the posterior 1/3 of the tongue, provides secretomotor innervation to the parotid gland, and provides motor innervation to the stylopharyngeus (essential for tactile, pain, and thermal sensation). Sensation is relayed to opposite thalamus and some hypothalamic nuclei. Located in jugular foramen

X Vagus nerve - supplies branchiomotor innervation to most laryngeal and pharyngeal muscles; provides parasympathetic fibers to nearly all thoracic and abdominal viscera down to the splenic flexure; and receives the special sense of taste from the epiglottis. A major function: controls muscles for voice and resonance and the soft palate. Symptoms of damage: dysphagia (swallowing problems). Located in jugular foramen

XI Accessory nerve (or cranial accessory nerve) - nucleus ambiguus, Spinal accessory nucleus; controls muscles of the neck and overlaps with functions of the vagus. Examples of symptoms of damage: inability to shrug, weak head movement, velopharyngeal insufficiency; Located in jugular foramen

XII Hypoglossal nerve - provides motor innervation to the muscles of the tongue and other glossal muscles; important for swallowing (bolus formation) and speech articulation. Located in hypoglossal canal."

CERVICAL NERVES

Dr. Schmidt continued, "In addition to the seven cervical vertebrae, cervical anatomy features eight cervical nerves (C1-C8) that branch off of the spinal cord and control different types of bodily and sensory activities.

"Each cervical nerve is named based on the lower cervical vertebra that it runs between. As an example, the nerve root that runs between the second cervical vertebra and the third cervical vertebra in the neck is described as the C3 nerve.

Branching off from the nerves in the spinal cord, the cervical nerves are responsible for relaying messages and ensuring functioning to different body parts. Here are the nerves and their functions:

C1 and C2 (the first two cervical nerves) control the head.

C3 and C4 help control the diaphragm (the sheet of muscle that stretches to the bottom of the rib cage and plays an important role in breathing and respiration).

C5 controls upper body muscles like the Deltoids (which form the rounded contours of the shoulders) and the Biceps (which allow flexion of the elbow and rotation of the forearm).

C6 controls the wrist extensors (muscles like the extensor carpi radialis longus, extensor carpi radialis brevis, and extensor carpi ulnaris that control wrist extension and hyperextension) and also provides some innervation to the biceps.

C7 controls the Triceps (the large muscle on the back of the arm that allows for straightening of the elbow).

C8 controls the hands."

THORACIC NERVES

She continued again, "T1-12 power the muscles that lie between the ribs (intercostal muscles). These also help you breathe by drawing the rib cage outwards and upwards, pulling the lungs in the same direction. The lungs expand, helping them fill with air.

"The diaphragm and the intercostal muscles are your major breathing muscles.

"Your lower thoracic spinal nerves T6-12 provide power to your abdominal muscles. These muscles help you cough and expel matter from your air passages. Abdominal muscles are also important in balance and posture."

Dr. Schmidt finalized, "Nerves in the lumbar (L) and sacral (S) sections of your spinal cord power leg muscles for walking, running and jumping.

L2 powers muscles that bend or flex your hip joint.

L3 powers your quadriceps muscle so your leg straightens at the knee.

L 4 powers muscles around your ankle, allowing your ankle to bend and draw the foot back towards your head (dorsi-flexion).

S1 powers muscles around your ankle, allowing it to bend and your foot and toes to point downwards (plantar flexion)

S2-S4 powers external sphincter muscles of your anal canal and urethra."

SMELL PHYSIOLOGY

Will questioned, "Doesn't the combination of smell and taste give food their flavor?"

Dr. Schmidt answered, " Yes - and our sense of smell becomes less acute with age and this can lead to a loss of appetite and sometimes weight loss."

She continued, "Most odors (smells), such as a perfume, are composed of a complex mixture of odorants. Each odorant has a distinctive molecular shape. These molecules enter the roof of the nasal cavity, dissolve in the moist mucosal protective layer and are recognized by receptors on the dendrites of olfactory cells. These cells project directly through the skull to mitral cells of the olfactory bulb, a part of the cortex. After a sudden head impact, these olfactory afferents can be sheared off, interrupting the sense of smell. But this is restored when the olfactory cells regrow in about a month."

She continued, "Olfactory cell axons are unmyelinated because the distance to the olfactory bulb is short and because speed is not important. It takes time for an odor molecule to diffuse across the mucus film and attach to a particular receptor site. When we sniff, different odor molecules diffuse across the mucus film at different rates. Because of this, the response at any particular time provides little information as to the odor. One needs to remember the temporal pattern of the whole sniff. For this reason memory is an important element in recognizing odors. Conversely, odors often elicit strong memories."

"Yeah, like somebody's cologne reminding me of my Dad's cologne" said Grace.

"Or the smell of homemade cookies reminding me of my grandma because she always made cookies for us" commented Will.

Dr. Schmidt continued, "The mitral cells in the olfactory bulb in turn project to the pyriform cortex which codes mixtures of odorants present in a particular smell (e.g. a particular perfume). The pyriform cortex then sends information to the amygdala and hippocampus and through the medial dorsal thalamus to the orbital frontal cortex.

"The amygdala is activated by the pleasant or unpleasant aspects of odors and the hippocampus facilitates the storage of odor memories.

"The orbital frontal cortex combines our sensations to odors with those of taste, texture (somatosensory), spiciness (pain) and vision. The combination of two or more of these modalities results in the perception of flavor. Cells here receive a multimodal input and can respond, for example, to the smell, sight, or taste of a banana. Patients with lesions of the orbitofrontal cortex are unable to discriminate odors."

Grace asked, "Are there basic smell qualities?"

Dr. Schmidt answered, "No. The sense of smell does not have a small number of basic receptor types (as there are 5 types of basic tastes, 3 types of cones, or 5 types of touch receptors). Instead humans have over 300 receptive types, and other species, such as dogs, have many more.

"Each odorant molecule can fit, unlock and thus activate several receptors because each end of the molecule can fit a different receptor. A molecular cascade amplifies the taste receptor's sensitivity, as in retina's receptors. To identify a particular odor, the cortex examines the pattern of afferents that are activated, the odor's combinatorial code. In Figure 7.26 the three colors signify three of the thousands of mitral cells that map smells. A particular odor is coded by a particular pattern of activated mitral cells. By comparing patterns, the number of odors that can be distinguished become enormous. This is like a combination lock in which 3 or 4 numbers can produce thousands of combination code possibilities. On average humans can discriminate between trillions of odors. But this is highly variable between subjects, ranging from a million trillion trillion to only 70 million. This is far more than the number of colors that can be distinguished. Genetic defects, such as those that produce various forms of color blindness, can also result in the absence of particular subtypes of olfactory cells. These can produce anosmia for specific odors.

"Each mitral cell receives input from many olfactory cells that expresses the same receptor subtype. As in the receptors, each odor activates a particular combination of mitral cells. A given mitral cell is activated by more than one odor. Similar odors stimulate adjacent mitral cells. Unlike other modalities, the nasal cavity is not mapped somatotopically onto the olfactory bulb. Rather the olfactory bulb is arranged in a topographical map of smells.

"When olfactory cells are damaged, due to a virus or toxic substance, they are replaced within about a month from basal cells. These then regrow to the same mitral cell in the same region of the olfactory bulb that was previously most sensitive to that particular odor. Thus the same map of odors in the olfactory bulb is maintained as is their memory."

SUMMARY OF TASTE AND SMELL

Will added, "So in summary, both olfactory cells, and taste cells in the taste bud, are constantly replaced."

"Right" confirmed Dr. Schmidt.

Grace spoke up, "And - both taste and smell project to the newer cerebral cortex, the neocortex for perception, and to the older cortex in the limbic system for automatic responses of hunger, pleasure, etc."

Dr. Schmidt added, "Smell is the only sensory system which projects from the periphery directly to the cortex (hippocampus and amygdala). Smell also has a projection through the thalamus to the cortex."

MORE ABOUT THE PNS

Dr. Schmidt continued, "Now that we have explored the CNS—brain and spinal cord—I'd like to revisit the peripheral (PNS). What do you know about it?" she asked Grace and Will.

"The PNS includes the nerves that branch from the spinal cord into the rest of the body including your limbs, as shown on the diagram below," Dr. Schmidt added.

Will asked, "And, doesn't it include your senses such as vision, taste, and hearing?"

"Yes it does. It's part of the PNS sensory afferent division. Have you ever studied the eye before?" asked Dr. Schmidt.

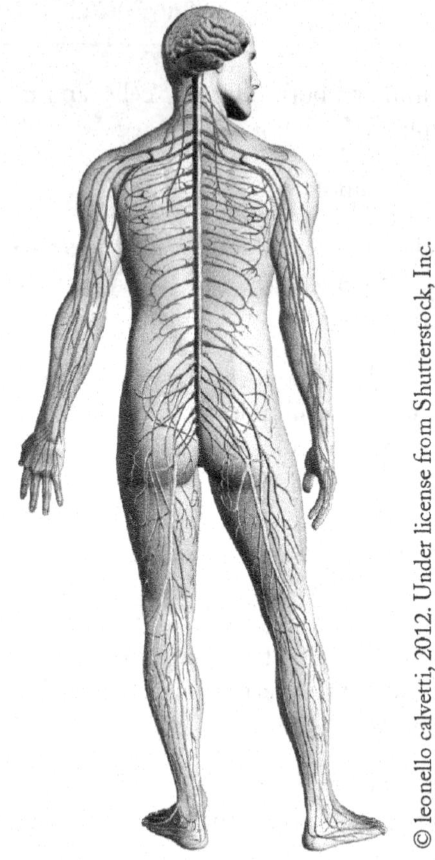

IMAGE 10.20 PNS

THE EYE

Dr. Schmidt explained, "It has been said that our dominant sense is vision. Our eyesight plays a huge role in our lives. It's easy to think of the individual parts of the eye working much like a camera. The **cornea**, behaving much like a lens cover, is the eye's main focusing element. The cornea takes widely diverging rays of light and bends them through the **pupil**, the dark, round opening in the center of the colored **iris**. The iris, a muscle, together with the pupil act like the aperture of a camera.

"Next in line is the **lens** which acts like the lens in a camera, helping to focus light to the back of the eye. The lens is made out of crystalline proteins that makes it hard like the lenses in reading glasses. The lens of your eye is held in place and moved by ciliary muscles (**ciliary body**) which can stretch it out more tightly or allow it to compress for better focusing power. The lens is the part that becomes cloudy and is removed/replaced during cataract surgery."

Dr. Schmidt continued, "The very back of the eye is lined with a layer called the **retina** which acts like film in a non-digital camera. The retina is a membrane containing photoreceptor nerve cells (rods and cones) that line the inside back wall of the eye. **Rods** are responsible for seeing in dim light and for black and white vision. **Cones** are helpful for seeing in bright light and for color vision. The photoreceptor nerve cells of the retina change the light rays into electrical impulses and send them through the **optic nerve** to the brain where an image is perceived. The center 10% of the retina is called the fovea centralis or **fovea**. This is responsible for your sharp vision and your reading vision. The peripheral retina is responsible for the peripheral vision. Your retina has a **blind spot** where there are no rods or cones. As with the camera, if the 'film' is bad in the eye (the retina), no matter how good the rest of the eye is, you will not get a good picture.

"Encircling the outer retina, there is another layer called the choroid. The **choroid** is responsible for supplying blood to the outer retina as well as nourishing the retina and absorbing scattered light. And there is an additional dense, white fibrous layer that along with the cornea forms the outer protective covering of the eyeball called the **sclera** or white of the eye."

Grace commented, "The human eye really is remarkable. It accommodates changing light conditions and focuses light rays originating from various distances from the eye. When all of the components of the eye function properly, light is converted to impulses and conveyed to the brain where an image is perceived."

The Physiology of How We See

Dr. Schmidt continued, "Light is the energy that our eyes detect when we see. When there's no light, we won't be able to see anything even if there is

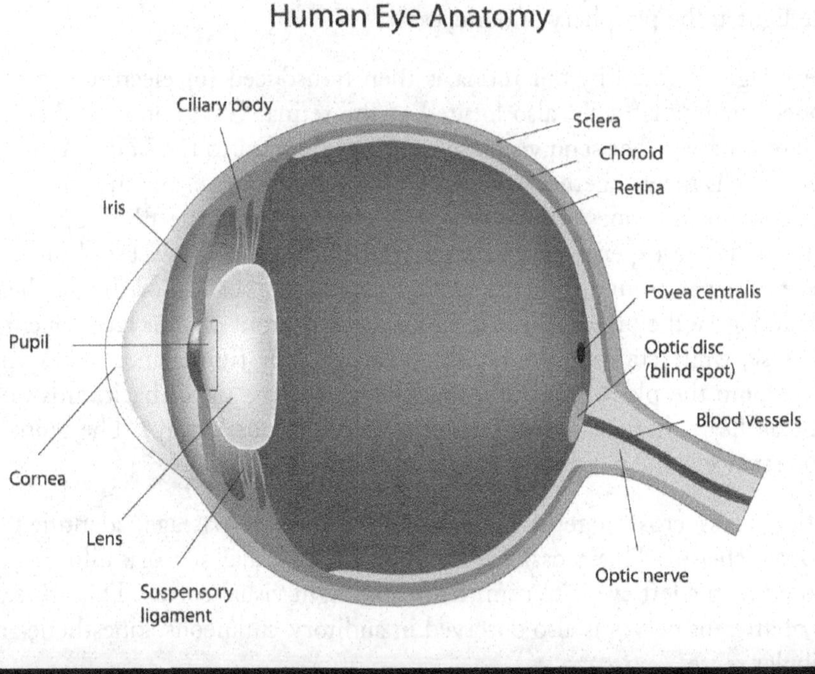

Human Eye Anatomy

IMAGE 10.21 Structures of the Eye

nothing physically wrong with us. Light occurs in waves, where the distance between one peak to another is called wavelength. Light is visible at 400-700 nm wavelength. (1 nm or nanometer is one-billionth of a meter.) Microwaves and infrared rays have longer wavelengths than visible light; while ultraviolet rays, x-rays and gamma rays have shorter wavelengths than visible light.

"Light may differ in color, brightness and purity. Color, or hue, is dependent upon the wavelength of light. Low-frequency light, or light with longer wavelengths, is reddish; while high-frequency light, or light with shorter wavelengths, is bluish. The brightness of light, on the other hand, is measured by the wave's amplitude (or height). High-amplitude light waves are brighter than low-amplitude light waves. Lastly, purity or saturation is measured by the amount of white light added. Pure light has lesser added white light than saturated light. The Color Tree is oftentimes used to demonstrate the different properties of light. Hue moves around the tree; saturation goes outward; and brightness moves upward."

Dr. Scmidt added, "The retina is considered as the primary mechanism of sight, because it is where visual sensory receptors are located. It may be compared to the film of a camera, where images are recorded. The retina contains roughly about 126 million visual sensory receptors of two kinds - rods and cones. Rods are thin and long, functions under low illumination, and can only register black and white images. There are around 120 million rods in the retina of a single eye. Cones are short and fat, function under high illumination, and can register colored images. There are around 6 million cones in the retina of a single eye. The main reason why dogs are said to be color-blind is that they lack cones in their retina. The fovea is a minute area at the center of the retina where most cones reside. It is where we register the best picture of the external world. Because there are too few cones scattered around the fovea, and because there are significantly more rods than cones, we can only register very few cones and white light at the periphery of our eyes.

"The image captured by our retina is then transduced (or electrochemically decoded) by bipolar cells, also located at the retina. Ganglion cells then act as afferent nerves that send visual sensory information to the brain. Axons of ganglion cells meet together and make up the optic nerve. Because the optic nerve contains no cones and no rods, it is also known as the Blind Spot. You can try a simple experiment to locate your blind spot. Draw two hearts on a piece of paper around 3 inches apart from each other. Color the left heart black and leave the other plain. Hold the paper in front of your face, touching your nose, while covering your right eye with your right hand. Focus your left eye onto the plain heart, and gradually move the paper backwards until the black heart disappears from the periphery of your left eye. The point of disappearance is the blind spot. Do the same to the other eye.

"Optic nerves cross together just behind the nasal passage, identified as the optic chiasm. This crossing over shows that visual sensory information recorded by the left eye is transmitted to the right visual cortex. This crossing over of afferent nerves is also observed in auditory, cutaneous, kinesthetic and vestibular senses."

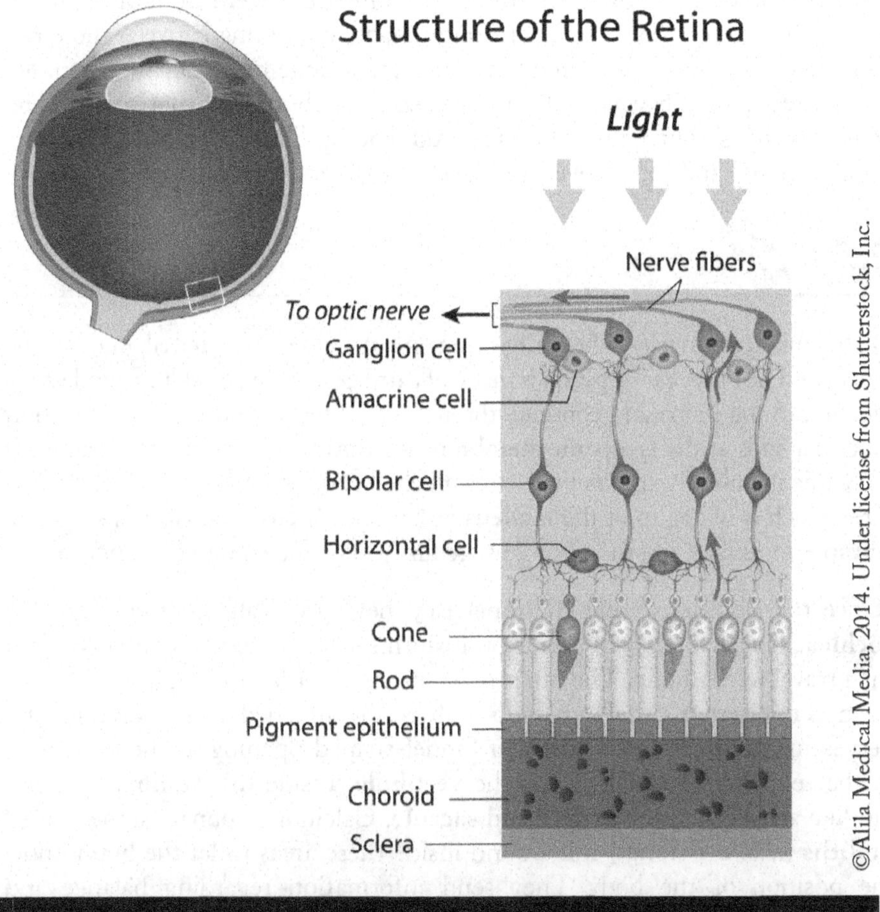

IMAGE 10.22 Rods and Cones in the Retina

Visual Processing in the Brain

She continued, "Visual sensory information is processed in the brain, particularly in the visual cortex, located at the occipital lobe. David Huber and Tolsten Wiesel (1965) won a Nobel Prize for demonstrating that visual stimulus features, such as size, shape, color, movement, lines and angles, are firstly detected in the visual cortex, hence the label of the cortex. Visual sensory information is also shown to move in two different pathways, after being registered in the visual cortex. The "what" pathway is primarily located in the temporal lobe. To illustrate the function of the "what" pathway, consider the case of Dr. P. who is popularly known as "the man who mistook his wife for a hat," thanks to the same-titled 1985 popular book of famous neurologist Oliver Sacks. Dr. P started having trouble recognizing his students until they speak to him, so he consulted an ophthalmologist, but the ophthalmologist found no direct physical problem with his eyes. Dr. P was subsequently referred to neurologist Oliver Sacks, who correctly diagnosed him of having brain damage, particularly in the "what" pathway. Because of this, he had difficulty even discriminating between a glove and a change purse. The "where" pathway, on the other hand, is located at the parietal lobe, where visual stimuli are judged against their relationship with other objects in the visual field, that is, by comparing them from other visible objects three-dimensionally. What is presumably interesting between these two pathways is that they work simultaneously, under same frequency vibration. For us to correctly identify what and where we see things, both the "what" and "where" pathways engage in parallel processing and binding. This means that they work simultaneously and connectively to process visual sensory information."

THE EAR

Dr. Schmidt continued, "The ear has three main areas—the external, middle, and inner ear. The outer ear or **pinna** is made of cartilage covered by skin. Sound waves funnel into the pinna and continue through the **external auditory canal,** a short tube that ends at the **tympanic membrane** (or eardrum). Sound waves cause the tympanic membrane and its three tiny attached ossicles (or bones) in the middle ear—which is made up of the **malleus** or hammer, **incus** or anvil, and **stapes** or stirrup—to vibrate, and the vibrations are further conducted into the inner ear.

"Once the vibrations enter the inner ear, they specifically continue into the **cochlea**. The snail-shaped cochlea transforms sound into nerve impulses that then travel to the brain. The other main structure in the inner ear is the fluid-filled **semicircular canals or ducts** (labyrinth) attached to the cochlea and nerves in the inner ear. There is a funnel-shaped opening at the beginning of the semicircular canals called the **vestibule**. Inside this vestibule are two sac-like areas called the **utricle** and **saccule**. Calcium carbonate stones called **otoliths** (aka: ear stones) roll around inside these areas to let the brain know the position of the body. They send information regarding balance and head position to the brain. The pharyngotympanic tube, also known as the **eustachian (or auditory) tube** drains fluid from the middle ear into the throat (pharynx) behind the nose. The middle ear typically contains only air while the inner ear contains fluid in the semicircular canals called **perilymph**."

Will stated, "So, the inner ear has two functions; the first is hearing and the second is balance."

"Yes, correct" echoed Dr. Schmidt.

She continued, " It has a lot of tubes filled with fluid encased within the temporal bone of the skull. The bony tubes also contain a set of cell membrane lined tubes. The bony tubes are called the bony labyrinth filled with perilymph fluid: the membranous labyrinth tubes are filled with endolymph. This is where the cells responsible for hearing are located (the hairy cells of the organ of Corti).

"The bony labyrinth itself has three sections. 1) The cochlea is responsible for hearing, 2) the semicircular canals have function associated with balance, and 3) the vestibule which connects the two and contains two more balance and equilibrium related structures, the saccule and utricle.

"The final structures of the inner ear are the round window and the eighth cranial nerve (cranial nerve VIII) which is composed of the vestibular nerve (balance) and the cochlear (hearing) nerve."

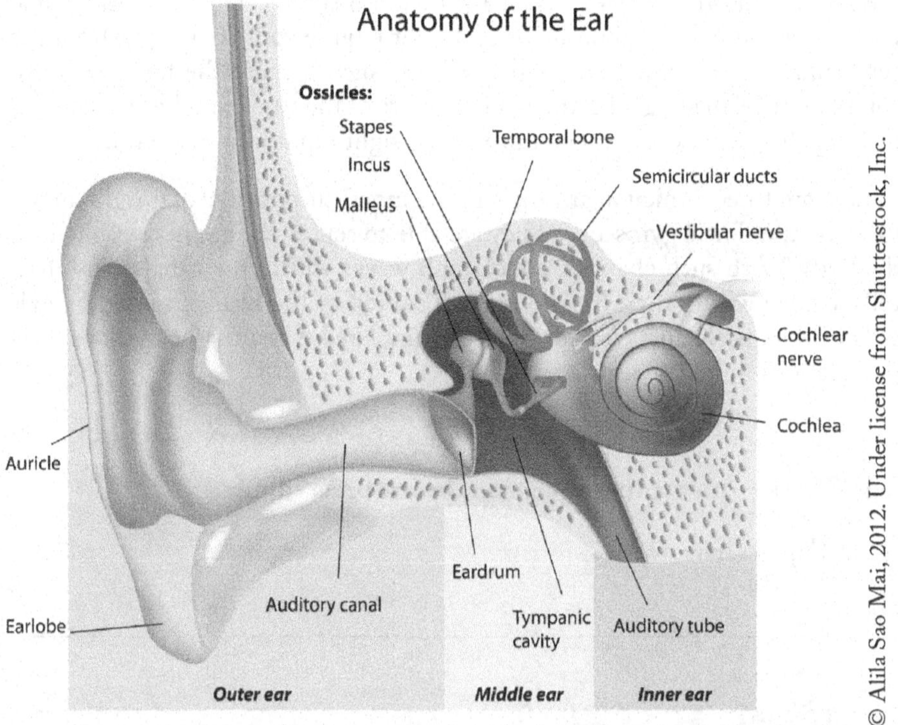

IMAGE 10.23 Structures of the Ear

The Physiology of Hearing

Dr. Schmidt continued, "We hear by funneling sound from the environment into the outer ear and causing the tympanic membrane to vibrate. Those sound wave vibrations are transferred into mechanical vibrations of the ossicles. Those mechanical vibrations cause the oval window to move back and forth causing the perilymph of the inner ear to begin wave-like motions. The perilymph fluid motion is transferred to the endolymph and the wave motion is transformed into electrical impulses picked up by the hairy cells of the organ of Corti and sent to the brain via the cochlear nerve. The round window is responsible for absorbing the fluid wave vibrations and releasing any increased pressure in the inner ear caused by the wave motion."

The Physiology of Balance

Dr. Schmidt continued, "Balance is a choreographed arrangement that takes sensory information from a variety of organs and integrates it to tell the body where it is in relation to gravity and the earth.

"Information from the vestibular system of the inner ear (semicircular canals, the saccule and the utricle) is sent to the brain stem, cerebellum, and spinal cord. Potential balance abnormalities do not require conscious input from the cerebrum of the brain. Abnormal vestibular signals cause the body to try to compensate by making adjustments in posture of the trunk and limbs as well as making changes in eye movement to adjust sight input into the brain.

"There are three semicircular canals in the inner ear positioned at right angles to each other like a gyroscope. They are able to sense changes in movement of the body. With such changes, endolymph waves within the canals cause hair cells located within their base to move. Position of the head is sensed by hair cells of the utricle and saccule which is stimulated when the head moves and the relationship to gravity changes."

Self-Check – Answer these questions before proceeding:

9. Explain how we see.

____

____

____

10. Explain how we hear.

____

____

____

____

____

TASTE

Grace asked, "Is taste part of the PNS also?"

Will commented, "Yeah, how does that work?"

Dr. Schmidt replied, "Well, like vision and hearing, the sense of taste is part of the PNS. The tongue not only detects gustatory (taste) sensations, but also helps sense the texture and temperature that give food flavor. Most people mistake the bumpy structures that cover the tongue's surface for taste buds. But these are actually papillae or small elevations that sometimes contain taste buds and help create contact with food. Taste buds are actually smaller structures, tucked away in the folds between papillae. Every taste bud is made up of cells that help maintain gustatory receptor cells. These specialized gustatory receptor cells are stimulated by the chemical makeup of solutions. They respond to tastes like sweet, salty, bitter, sour, umami (savory), and fat. When a stimulus activates a gustatory cell, the receptor will synapse with neurons and send an electrical impulse to the gustatory region of the cerebral cortex. The brain interprets the sensation as taste. Each gustatory receptor cell has a long, hair-like protrusion called a **gustatory hair** that comes into contact with the outside environment. The hair extends from a small opening or taste pore, and mingles with molecules of food introduced by saliva."

THE PHYSIOLOGY OF TASTE

She continued, "The sense of taste is mediated by taste receptor cells which are bundled in clusters called taste buds. Taste receptor cells sample oral concentrations of a large number of small molecules and report a sensation of taste to centers in the brain stem.

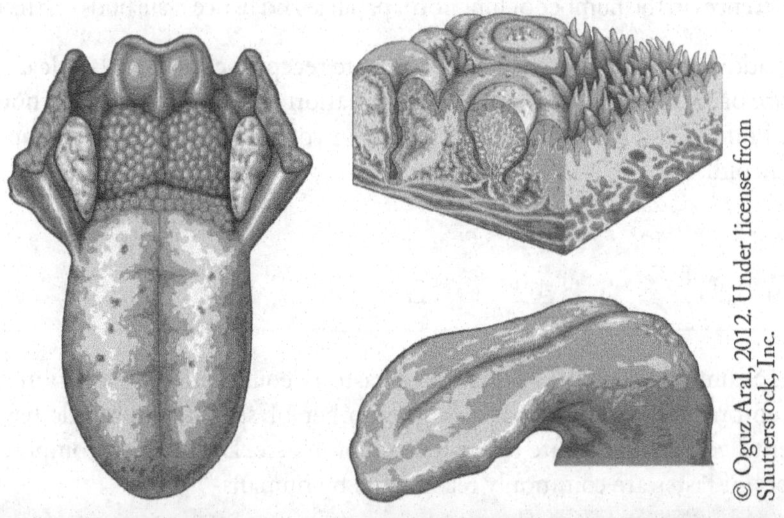

© Oguz Aral, 2012. Under license from Shutterstock, Inc.

IMAGE 10.24 Tongue with Papillae, Taste Buds, and Gustatory Hairs

"In most animals, including humans, taste buds are most prevalent on small pegs of epithelium on the tongue called papillae. The taste buds themselves are too small to see without a microscope, but papillae are readily observed by close inspection of the tongue's surface. To make them even easier to see, put a couple of drops of blue food coloring on the tongue of a loved one, and you'll see a bunch of little pale bumps - mostly fungiform papillae - stand out on a blue background."

"Oh wow - that would be fun to try. So food coloring, eh?" commented Will.

Dr. Schmidt continued, "Taste buds are composed of groups of between 50 and 150 columnar taste receptor cells bundled together like a cluster of bananas. The taste receptor cells within a bud are arranged such that their tips form a small taste pore, and through this pore extend microvilli from the taste cells. The microvilli of the taste cells bear taste receptors.

"Interwoven among the taste cells in a taste bud is a network of dendrites of sensory nerves called "taste nerves." When taste cells are stimulated by binding of chemicals to their receptors, they depolarize and this depolarization is transmitted to the taste nerve fibers resulting in an action potential that is ultimately transmitted to the brain. One interesting aspect of this nerve transmission is that it rapidly adapts - after the initial stimulus, a strong discharge is seen in the taste nerve fibers but within a few seconds, that response diminishes to a steady-state level of much lower amplitude.

"Once taste signals are transmitted to the brain, several efferent neural pathways are activated that are important to digestive function. For example, tasting food is followed rapidly by increased salivation and by low-level secretory activity in the stomach.

"Among humans, there is substantial difference in taste sensitivity. Roughly one in four people is a 'supertaster' that is several times more sensitive to bitter and other tastes than those that taste poorly. Such differences are heritable and reflect differences in the number of fungiform papillae and hence taste buds on the tongue.

"In addition to signal transduction by taste receptor cells, it is also clear that the sense of smell profoundly affects the sensation of taste. Think about how tastes are blunted and sometimes different when your sense of smell is disrupted due to a cold."

TASTE SENSATIONS

Dr. Schmidt opened, "The sense of taste is equivalent to excitation of taste receptors, and receptors for a large number of specific chemicals have been identified that contribute to the reception of taste. Despite this complexity, five types of tastes are commonly recognized by humans."

"Five?" questioned Grace.

"I only learned four previously in school" added Will.

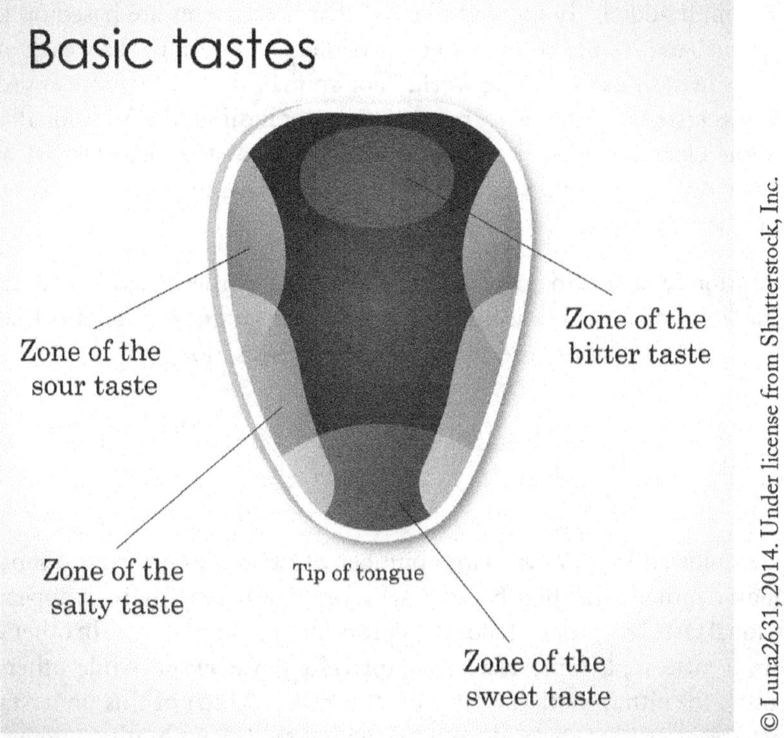

IMAGE 10.25 Zones of Taste

"Yes" answered Dr. Schmidt. She continued, "They are:

> Sweet - usually indicates energy-rich nutrients
> Umami - the taste of monosodium glutamate (e.g., a salty preservative)
> Salty - allows modulating diet for electrolyte balance
> Sour - typically the taste of acids
> Bitter - allows sensing of diverse natural toxins"

"Ok, umami is the one I had not heard of before" said Will.

Dr. Schmidt continued, "None of these tastes are elicited by a single chemical. Also, there are thresholds for detection of taste that differ among chemicals that taste the same. For example, sucrose, 1-propyl-2 amino-4-nitrobenzene and lactose all taste sweet to humans, but the sweet taste is elicited by these chemicals at concentrations of roughly 10 mM, 2 uM and 30 mM respectively - a range of potency of roughly 15,000-fold. Substances sensed as bitter typically have very low thresholds."

Examples of some human thresholds

Taste	Substance	Threshold for tasting
Salty	NaCl	0.01 M
Sour	HCl	0.0009 M
Sweet	Sucrose	0.01 M
Bitter	Quinine	0.000008 M
Umami	Glutamate	0.0007 M

Dr. Schmidt added, "It should be noted that these tastes are based on human sensations and some comparative physiologists caution that each animal probably lives in its own 'taste world.' For animals, it may be more appropriate to discuss tastes as being pleasant, unpleasant or indifferent. Additionally, there are some clear differences among animals in what they can taste. Cats, for example, do not respond to sweets due to a deletion in the gene that encodes one of the sweet receptors.

"Perception of taste also appears to be influenced by thermal stimulation of the tongue. In some people, warming the front of the tongue produces a clear sweet sensation, while cooling leads to a salty or sour sensation."

TASTE RECEPTORS

Dr. Schmidt added, "A very large number of molecules elicit taste sensations through a rather small number of taste receptors. Furthermore, it appears that individual taste receptor cells bear receptors for one type of taste. In other words, within a taste bud, some taste receptor cells sense sweet, while others have receptors for bitter, sour, salty and umami tastes. Much of this understanding of taste receptors has derived from behavioral studies with mice engineered to lack one or more taste receptors.

"The pleasant tastes (sweet and umami) are mediated by a family of three T1R receptors that assemble in pairs. Diverse molecules that lead to a sensation of sweet bind to a receptor formed from T1R2 and T1R3 subunits. Cats have a deletion in the gene for T1R2, explaining their non-responsiveness to sweet tastes. Also, mice engineered to express the human T1R2 protein have a human-like response to different sweet tastes. The receptor formed as a complex of T1R1 and T1R3 binds L-glutamate and L-amino acids, resulting the umami taste.

"The bitter taste results from binding of diverse molecules to a family of about 30 T2R receptors. Sour tasting itself involves activation of a type of TRP (transient receptor potential) channel. Surprisingly, the molecular mechanisms of salt taste reception are poorly characterized relative to the other tastes."

THE ANS

Dr. Schmidt said, "And finally, lets discuss the ANS. Your autonomic nervous system (ANS), is part of the peripheral nervous system, and consists of two divisions: parasympathetic and sympathetic, that work in a complementary manner.

"The ANS is responsible for bodily functions such as digestion, urination, changing the size of blood vessels in order to regulate blood pressure, regulating body temperature and keeping your heart beating.

"Your parasympathetic and sympathetic nerves receive information from your brain down through your spinal cord. This information is then passed on to organs, glands and blood vessels."

PARASYMPATHETIC NERVOUS SYSTEM

Dr. Schmidt continued, "Your parasympathetic division is subdivided into the cranial and sacral sections.

"The cranial section consists of the cranial nerves III, VII, IX, X which are located in the lower part of your brain, your brain stem.

"The sacral section is made up of autonomic nerves from the sacral section of your spinal cord, the S2, S3 and S4 levels.

"The parasympathetic system is responsible for slowing your heart rate, bronchial or air passage constriction, increasing gastric secretions, bladder function (e.g. bladder muscle contraction, release of urine), bowel function, and sexual function (e.g. erectile function and lubrication)."

SYMPATHETIC NERVOUS SYSTEM

She continued, "The sympathetic system consists of nerves that are located in the thoracic and lumbar region of the spinal cord between T1–L2 levels.

"The sympathetic system is responsible for increasing your heart rate, increasing blood pressure, increasing respiratory or breathing rate, regulating your temperature, pupil dilation (enlargement), bronchial or air passage dilation, decreasing gastric secretions, bladder function (e.g. bladder muscle relaxation, storage of urine), and sexual function."

SPINAL REFLEXES

Dr. Schmidt said, "Now lets talk about reflexes. Spinal nerves provide a pathway for reflex activity. Reflexes are fast and automatic responses (e.g. when an area of your skin contacts a hot surface, you quickly withdraw that part of your body away from the heat).

"A variety of reflexes occur through your spinal cord.

"Reflex activity happens between your spinal cord and skeletal muscles via your spinal nerves (e.g. when the tendon of your quadriceps muscle is tapped below the knee joint, your knee jerks upwards).

SPINAL REFLEX ARC

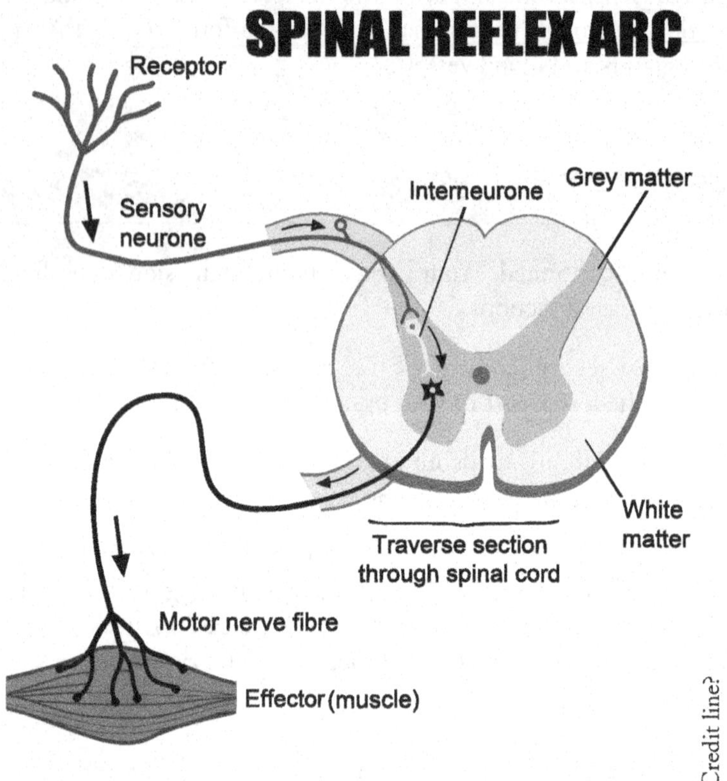

IMAGE 10.24 Reflex Arc

"Reflex activity also happens between the spinal cord and organs and glands through autonomic nerves. These allow the many systems of your body to function without conscious control.

"Reflexes assist movement of food through the digestive tract, in emptying your bladder, in emptying your bowel, in erection of the penis, and in stimulating the flow of secretions from sex glands that lubricate the vagina."

WRAP-UP

Grace looked at her watch thinking time flies when you're learning!

Will commented, "We are not required by our instructor, Mrs. Lanoue, to delve into the ANS, so we want to thank you for your time and help, Dr. Schmidt."

Dr. Schmidt shook both their hands, smiled, and walked away.

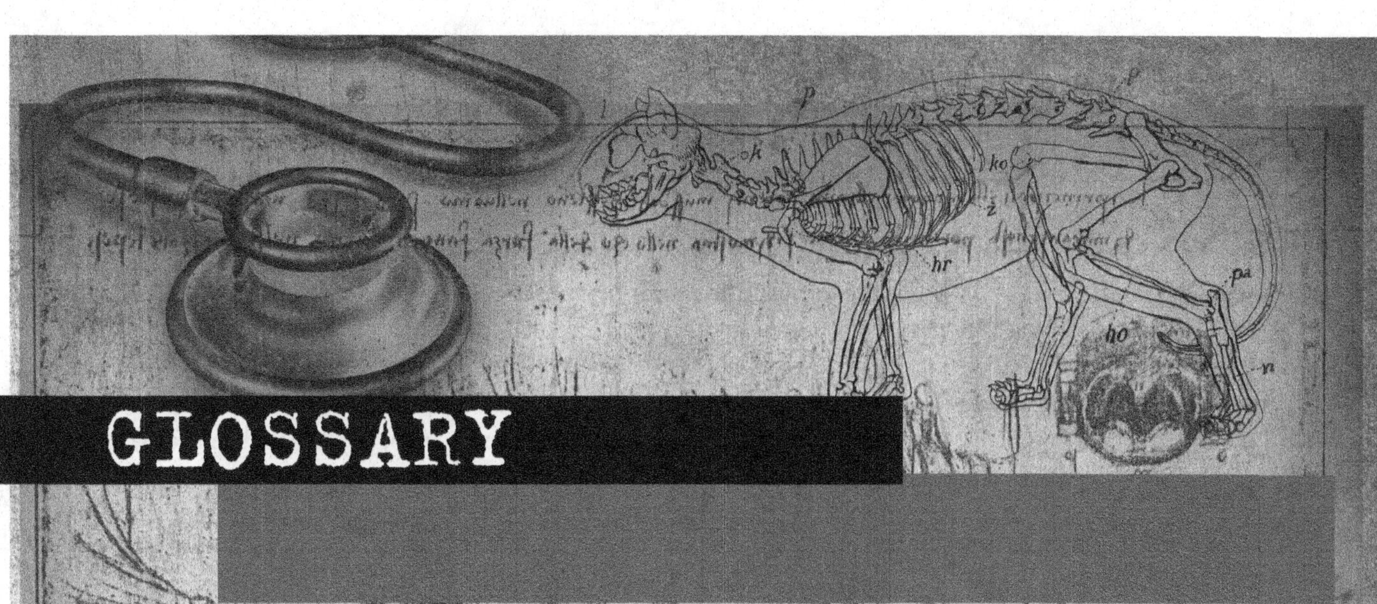

GLOSSARY

abduction the movement of a limb away from the midline or axis of the body

acetabulum the 'indented' hip socket; articulates with the head of the femur

Achilles tendon the common distal tendon of the soleus and gastrocnemius muscles of the leg; it is the thickest and strongest tendon in the body and connects the triceps surae to the heel bone; in an adult, it is about 15 cm long

acid a substance that yields hydrogen ions in solution and from which hydrogen may be displaced by a metal to form a salt; sour

acne a common skin disease characterized by pimples on the face, chest, and back; it occurs when the pores of the skin become clogged with oil, dead skin cells, and bacteria

acromion the region known as the shoulder

adduction the movement of a limb toward the midline or axis of the body

adipose fat tissue

albinism an inherited condition present at birth, characterized by a lack of pigment that normally gives color to the skin, hair, and eyes

amino acid any of various organic acids containing both an amino group and a carboxyl group, especially any of the 20 or more compounds that link together to form proteins

amphiarthroses joints that are known as 'somewhat' or having slight mobility (i.e., the fibrocartilage between vertebrae in the spine)

anabolic reactions in the human body that build, repair, etc. (i.e., the creation of a protein from amino acid molecules)

anaphase phase in mitosis where chromosomes pull away from one another (form chromatids)

anterior the front (or ventral surface) of the human body

apocrine sweat glands a type of large, branched, specialized sweat gland, after puberty producing a viscous secretion that is acted on by bacteria to produce a characteristic acrid odor; among other places found in the groin, hands, feet, and underarms

appendicular skeleton upper limbs, lower limbs, pectoral girdle (shoulder), and pelvis

arrector pili muscle provide insulation whereby they help to retain heat by trapping air between erect hairs; these are tiny muscles attached to hair

articulation where two bones touch; usually also considered a joint (also known as a facet)

associative neurons a nerve cell that is within the central nervous system and that links sensory and motor neurons

astrocytes the most common type of glial cell (neuroglia) in the nervous system

atlas also known as cervical vertebrae one (C1)

atoms the smallest particle of an element with all the properties of the element; it consists of a positively charged nucleus (made up of protons and neutrons) and negatively charged electrons

ATP adenosine triphosphate; supplies large amounts of energy to cells for various biochemical processes, including muscle contraction and sugar metabolism, through its hydrolysis to ADP

auditory tube the narrow channel connecting the middle ear and the nasopharynx; also known as the pharyngotympanic tube (or Eustachian tube)

axial skeleton skull, hyoid bone, vertebral column, thoracic cage (sternum and ribs)

axis the second vertebrae of the neck (C2)

axon the 'tail' of a neuron (nerve cell); conducts impulses

axon terminals the ending of an axon which releases neurotransmitters onto a synaptic space near another neuron, muscle cell, or gland

basal cell carcinoma the most common and treatable form of skin cancer; arises from the stratum basal of the epidermis

base in chemistry, a substance that combines with acids to form salts; a substance that dissociates to give hydroxide ions in aqueous solutions; a substance whose molecule or ion can combine with a proton (hydrogen ion); a substance capable of donating a pair of electrons (to an acid) for the formation of a coordinate covalent bond

biceps brachii the main anterior upper arm muscle

biceps femoris one of the posterior femoral muscles. It has two heads at its origin; the biceps femoris flexes the leg and rotates it laterally and extends the thigh, rotating it laterally; one of the 'hamstrings'

biochemistry the chemistry dealing with living things (i.e., chemical reactions in the body)

bipennate referring to a muscle with a fiber architecture coursing obliquely on both sides of a tendon, as seen in hand muscles

body of a vertebrae ventral, rounded area on a vertebrae in the spine (i.e., as in a lumbar vertebrae)

bone marrow soft bone that produces blood

brachialis a muscle of the upper arm, covering the distal half of the humerus and the anterior part of the elbow joint; it functions to flex the forearm

brachioradialis muscle of posterior (extensor) compartment of forearm; flexes elbow and assists in returning the pronated or supinated limb to the neutral position

brainstem the portion of the brain, consisting of the medulla oblongata, pons, and mesencephalon, that connects the spinal cord to the forebrain and cerebrum

brown fat a type of fat present in newborns and rarely found in adults; brown fat is a unique source of heat energy for the infant because it has greater thermogenic activity than ordinary fat; deposits occur around the kidneys, neck, and upper chest

bursae a fibrous sac between certain tendons and the bones beneath them. Lined with a synovial membrane that secretes synovial fluid, the bursa acts as a small cushion that allows the tendon to move over the bone as it contracts and relaxes

calcaneus 'heel' bone in the foot

canaliculi a very small tube or channel, such as the microscopic haversian canaliculi throughout bone tissue

cancer a large group of almost 100 diseases; its two main characteristics are uncontrolled growth of the cells in the human body and the ability of these cells to migrate from the original site and spread to distant sites; can be fatal

carbohydrate any of a group of organic compounds that includes sugars, starches, celluloses, and gums and serves as a major energy source in the diet of animals; they are produced by photosynthetic plants and contain only carbon, hydrogen, and oxygen, usually in the ratio 1:2:1

cardiac muscle heart muscle

carpal referring to the wrist or wrist bones

cartilaginous joint a joint that has hyaline cartilage (i.e., knee, shoulder)

catabolic a chemical reaction in the human body that breaks down or tears apart something (i.e., digestion)

catalyst a substance, usually used in small amounts relative to the reactants, that modifies and increases the rate of a reaction without being consumed in the process

cell the fundamental, structural, and functional unit of living organisms

cell cycle the cycle of biochemical and morphological events occurring in a reproducing cell population; it consists of the *S phase*, occurring toward the end of interphase, in which DNA is synthesized; the *G2 phase*, a relatively quiescent period; the *M* phase, consisting of the four phases of mitosis; and the *G1 phase* of interphase, which lasts until the *S phase* of the next cycle

cell (plasma) membrane the semipermeable membrane that encloses the cytoplasm of a cell

central canals small tunnels in bone that run 'north and south'; areas for blood vessels and nerves

centrioles either of the two cylindrical organelles located in the centrosome and containing nine triplets of microtubules arrayed around their edges; centrioles migrate to opposite poles of the cell during cell division and serve to organize the spindles; they are capable of independent replication and of migrating to form basal bodies

centromere the most condensed and constricted region of a chromosome to which the spindle fiber is attached during mitosis; also called the *kinetochore*

centrosome a specialized area of condensed cytoplasm containing the centrioles and playing an important part in mitosis

cerebellum the part of the brain involved in coordination of movement, walking, and balance

cerebral hemispheres either of the lateral halves of the cerebrum; a cerebral hemisphere

cerebrospinal fluid the serumlike fluid that circulates through the ventricles of the brain, the cavity of the spinal cord, and the subarachnoid space, functioning as a shock absorber; also called *spinal fluid*

cerebrum the main portion of the brain, occupying the upper part of the cranial cavity; its two hemispheres, united by the corpus callosum, form the largest part of the central nervous system in humans

ceruminous glands wax-secreting glands in the skin of the external auditory canal

cervical vertebrae referring to the seven vertebrae in the neck

channels types of gates that open and close to allow certain materials (like ions) to pass into or out of a cell's membrane

chondrocytes cartilage cells

choroid the dark brown vascular coat of the eye between the sclera and the retina

chromatin the substance of chromosomes, the portion of the cell nucleus that stains with basic dyes

chromosomes contain the genetic information necessary to direct the development and functioning of all cells and systems in the body; they pass on hereditary traits from parents to child (like eye color) and determine whether the child will be male or female

cilia small, hairlike projections that 'sweep' and move materials like mucus across cells (i.e., trachea); structures of motility (movement)

ciliary body the thickened part of the vascular tunic of the eye, connecting the choroid and iris

circular muscle fascicles that form a round shape (i.e., muscle surrounding the eyes, mouth)

circumduction circular movement of a limb or the eye

clavicle collarbone

CNS stands for central nervous system; the brain and spinal cord

coccyx tailbone

cochlea snail-like organ of hearing in the inner ear

collagen any of a family of extracellular, closely related proteins occurring as a major component of connective tissue, giving it strength and flexibility

columnar one of three shapes that epithelial tissue grows; long, tall cells

compact bone hard outer bone

compounds a substance comprised of two or more elements

conductivity the capacity of a body to transmit a flow of electricity or heat; the conductance per unit area of the body

cones help us to see colors, and images in bright light

connective tissues that bind together and comprise the ground substance of the various parts and organs of the body

convergent tending toward a common point

coracoid process a curved process arising from the upper neck of the scapula and overhanging the shoulder joint

cornea the transparent anterior part of the eye

coronal suture the line of junction of the frontal bone with the two parietal bones in the skull

corpus callosum an arched mass of white matter in the depths of the longitudinal fissure, composed of transverse fibers connecting the cerebral hemispheres

covalent relating to a chemical bond characterized by one or more pairs of shared electrons

crista galli the very tip (point) of the ethmoid bone; best seen from inside the skull

cuboid bone a basic bone shape that is square; as in the wrist bones

cutaneous glands any of the skin glands

cuticle cortex the strip of hardened skin at the base and sides of a fingernail or toenail

cyanosis bluish color; typically caused by lack of oxygen

cytokinesis the 'pinching in' (division) of the cytoplasm of a cell following the division of the nucleus; happens during mitosis

cytoplasm the protoplasm outside a cell nucleus

cytoskeleton the internal framework of a cell, composed largely of actin filaments and microtubules

daughter cells one formed by division of a mother cell

decomposition reaction (chemistry) separation of a substance into two or more substances that may differ from each other and from the original substance

deep found deep within the body (i.e., the heart, the lungs rather than the skin)

deltoid shoulder muscles

dendrites treelike branches on a neuron

dens bulgelike process unique to C2 in the neck

dense connective forms strong, ropelike structures such as tendons and ligaments

depression a scooped-out area (i.e., hole, dip, rut)

dermatitis eczema a form of chronic inflammation of the skin

diaphysis the shaft of a long bone

diarthroses the most common and most movable type of joint in the body; a synovial joint

diploe spongy bone structure (or tissue) of the internal part of short, irregular, and flat bones

disaccharide two monosaccharide together (i.e., sucrose)

distal far away from the joint

division to divide

DNA deoxyribonucleic acid; a molecule that encodes the genetic instructions used in the development and functioning of all known living organisms and many viruses

DNA polymerase synthesizes DNA molecules from their nucleotide blocks; DNA polymerases are essential for DNA replication, and usually function in pairs while copying one double-stranded DNA molecule into two double-stranded DNAs in a process termed semiconservative replication

dorsal posterior

dorsiflexion a movement where you point the toes upward

eccrine sweat glands controlled by the sympathetic nervous system, regulates body temperature; when internal temperature rises, the eccrine glands secrete water to the skin surface, where heat is removed by evaporation

Ehlers-Danlos syndrome an inherited syndrome; abnormal collagen renders structures like skin, joints, muscles, ligaments, blood vessels, and visceral organs more elastic; depending on the severity of the gene mutations, it can be lethal

eicosanoid in biochemistry, signaling molecules made by oxidation of 20-carbon fatty acids; they exert complex control over many bodily systems, mainly in inflammation, and as messengers in the CNS

elastic cartilage the most elastic of the three cartilage types (i.e., found in the ear lobes, epiglottis)

electrochemical combination of both electrical impulses and chemicals (like neurotransmitters); the basis for how our nervous system operates

electrolytes a compound that ionizes when dissolved in suitable ionizing solvents such as water (i.e., dissolved salts in solution)

electrons a subatomic particle with a negative charge

element a substance that is pure; all the same type of atoms

elevation higher than normal; found above

endergonic the system absorbs energy from the surroundings; as a result, during an endergonic process, energy is *put into* the system

endomysium a wispy layer of areolar connective tissue that ensheaths each individual muscle fiber. It also contains capillaries, nerves, and lymphatics; it overlies the muscle fiber's cell membrane

endoplasmic reticulum (ER) organelle in eukaryotic cells; two types: smooth and rough; the rough is studded with ribosomes (sites of protein synthesis); the smooth has no ribosomes and is concerned with lipid metabolism, carbohydrate metabolism, and detoxification

endosteum thin layer of connective tissue that lines the surface of the bony tissue that forms the medullary cavity of long bones

enzyme a substance that acts as a catalyst in living organisms, regulating the rate at which chemical reactions proceed without itself being altered in the process

ependymal cells a type of neuronal support cell (neuroglia) that forms the epithelial lining of the ventricles (cavities) in the brain and the central canal of the spinal cord

epicranius occipitofrontalis muscle

epidermis upper region of the skin; superficial

epimysium the external connective-tissue sheath of a muscle

epiphyseal line the line marking the site of the epiphyseal plate

epiphysis the end(s) of a long bone

epithelial refers to layers of cells that line hollow organs and glands; it also is those that make up the outer surface of the body; help to protect or enclose organs. Most produce mucus or other secretions

erythema redness of the skin; can be due to superficial capillaries

erythrocytes red blood cells

ethmoid bone bone that separates the nasal cavity from the brain

eversion turning the sole of the foot outward and upward

exchange reaction are those in which cations and anions that were partners in the reactants are interchanged in the products; the products must remain electrically neutral

excitability neurons respond to stimuli (i.e., electricity)

exergonic relating to a chemical reaction that releases energy; the breakdown of sugars by cellular metabolism is an exergonic reaction that releases energy used to drive endergonic reactions

extension a physical position that decreases the angle between the bones of the limb at a joint

extensor carpi radialis longus one of the five main muscles that control movements at the wrist

external auditory canal opening to the ear

external oblique the largest and most superficial muscle on the side of the body; one of the three flat muscles of the lateral anterior abdomen

extracellular matrix part of multicellular structure (e.g., organisms, tissues) that typically provides structural and biochemical support to the surrounding cells

false ribs a rib that does not attach directly to the sternum; the upper three false ribs connect to the costal cartilages of the ribs just above them. The last two false ribs usually have no ventral attachment to anchor them in front and so are called floating, fluctuating, or vertebral ribs

fascicle the shape of a muscle (round, convergent, etc.)

fatty acid an essential aspect to human health, but cannot be manufactured in the body

femur upper thing bone

fibroblasts a type of cell found in connective tissue throughout the body that produces collagen and other proteins found in the extracellular spaces

fibrocartilage the strongest type of cartilage; found in the discs between the vertebrae in the spine

fibrous connective tissue made up of high-strength, slightly stretchy fibers; these fibers consist mainly of collagen, water, and complex strands of polysaccharides; they provide support and shock absorption to surrounding organs and bones

fibrous joint a type of synarthrosis; the bones are united by continuous intervening fibrous tissue

fibula the thinnest of the lower leg bones (located laterally)

fibularis longus also referred to as the peroneus longus; a muscle inside the lateral area of the human leg that everts and flexes the ankle

first gap phase (G1) the first of four phases of the cell cycle that takes place in eukaryotic cell division

first–third degree burns a scale to evaluate burned patients (third degree is the most severe)

flagella a whiplike structure used for motility

flat bones a type of bone classification by shape; referring to skull bones (flat) and breastbone (flat)

flexion a bending movement around a joint in a limb (as the knee or elbow) that decreases the angle between the bones of the limb at the joint

flexor carpi radialis a muscle of the human forearm that acts to flex and (radial) abduct the hand

flexor carpi ulnaris a muscle of the human forearm that acts to flex and adduct the hand

floating ribs a rib that is attached to the spine, but not the sternum; there are two pairs

follicle a small spherical group of cells containing a cavity, such as a hair follicle

foramen a bone marking that presents as a hole through a bone (i.e., the magnum foramen at the base of the skull is the largest foramen in the skull)

fovea a small rodless area of the retina that affords acute vision

freckles accumulations of melanin

free radicals an especially reactive atom or group of atoms that has one or more unpaired electrons; one that is produced in the body by natural biological processes or introduced from an outside source (as tobacco smoke, toxins, or pollutants) and that can damage cells, proteins, and DNA by altering their chemical structure

frontal bone bone of the forehead

frontalis muscle covering the frontal bone

frontal lobe front lobe of the brain

fusiform a muscle shape tapering toward each end

galea aponeurotica top of the scalp

gastrocnemius calve muscle

gates openings (pores) in a cell membrane that allow substances to enter or leave a cell

gene a specific sequence of nucleotides in DNA or RNA that is located usually on a chromosome and that is the functional unit of inheritance controlling the transmission and expression of one or more traits by specifying the structure of a particular polypeptide and especially a protein or controlling the function of other genetic material

genome the genetic material of an organism. It is encoded either in DNA or, for many types of viruses, in RNA; the genome includes both the genes and the non-coding sequences of the DNA/RNA

genomic medicine an emerging field of medicine that deals with understanding the predictive power of patients' genomes and the products of those genomes; it should be possible to identify individuals at risk of disease and to create smarter, more effective treatments for those who are already ill

glenoid fossa a part of the shoulder; a shallow articular surface, which is located on the lateral angle of the scapula. It is directed laterally and forward and articulates with the head of the humerus; it is broader below than above and its vertical diameter is the longest

glial cells cells in the nervous system that support and nourish neurons; also known as neuroglia

gluteus maximus the largest and most superficial of the three gluteal muscles; the buttocks

gluteus medius smaller of the three gluteal muscles; located high on the hip

glycemic index a measurement carried out on carbohydrate-containing foods and their impact on our blood sugar

glycerol simple polyol compound. It is a colorless, odorless, viscous liquid that is widely used in pharmaceutical formulations. Glycerol has three hydroxyl groups that are responsible for its solubility in water and its hygroscopic nature

glycocalyx an extracellular polymeric material produced by some bacteria, epithelia, and other cells. The slime on the outside of a fish is an example of glycocalyx; it is unique to each individual

goblet cells columnar epithelial cells

Golgi apparatus considered the distribution and shipping department for the cell's chemical products; it modifies proteins and lipids (fats) that have been built in the endoplasmic reticulum and prepares them for export outside of the cell or for transport to other locations in the cell. Proteins and lipids built in the smooth and rough endoplasmic reticulum bud off in tiny bubblelike vesicles that move through the cytoplasm until they reach the Golgi complex. The vesicles fuse with the Golgi membranes and release their internally stored molecules into the organelle

gomphoses a fibrous, mobile peg-and-socket joint. The roots of the teeth (the pegs) fit into their sockets in the mandible and maxilla and are the only examples of this type of joint

gracilis a muscle in the inner thigh

greater wings part of the sphenoid bone

gyri the crest of a single convolution on the brain's cerebral cortex

hair bulb contains the hair follicle

hair follicle grows hair

hair root anchor for the hair

hair shaft tubelike area where the hair pushes through the scalp

hemangiomas small benign tumor made up of new capillaries near the skin's surface

hematoma Large, thin-walled venous sinuses lie within the cranial cavity. While they are thus protected by the cranium, in many places they are so close beneath the bones that a fracture or a penetrating wound may tear the sinus wall and lead to bleeding. The blood frequently is trapped beneath the outermost and toughest brain covering, the dura mater, in a mass called a subdural hematoma

histology the study of cells and tissues

homeostasis the body's ability to maintain a stable internal environment

Human Genome Project the study of human genetics; mapping of all human chromosomes

humerus upper arm bone

hyaline cartilage most common of the cartilage types; usually covers the ends of long bones

hydrogen a pure element

hydrophilic a substance with an attraction for water

hydrophobic having little or no affinity for water

hyperextension extension of a body part beyond normal extension

hypodermis the region below the dermis of the skin; contains a high percentage of adipose tissue

hypothalamus directs a multitude of important functions in the body. It is the control center for many autonomic functions of the PNS; connections with structures of the endocrine and nervous systems enable the hypothalamus to play a vital role in maintaining homeostasis. For example, blood vessel connections between the hypothalamus and pituitary allow hypothalamic hormones to control pituitary hormone secretion. As a limbic system structure, the hypothalamus also influences various emotional responses

iliac crest prominent ridgelike bone marking on ilium

iliopsoas major thigh muscle

iliotibial tract large tendon holding down gluteus maximus to lateral leg

ilium top bone of the pelvis; shaped like an ear

incus second bone in the ear

inferior below

inorganic in chemistry, not containing carbon

integumentary system body system dealing with hair, skin, nails

intercalated discs areas where cardiac muscle cells (among a few other cell types) are joined together. These discs contain 'gap junctions,' which are small pores that connect the cytoplasm of one cell to the next. In other words, the membranes of the cells are fused together, and there are tiny holes that connect the interior of the adjacent cells

intermediate between two points

intermediate cuneiform bone in the foot

interphase first portion of the cell cycle; DNA replicates itself

inversion pointing the sole of the foot medially

ionic bond one that involves the electrostatic attraction between oppositely charged ions

ions an atom or molecule in which the total number of electrons is not equal to the total number of protons, giving the atom a net positive or negative electrical charge

irregular bones a type of bone classification by shape; odd-shaped bones that defy description (e.g., pelvis)

ischium lower bones of the pelvis; the ones that look like stirrups

jaundice yellowing of the skin

keratin protein

keratinocytes cells containing protein

keratinohyaline cells that contain both keratin and hyaline cartilage

kinetochore same as a centromere; see *centromere*

lacrimal bone tear bone near the eye

lacunae areas in bone where osteocytes live

lambdoid suture immovable joint in the skull between the occipital bone and the parietal bones

lamellae a thin layer, plate, or membrane, especially any of the calcified layers of which bone is formed

Langerhans cells dendritic cells (antigen-presenting immune cells) of the skin and mucosa

lanugo very fine, soft, and usually unpigmented, downy hair as can be found on the body of a fetus or newborn baby

lateral toward the sides of the body

lateral cuneiform a foot bone

latissimus dorsi 'lats'; large back muscles

lens hard, crystalline piece that focuses light in the eye

leukocytes white blood cells

levator labii superioris a muscle whose purpose is to dilate the nostril and elevate the upper lip

ligament binds bone directly to bone

lipid fat (adipose)

long bones a type of bone classification by shape; bones that are longer than they are wide (e.g., humerus and femur)

longitudinal fissure elongated bone marking that looks like a deep 'crack'

loose connective a category of connective tissue which includes areolar tissue, reticular tissue, and adipose tissue

lumbar vertebrae five vertebrae in the lower back

lysosome a cell organelle that contains acid to digest old cells (those that lyse) or destroy bacteria and unwanted agents

macrophages white blood cells that are mobile warriors; chase, surround, and engulf harmful organisms

malleus the first bone of the middle ear; the hammer

mammary glands glands in the breast; in females they secrete milk to nurse the young

mandible bone lower jaw

masseter a muscle that runs along the 'sideburn' area of the face; assists with chewing

mast cells mediate inflammatory responses such as hypersensitivity and allergic reactions

mastoid process large process projecting down behind the ear; conical prominence projecting from the undersurface of the mastoid portion of the temporal bone

maxilla bone upper jaw

medial tending toward the midline of the body

medial cuneiform a foot bone

medulla a general term that refers to the center of an organ

medulla oblongata the lower half of the brainstem, which is continuous with the spinal cord

Meissner's corpuscles a type of mechanoreceptor; a type of nerve ending in the skin that is responsible for sensitivity to light touch

melanin a natural substance that gives color (pigment) to hair, skin, and the iris of the eye. It is produced by cells in the skin called melanocytes

melanocytes cells that contain melanin

melanoma a very serious type of skin cancer; often resistant to chemotherapy and radiation

Merkel cell a sensory touch function within the skin; highly concentrated in the lips, fingertips, etc.

messenger RNA a large family of RNA molecules that convey genetic information from DNA to the ribosome, where they specify the amino acid sequence of the protein products of gene expression

metacarpals the five bones of the hand

metaphase a phase in mitosis where chromosomes are lined up down the middle of the cell

metastasis the spreading of cancer

metatarsals the five bones of the middle foot

microglia also called glial cells; they provide support, nourishment, and protection to neurons

microvilli small projections protruding off the walls of villi in the small intestine; increase nutrient absorption

mineralization the hardening process of bones

mitochondria cell organelles that make ATP

mitosis a part of the cell cycle which includes prophase, metaphase, anaphase, telophase, and cytokinesis; results in two daughter cells

molecules an electrically neutral group of two or more atoms held together by chemical bonds

moles clusters of melanocytes

monosaccharide a simple sugar (i.e., like glucose)

monounsaturated a heart-healthy type of fat (i.e., olive oil, avocado)

motor (efferent) division motor nervous system cells carry information from the CNS to organs, muscles, and glands; the motor nervous system is divided into the somatic nervous system and the autonomic nervous system

motor neurons efferent neurons that originate in the spinal cord and synapse with muscle fibers to facilitate muscle movement and response

multipennate a type of pennate muscle wherein the diagonal muscle fibers are in multiple rows with the central tendon branching into two or more tendons

muscle the body's 'engine' (i.e., skeletal muscle for movement; smooth muscle for digestion, vasodilatation, and vasoconstrictions; cardiac muscle to keep our hearts beating)

muscle fibers cylindrical, multinucleate cell composed of numerous myofibrils that contracts when stimulated

mutations a change of the nucleotide sequence of the genome of an organism, virus, or extrachromosomal genetic element; some are harmful, many are not

myelin sheath an insulating covering of fat (i.e., found on the tails of neurons)

myocytes muscle cells

nasal bone bone of the nose (i.e., near bridge of nose)

navicular one of the tarsal bones (in the ankle)

negative feedback occurs when the result of a process influences the operation of the process itself in such a way as to reduce changes

nervous pertaining to the nervous system in the human body

neurobiology the special study of biology focused on the nervous system

neuroglia glial cells

neurons an electrically excitable cell that processes and transmits information through electrical and chemical signals

neurotransmitters endogenous chemicals that transmit signals across a synapse from one neuron to another 'target' neuron

neutrons particles within the nucleus of an atom that carry a neutral charge

nodes of Ranvier a gap occurring at regular intervals between segments of myelin sheath along a nerve axon

nuclear envelope the double-lipid bilayer membrane that surrounds the genetic; also known as the nuclear membrane

nucleic acid naturally occurring chemical compound that is capable of being broken down to yield phosphoric acid, sugars, and a mixture of organic bases (purines and pyrimidines); nucleic acids are the main information-carrying molecules of the cell, and, by directing the process of protein synthesis, they determine the inherited characteristics of every living thing. The two main classes of nucleic acids are deoxyribonucleic acid (DNA) and ribonucleic acid (RNA)

nucleoli dark, spherical area within the nucleus; serves as the site of protein synthesis

nucleosomes a section of DNA that is wrapped around a core of proteins

nucleotides any member of a class of organic compounds in which the molecular structure comprises a nitrogen-containing unit (base) linked to a sugar and a phosphate group

nucleus an organelle that is the command center of a cell; contains DNA

obturator a large foramen located in the ischium

occipitalis the posterior belly of the occipitofrontalis that arises from the lateral two-thirds of the superior nuchal lines and from the mastoid part of the temporal bone, inserts into the galea aponeurotica, and acts to move the scalp

occipital lobe the lobe of the brain located at the posterior (back) of the skull

odd-shape bones or irregular bones; a type of bone classification by shape; little bones, squarish in shape (e.g., wrist and ankle bones)

oligodendrocytes a type of glial cell that provides support and insulation to neurons

opposition moving your thumb diagonally across your palm to get to the other fingers

optic nerve a nerve that carries information from the eye to the brain

orbicularis oculi a round fascicle muscle surrounding the eye; allows the eye to squint in bright light

orbicularis oris a round fascicle muscle surrounding the mouth; allows the mouth to pucker among other things

organelles cell parts (i.e., structures)

organic in biochemistry, it means 'containing carbon'

osteoblasts responsible for bone formation

osteoclasts responsible for the reabsorption of bone

osteocytes bone cells

osteon (Haversian system) central canal and the concentric osseous lamellae encircling it, occurring in compact bone; also called *Haversian system*

osteoporosis a degenerative disease where bone loses its minerals (primarily calcium) and becomes hollowed out and brittle

otoliths tiny 'rocks' of calcium carbonate found in the saccule and utricle of the inner ear

pacinian corpuscles cutaneous mechanoreceptors that sense pressure and stretch

Paget's disease disrupts the body's normal bone recycling process, in which old bone tissue is gradually replaced with new bone tissue. Over time, the affected bones may become fragile and misshapen. Paget's disease of bone most commonly occurs in the pelvis, skull, spine, and legs

pallor a very 'washed out' white look

papillary layer top layer within the dermis; superficial to the reticular layer

parallel muscles that run side by side

parasympathetic division the part of the autonomic nervous system originating in the brainstem and the lower part of the spinal cord that, in general, inhibits or opposes the physiological effects of the sympathetic nervous system, as in tending to stimulate digestive secretions, slow the heart, constrict the pupils, and dilate blood vessels

parietal bone two primary bones on the top of the skull

parietal lobe the upper middle lobe of each cerebral hemisphere, located above the temporal lobe. Complex sensory information from the body is processed in the parietal lobe, which also controls the ability to understand language

parietal serosa most of the organs within the ventral cavity are lined by a pair of membranes known as the *serous membranes*. In each case, the outer membrane (known as the *parietal membrane* or *parietal serosa*) lines the cavity within which the organ is contained. For example, the *parietal pericardium* (or *parietal pericardial membrane*) lines the pericardial cavity

patella also known as the knee cap

pectineus a flat, quadrangular muscle, situated at the front part of the upper and middle part of the thigh

pectoral girdle the shoulder (i.e., clavicle and scapula)

pectoralis major large muscle in the upper chest

pedicles thick, stalklike pieces on the lateral sides of vertebrae

pelvic girdle the hip (ilium, ischium, etc.)

perilymph a thin liquid that fills the space within the bony labyrinth surrounding the membranous labyrinth

perimysium the connective tissue surrounding bundles of skeletal muscle fibers

periosteum the outer membrane of compact bone

peroxisome a cell organelle that contains peroxides that handle formaldehydes, etc.

pH a scale that serves as a measure to determine whether a substance is an acid or base; 7 is neutral (like water); 0–6 indicates an acid; 8–14 indicates a base

phalanges digits (i.e., fingers, toes)

phalanx a single bone piece in the digits (i.e., the thumb has two phalanxes, the other fingers have three phalanxes

phospholipids

pineal gland also known as the epithalamus; center that helps to regulate sleep; secretes the hormone melatonin

pinna fleshy outer ear

pituitary gland the master gland; located in the sella tursica of the sphenoid bone within the skull

plantar flexion movement of the foot that flexes the foot or toes downward toward the sole

platelets small fragment-type cells that clot blood

platysma muscle under the chin and neck; helps open and close mouth

PNS peripheral nervous system

polypeptide chain a chain of amino acids held together by peptide bonds

polysaccharide a long chain of monosaccharides; examples include cellulose, glycogen, and starch

pons part of the brainstem that links the medulla oblongata and the thalamus

positive feedback a process in which an initial change will bring about an additional change in the same direction

posterior dorsal surface

primary proteins the most basic of protein structures; like beads on a necklace

pronation rotation of the hand and forearm so that the palm faces backwards or downwards

prone laying flat or prostrate

prophase first stage of mitosis; chromatin condenses and uncoils; nucleus disappears

protein a macromolecule composed of amino acids

protons a very small particle of matter that is part of the nucleus of an atom and that has a positive electrical charge

protraction the act of moving an anatomical part forward (i.e., project the lower mandible forward)

proximal on a limb, the area closest to the origin of the joint

pseudostratified appearing as two distinct layers when it is actually one layer

psoriasis a chronic skin disease characterized by circumscribed red patches covered with white scales

pubis anterior bone of the pelvis

pupil small opening that light passes through in the eye

quaternary protein the arrangement of a multiple folded protein or coiling protein molecules in a multi-subunit complex

radius one of two bones in the lower forearm; the bone that rotates over the other

range of motion (ROM) the range of movement; how a joint normally moves

rectus abdominis abdominal muscles

rectus femoris anterior quadriceps muscle; thigh area

remodeling normal bone repair and growth

reticular layer part of the dermis made up of irregular connective tissue; this layer of the skin gives the skin its overall strength and elasticity

reticular tissue irregular connective tissue

retina the photoreceptor layer of the eye; contains rods and cones

retraction pulling something inward

ribosomal RNA the RNA component of the ribosome, and is essential for protein synthesis in all living organisms

ribosome tiny organelle that is present in large numbers in all living cells and serves as the site of protein synthesis

rickets lack of calcium in infants and toddlers resulting in softened bones

RNA ribonucleic acid

rods photoreceptor cells; allow us to see in dim light; responsible for black and white vision

rosacea a chronic skin condition that makes your face turn red and may cause swelling and skin sores that look like acne

rough endoplasmic reticulum (ER) involved in some protein production, protein folding, quality control and dispatch; it is called 'rough' because it is studded with ribosomes

saccule one of two chambers in the inner ear; the smaller chamber of the membranous labyrinth of the ear

sacrum posterior bone of the pelvis; forms rear of pelvis

sagittal suture a fibrous and immovable interlocking joint between the right and left parietal bones of the skull

satellite cells glial cells that cover the surface of nerve cell bodies in sensory, sympathetic, and parasympathetic ganglia

scapula shoulder blade

Schwann cells supporting cells of the PNS; wrap themselves around nerve axons

sclera the membrane that is the 'white' of the eye

sebaceous glands oil glands in the skin

sebum oil secreted from sebaceous glands of the skin; especially in the face region

secondary proteins there are two types of secondary structures observed in proteins. One type is the alpha (α) helix structure; this structure resembles a coiled spring and is secured by hydrogen bonding in the polypeptide chain. The second type of secondary structure in proteins is the beta (β) pleated sheet. This structure appears to be folded or pleated and is held together by hydrogen bonding between polypeptide units of the folded chain that lie adjacent to one another

second gap phase (G2) the period after DNA synthesis has occurred but prior to the start of prophase; the cell synthesizes proteins and continues to increase in size. (Note that the G in G2 represents gap and the 2 represents second, so the G2 phase is the second gap phase.)

secretion the giving forth of a product (mucus, sweat, etc.)

sella tursica a protective cradle in the sphenoid bone that contains the pituitary gland

semicircular canals structures in the inner ear that assist with balance

semimembranosus one of the hamstrings (i.e., posterior knee)

semitendinosus one of the hamstrings (i.e., posterior knee)

sensory (afferent) division a division of the PNS that acquires information from the external and internal environments of the body, and transmits impulses from sense organs to the CNS

sensory neurons nerve cells that transmit sensory information (sight, sound, feeling, etc.). They are activated by sensory input, and send projections to other elements of the nervous system, ultimately conveying sensory information to the brain or spinal cord

Sharpey's fibers a matrix of tiny 'bone fibers'; a type of connective tissue consisting of bundles of strong collagenous fibers connecting the outer membrane of compact bone (the periosteum) to the rest of the bone

simple epithelial one of the three most basic tissue types found in the body; grows in sheets

sister chromatid are generated when a single chromosome is replicated into two copies of itself, these copies being called sister chromatids

skeletal muscle muscle that covers bones; allows movement; under voluntary control

skeletal system system that includes bones, cartilages, and ligaments

smooth endoplasmic reticulum (ER)

smooth muscle involuntary muscle; located in many places within the human body including the small intestine, lining of blood vessels

soleus a broad muscle in the lower calf, below the gastrocnemius, that flexes the foot to point the toes downward

somatic motor division the part of the PNS associated with the voluntary control of body movements via skeletal muscles; it consists of afferent and efferent nerves. Afferent nerves are responsible for relaying sensation to the central nervous system; efferent nerves are responsible for stimulating muscle contraction, including all the non-sensory neurons connected with skeletal muscles and skin

sphenoid bone a compound bone that forms the base of the cranium, behind the eye and below the front part of the brain; it has two pairs of broad lateral "wings" and a number of other projections, and contains two air-filled sinuses

spinous process a bone marking that resembles a sharp spine or dagger (i.e., the spinous process on vertebrae)

spongy bone the soft bone marrow inside compact bone

spray-on skin a solution of a person's stem cells that are sprayed on third degree burns; heals skin without scarring in about 3 days; administered at a hospital's burn unit

squamous cell carcinoma a type of skin cancer; curable with surgical removal if caught early; can spread and become lethal if not treated early

squamous cuboidal a common type of epithelial tissue; cells are shaped like small cubes ('box cars'); tissue from the ovary is an example of where this tissue type is found in the body

squamous suture joint/'seam' in the skull that occurs between the parietal and temporal bones

stapes the third ear bone of the middle ear; also known as the stirrup

sternocleidomastoid each of a pair of long muscles that connect the sternum, clavicle, and mastoid process of the temporal bone and serve to turn and nod the head

sternohyoid this muscle's origin is from the episternum and the first costal cartilage, with insertion into the hyoid bone, with nerve supply from the upper cervical nerve through the cervical ansa, and whose action depresses the hyoid bone

steroid a natural cholesterol-based hormone produced by the body; examples include progesterone, estrogen, testosterone

stratified columnar epithelial composed of column-shaped cells arranged in multiple layers; found in the ocular conjunctiva of the eye, in parts of the pharynx and anus, the female's uterus, the male urethra and vas deferens

stratified cuboidal epithelial composed of multiple layers of cube-shaped cells; only the most superficial layer is made up of cuboidal cells, and the other layers can be cells of other types

stratified epithelial consists of squamous (flattened) epithelial cells arranged in layers upon a basal membrane. Only one layer is in contact with the basement membrane; the other layers adhere to one another to maintain structural integrity

stratified squamous epithelial consists of squamous (flattened) epithelial cells arranged in layers upon a basal membrane. Only one layer is in contact with the basement membrane; the other layers adhere to one another to maintain structural integrity

stratum basale 'bottom' or basal layer of the epidermis; it is an active mitotic layer; youngest layer of the epidermis

stratum corneum 'topmost', horny layer of the epidermis; oldest layer; it is dead and typically sheds off

stratum granulosum a very 'grainy' appearing layer within the epidermis; found immediately inferior to the stratum corneum

stratum lucidum a clear, thicker layer within the epidermis found in thick areas like palms, soles of feet

stratum spinosum layer inferior to the stratum granulosum; very spiny in appearance

striae a linear mark, slight ridge, or groove on a surface, often one of a number of similar parallel features

striated striped appearance in skeletal muscle

styloid process one of two very tiny, sharp daggerlike spines found underneath the skull near the mastoid processes

sulci the 'dip/rut' of a single convolution on the brain's cerebral cortex

superficial near the surface (i.e., the skin is superficial to the muscles)

superior located above something else (i.e., the head is superior/above the shoulders)

superior nasal concha a long, narrow and curled bone shelf (shaped like an elongated seashell) that protrudes into the breathing passage of the nose; there are two sets—superior is the uppermost pair

supination rotation of the arm or leg outward; in the case of supination of the arm, the palm of the hand faces forward

supine laying on your dorsal surface (i.e., back), ventral surface (i.e., human abdomen) up

surface tension the tendency for water to hold together due to the hydrogen bonds in water molecules (i.e., puddles of water on a freshly waxed car hood form, become larger, and pool together into larger puddles

sutures immovable seams/joints in the skull

sweat glands glands that secrete sweat to the surface of the skin

sympathetic a part of the nervous system that serves to accelerate the heart rate, constrict blood vessels, and raise blood pressure; the sympathetic nervous system and the parasympathetic nervous system constitute the autonomic nervous system

symphyses a place where two bones are closely joined, either forming an immovable joint (as between the pubic bones in the center of the pelvis) or completely fused (as at the midline of the lower jaw)

synapse the space between neurons; location of some neurotransmitters

synarthroses an immovably fixed joint between bones connected by fibrous tissue (e.g., sutures of the skull)

synchondroses an almost immovable joint between bones bound by a layer of cartilage, as in the vertebrae

syndesmoses an immovable joint in which bones are joined by connective tissue (e.g., between the fibula and tibia at the ankle)

synovial fluid a thin, lubricating fluid found in synovial joints (e.g., knees)

synovial joint a joint that has a capsule that contains synovial fluid (i.e., knee)

synthesis phase (S) the period during which DNA is synthesized; in most cells, there is a narrow window of time during which DNA is synthesized

synthesis reaction one of the most common types of chemical reactions. In a synthesis reaction two or more chemical species combine to form a more complex product

talus ankle bone

tarsal ankle region

telophase last phase of mitosis; nuclei reform; organelles reappear; cytokinesis occurs and results in two new daughter cells

temporal bone two bones; located at temple areas of skull

temporalis temporal muscles

temporal lobe lobes of the brain (2) located just under the temporal bones

tendon hold skeletal muscle to bone

tendon sheath a protective covering wrapping the tendon and adding strength

teres major the largest of two muscles in the shoulder region that move the shoulders and arms

teres minor the smallest of two muscles in the shoulder region that move the shoulders and arms

terminal hair hair that is coarse and dark (hair on our head, armpits, pubic hair, etc.)

tertiary protein a single polypeptide chain 'backbone' with one or more protein secondary structures

thalamus a large, dual-lobed mass of grey matter buried under the cerebral cortex; it is involved in sensory perception and regulation of motor functions. The thalamus is a limbic system structure and it connects areas of the cerebral cortex that are involved in sensory perception and movement with other parts of the brain and spinal cord that also have a role in sensation and movement. As a regulator of sensory information, the thalamus also controls sleep and awake states of consciousness; its function includes motor control, receiving auditory, somatosensory, and visual sensory signals; relaying sensory signals to the cerebral cortex; controlling sleep and awake states

thoracic cage a protective area formed by the ribs in the chest to protect the heart and lungs

thoracic vertebrae vertebrae in the mid-chest region; T1–T12; each thoracic vertebrae articulates with a rib

tibia the shin bone; the largest of the two bones in the lower legs

tibialis anterior muscles on the anterior shin

tissue cells working together to form similar functions

trans fat known as 'unhealthy fat'

transfer RNA an adaptor molecule composed of RNA, typically 73 to 94 nucleotides in length, that serves as the physical link between the nucleotide sequence of nucleic acids and the amino acid sequence of proteins

transitional epithelial a type of tissue consisting of multiple layers of epithelial cells which can contract and expand; these cells, part of the epithelium, are found in the urinary bladder, in the ureters, and in the superior urethra and gland ducts of the prostate

translation in molecular biology and genetics, the process in which cellular ribosomes create proteins; part of the process of gene expression

transverse running horizontally

transverse process bone markings that typically form the transverse/lateral sides of vertebrae in the back

trapezius shoulder muscles

triceps brachii muscle forming the back of the upper arm

triglycerides a type of fat produced by the body; they can elevate artificially from a diet of high-fat foods (especially fried foods) and become troublesome by contributing to premature atherosclerosis of the blood vessels

true ribs ribs that attach directly to the sternum

tympanic membrane eardrum

ulna one of two bones in the forearm

unipennate having the fibers arranged obliquely and inserting into a tendon only on one side in the manner of a feather barbed on one side

utricle one of two tiny sacs found in the vestibule of the semicircular canals

vastus lateralis knee muscle toward the side of the knee

vastus medialis knee muscle toward the inner knee

vellus hair very fine hair

ventral anterior

ventricles the two pumping chambers of the heart

vertebrae individual bones in the spine

vertebral foramen a large holelike opening found within the vertebrae of the spine; area for the spinal cord

vestibule entrance to the semicircular canals

vitiligo a condition resulting in areas characterized by the absence of melanin (pigment)

Volkmann's canals small canals that run 'east and west' through bone; areas for blood vessels and nerves

vomer bone small bone at the base of the nose/end of the nasal septum closest to the maxilla; the vomer together with the ethmoid form the nasal septum

Wolff's law of bone stress on bone (i.e., pounding-type exercises) helps it to maintain its girth, width, and strength

wrist bones eight bones comprise the wrist: scaphoid, lunate, triquetral, pisiform, trapezium, trapezoid, capitate, hamate

zygomatic bone one of two bones in the cheek; forms the front of the cheek

zygomatic process a piece of the temporal bone that together with the zygomatic bone form the cheek

zygomaticus major one of two 'smile' muscles; runs from the zygomatic bone of the cheek to the mouth; the major is the larger of the two

zygomaticus minor one of two 'smile' muscles; runs from the zygomatic bone of the cheek to the mouth; the minor is the smaller of the two

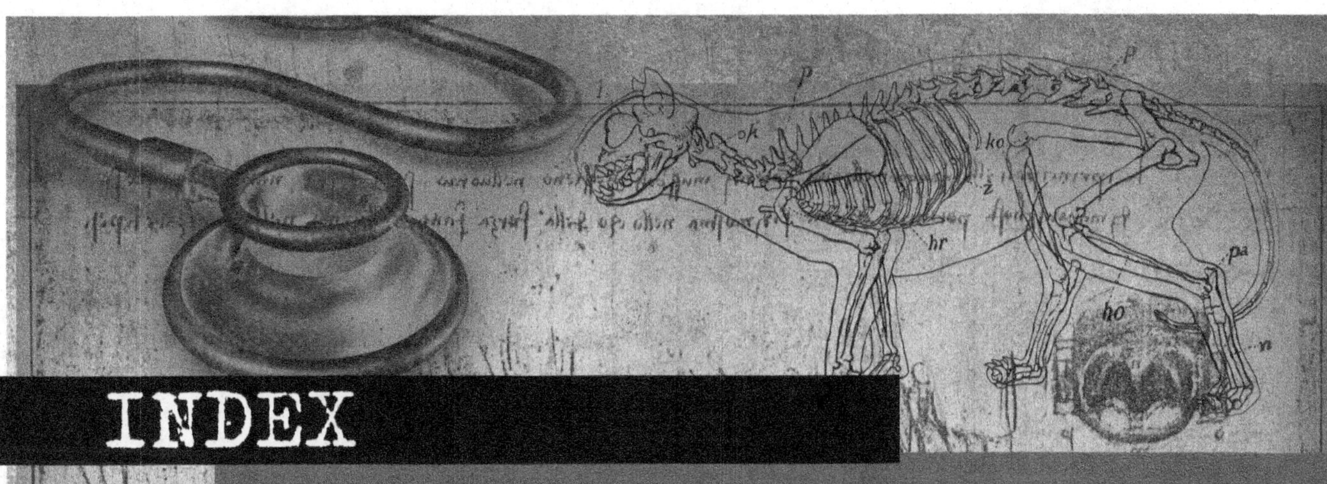

INDEX

Note: the italic numbers indicate a figure

A

abdominopelvic cavity, 22

abdominopelvic quadrants, 22, *22*

abdominopelvic regions, *23*

abduction, 212, *212*

acetabulum, 191

acids, *47*, 47–49

acne, 123

acromion, 186

actin, 232

action potential, 233, 255, 259, 261–262, *262*

adduction, 212, *212*

adensosine triphosphate (ATP), 63

adipose tissue, 96, 98, *98*, 98–99, 130

afferent (sensory) division, 252

Alzheimer research, 251

amino acids, 33, 72

amphiarthrosis, 202

anabolic reactions, 44

anagen phase, of hair, 124

anaphase, 75

anatomical position

    abdominopelvic quadrants, 22, *22*

    abdominopelvic regions, *23*

    body cavities and membranes, *20*, 20–21, *21*

    defined, 6

    directional terms, 12–13

    regional terms, *15*, 15–17, *16*, *17*, *18*, *19*

angiogram, *9*

ankles

    bones of, 195–197, *196*

    joints of, *207*

anterior terms, defined, *13*, *17*

apocrine sweat glands, 120

appendicular skeleton, 140, 167, *167*

arms, major muscles of, 238, *239–240*, *241–242*

arrector pili muscles, 123

articular cartilage, 137, *137*

associative neurons, 262

atlas (C1), 178

atoms, 41–42, *42*

autonomic nervous system (ANS), 292–293

axial skeleton, 140, *140*, 167, *167*

axis (C2), 178

axon, 254

axon terminals, 254

B

balance, physiology of, 287–288

ball-and-socket joints, *205*

basal cell carcinoma, 114, *114*

basement membrane, 89, *89*

bases, *47*, 47–49

biochemistry, 25–49

    acids, bases, and pH concentration, *47*, 47–49

    atoms and molecules, 41–42, *42*

    carbohydrates, 30–33, *31*, *32*, *33*

    chemical bonds and reactions, 42–45, *43*, *44*

    compounds, 41

    elements in human body, 39, *39–40*

central nervous system (CNS), 265–273, *267, 268,*
      269, 270, 272, 273
centrioles, 64, *65*
centromere, 69
centrosomes, 64, *65*
cerebellum, 266
cerebral hemispheres, 266
cerebrospinal fluid, 274
cerebrum, 266
ceruminous glands, 119
cervical nerves, 278
cervical vertebrae, 178, *178*
channels, 58
cheek cells, *93*
chemical bonds
      of molecules, 41–42, *42*
      reactions, 42–45, *43, 44*
chondrocytes, 138
choroid, 283
chromatin, 69–71
chromosomes, 69–71, *70*
chrondocytes, 100
cilia, 59, *59*
ciliary body, 282
circular muscle, 226, *227*
circumduction, 212, *212*
clavicle, 186
Clostridium tetani, 235
coccyx, 181–182, *182*
cochlea, 286
collagen, *35,* 96, *97*
collagen fibers, 86
columnar epithelial tissue, 91
comminuted (bone fracture), *159*
compact bone, 101, 155, *155, 156*
complimentary base pairing, 68
compounds, 41
compression (bone fracture), *159*
conductive neurons, 258
condylar joint, *206*
condyle, *148*
cones, 283, *285*
connective tissue, *85, 86,* 95–102, *96, 97, 98, 99, 100,*
      101, 102
convergent muscle, 226, *228*
coracoid process, 186
cornea, 282
coronal plane, 5
coronal suture, 170
corpus callosum, 269
cortex, 123
covalent bonds, 42, *43*
cranial cavity, 21, *21*

cranial nerves, 276–277
crest, *144*
cribiform plate, 172
crista galli, 172
CT (computer tomography) scan, *4,* 4–8, *5*
cuboid, 195
cuboidal epithelial tissue, 91
cuticle, 123, 129
cyanosis, 118
cytology, 52. *see also* cells
cytoskeletons, 64

D

daughter cells, 75
decomposition, 44
deep, *13*
deletion, 79
dendrites, 254
dens, 178
dense connective tissue, 97, *98*
depression (bone fracture), *159*
depression (of joints), 212–213, *213*
dermatitis, 118
dermis, *110,* 110–111, *119,* 119–122, *120, 121, 126*
diaphysis, 151
diarthroses, 204
diet. *see also* biochemistry
      dietary fats, *28*
      effect of soda on bones, 160, *161*
diploe, 152
directional terms, 12, *13*
disaccharides, 32
distal, *13*
DNA (deoxyribonucleic acid)
      polymerase, 68
      replication and cell cycle, 73–78, *74, 75, 76*
      RNA and, 72–73, *73*
      structure, *67,* 67–68
dorsal, *13*
dorsal cavity, 21
dorsiflexion, 215, *215*
duplication, 79

E

ear, 286–288, *287*
eccrine sweat glands, 120
eczema, 118
efferent (motor) division, 252

H

L

lacrimal bone, 169

lacunae, 154, *154*

lambdoid suture, 170

lamellae, 155, *156*

Langerhans cells, 112

lanugo, 122

lateral, defined, *13*

lateral cuneiform, 195

lateralization, 268, *268*

legs, major muscles of, 238, *239–240, 241–242*

lens, 282

lesser wings, 172

leukocytes (white blood cells or WBCs), 96, 102, 119

levator labii superioris, 237

ligaments, 137–138, *138*

line (on bone), *147*

lipids, 27

long bones, 140, *141*

longitudinal fissure, 269

loose connective tissue, 97, *98*

lower limbs, bones of, 191–192, *192, 193, 194,* 194–195

lumbar spinal nerves, 279

lumbar vertebrae, 180–181, *181*

lunate, 190

lysosomes, 62, *63*

M

macrophages, 96, *96,* 119

magnum foramen, 171

malleus, 286

mammary glands, 119

mandible, 169

markings, bones and, *143–148,* 143–150

marrow, *150,* 150–151

masseter, 237

mast cells, 96, 119, *120*

mastoid process, 171

maxilla, 169

meatus, *146*

medial, *13*

medial cuneiform, 195

median planecoronal plane, 45

medulla, 123

Meissner's corpuscles, 121

melanin, 117

melanocytes, 112

melanoma, *113,* 114

membrane potentials, 258–260, *259, 260*

membranes, *20,* 20–21, *21*

mental foramen, 169

Merkel cells, 112

messenger RNA (mRNA), 72

metacarpals, 190

metaphase, 75

metastasis, 109

methane, *43*

microanatomy, of bone, 154

microphage, *120*

microscopes, *53,* 53–54, *56*

microvilli, 59

mineralization, 139

minerals, 40, *40–41*

minerals—inorganic, 40

missense mutation, 78

mitochondria, 63, *64*

mitochondrial DNA, 63

mitosis, 66

mitotic spindle, 64

molecules

    chemical bonds and, 41–42, *42*

    organic molecules of life, 27–29, *28, 29, 30*

moles, 117

monosaccharides, 31

motor cortex, 270

motor neurons, 233, *233, 254,* 262

motor unit, 233

MRI (magnetic resonance imaging), 10

multipennate muscle, 226, *228*

muscles, 219–248

    arrector pili muscles, 123

    cardiac muscle tissue, *221,* 247–248, *248*

    fascicles and muscle shapes, *225,* 225–230, *227–229*

    functions of, 223–224

    major muscles of head and neck, 236–237, *238*

    major muscles of torso, arms, legs, 238, *239–240, 241–242*

    muscle fibers, 103

    muscular tissue (muscle), *85, 86, 103,* 103–106, *104,* 221–223

    origins and attachments, 230–231

    skeletal muscle, 103, *225*

    skeletal muscle fiber, microscopic anatomy, *221, 231,* 231–236, *232, 233*

    smooth muscle, *221, 246,* 246–247

mutations, 78

mutations, cells, 78–81, *80*

myelin sheath, 254

myocytes, 103

myofibrils, 232

myosin, 232

myosin light chain kinase, 247

T